Two Manuals Combined into One Book

AF801430

Tow Weapon System

And

Army M47 Dragon Med Antitank Weapon

Department of Defense

Pentagon
Publishing

FIELD MANUAL
NO. 3-22.34(23-34)

HEADQUARTERS
DEPARTMENT OF THE ARMY
Washington, DC, 28 November 2003

TOW WEAPON SYSTEM

CONTENTS

DISTRIBUTION RESTRICTION: Approved for public release; distribution is unlimited.

*This publication supersedes FM 23-34, 17 August 1994.

PREFACE

This manual discusses the many changes in the TOW missile, TOW training, and OPFOR armored vehicles and countermeasures. It includes training information on the M220A1 (basic) and M220A2 tube-launched, optically tracked, wire-guided (TOW) weapon systems and carrier M966. Its intended audience includes leaders and members of mechanized infantry, light infantry, airborne, and airmobile infantry organizations who train and test individuals and crews. Units with M2/M3 Bradley fighting vehicle systems should also refer to FM 23-1 for additional training and testing requirements.

The proponent of this publication is US Army Infantry School. Send comments and recommendations on DA Form 2028 directly to Commandant, US Army Infantry School, ATTN: ATSH-INB-B, Fort Benning, Georgia 31905-5595, or email to doctrine@benning.army.mil.

Unless otherwise stated, masculine nouns and pronouns do not refer exclusively to men.

CHAPTER 1

INTRODUCTION

The TOW is a crew-portable heavy antitank missile weapon system that can be operated from armored, lightly armored, and unarmored multipurpose vehicles and helicopters, and from the ground mount. The system consists of a guided missile and a launcher to engage tanks, fortifications, and other materiel targets. The term "TOW" is an acronym derived from the words that describe the principal operating features of the missile—tube-launched, optically tracked, wire-guided. The launcher initiates, tracks, and controls the flight of the missile by guidance signals transmitted over a command-link wire that connects the missile with the launcher.

1-1. CHARACTERISTICS

The TOW weapon system consists of a launcher, which has tracking and control capabilities, and the missile, which is encased in a launch container. The launcher is equipped with self-contained, replaceable units.

a. **Capabilities.** The TOW is mainly an antitank weapon used for long-range engagement of armored targets. It can be employed in all weather conditions as long as the gunner can see his target through the daysight tracker or nightsight. The TOW also provides a long-range assault capability against heavily fortified bunkers, pill boxes, and gun emplacements.

(1) The ITOW missile can destroy targets at a minimum range of 65 meters and a maximum range of 3,750 meters. The TOW 2B missile can destroy targets at a minimum range of 288 meters when fired from the ground mount and 200 meters when fired from the HMMWV or BFV. The TOW 2B has a maximum range of 3,750 meters whether ground- or vehicle-mounted.

(2) The automatic missile tracking and control capabilities of the TOW weapon system provide a high first-round-hit probability. To operate the system, the gunner places the crosshairs of the sighting system (either the daysight tracker or the nightsight) on the target, fires the missile, and centers the crosshairs on the target image until missile impact. The optical tracking and command functions within the system guide the missile along the gunner's line of sight. The gunner does not apply lead, windage, or elevation.

b. **Mobility.** The TOW weapon system can be vehicle-mounted or ground emplaced (tripod-mounted) for operation. Missiles can be launched from either operational mode. (Detailed operating procedures for each configuration are provided in Chapters 2 and 3.)

(1) The entire system can be hand-carried by a weapon crew; therefore, emplacement sites can be changed quickly to minimize detection or to engage targets that are out of range.

(2) The vehicle-mounted launcher is more mobile and can be quickly prepared for use. The launcher can be assembled and disassembled without the use of tools.

(3) Built-in self-test circuits can be used to check the operational condition of the assembled launcher whether ground- or vehicle-mounted.

1-2. SYSTEM CONFIGURATIONS

The TOW weapon system comes in two configurations—the M220A1 (TOW) and the M220A2 (TOW 2).

a. **M220A1 TOW Weapon System.** The M220A1 TOW (Figure 1-1) consists of a tripod, a traversing unit, a launch tube, a daysight tracker, an AN/TAS-4 nightsight, a missile guidance set (MGS), a battery assembly housed in the compartment of the MGS, and an encased missile.

(1) The M220A1 TOW weighs about 265.5 pounds with all of its components and carrying cases; with encased missile (BGM-71A), it weighs 320 pounds. (For a detailed description of the M220A1 TOW, see TM 9-1425-472-12.)

(2) The M220A1 TOW can fire all four configurations of TOW missiles, which include many types. However, it cannot take full advantage of the BGM-71D TOW 2 and BGM-71E TOW 2A or TOW 2B counter countermeasures when tracking through obscurants.

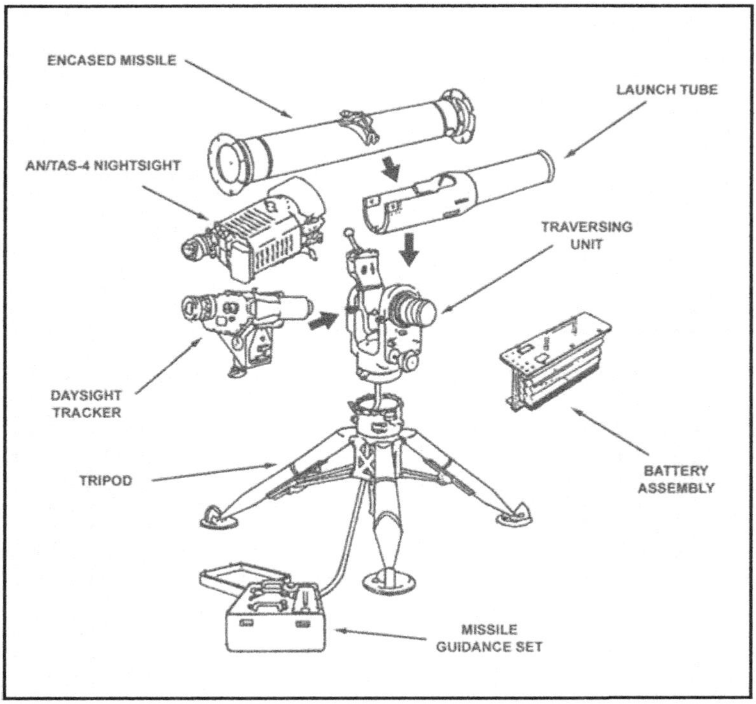

Figure 1-1. M220A1 TOW components.

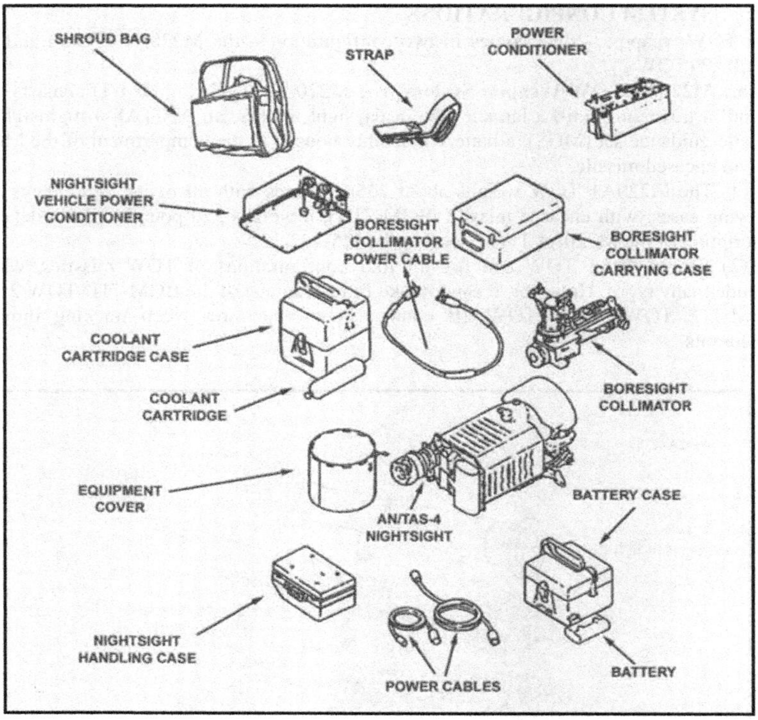

Figure 1-1. M220A1 TOW components (continued).

b. **M220A2 TOW 2 Weapon System.** The M220A2 TOW 2 (Figure 1-2, page 1-4) is a combination of a modification to the M220A1 TOW launcher and the addition of a new missile. The traversing unit, the digital MGS, and the AN/TAS-4A thermal nightsight of the M220A1 TOW launcher were modified to form the M220A2 TOW 2 launcher. (For a detailed description of the M220A2 TOW 2, see TM 9-1425-450-12.)

(1) The M220A2 TOW 2 weapon system weighs about 256.5 pounds with all of its components and carrying cases; with encased missile BGM-71D, it weighs about 318.5 pounds.

(2) The TOW 2 missile (BGM-71D) improvements include a new guidance link, a full-caliber 6-inch warhead, a reloaded flight motor, and a longer warhead probe.

(3) The M220A2 TOW 2 launcher is compatible with all four missile configurations. It can achieve a higher probability of hit against all types of targets through improved microprocessor-based electronics that use digital design techniques.

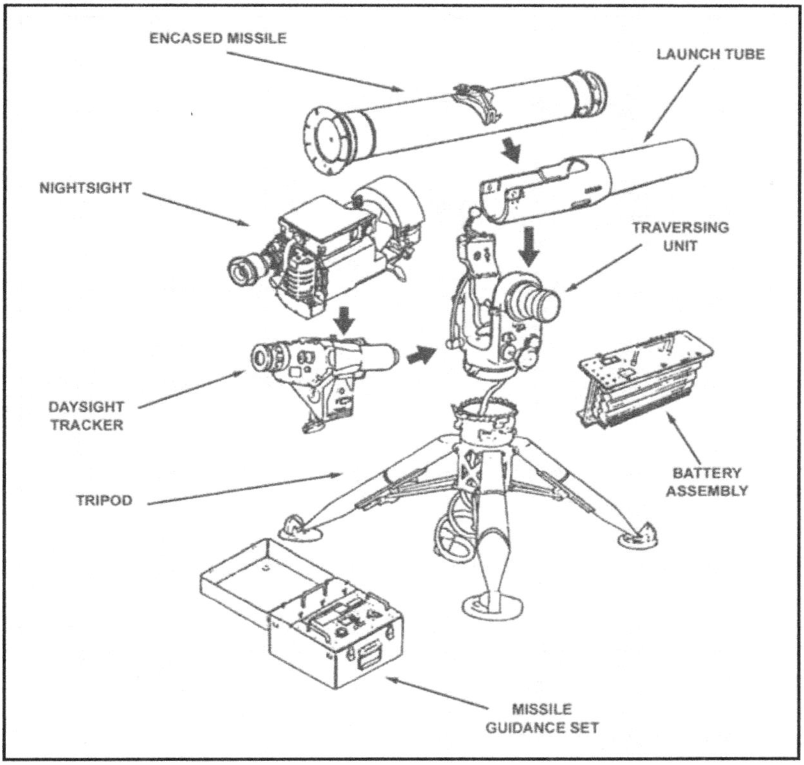

Figure 1-2. M220A2 TOW components.

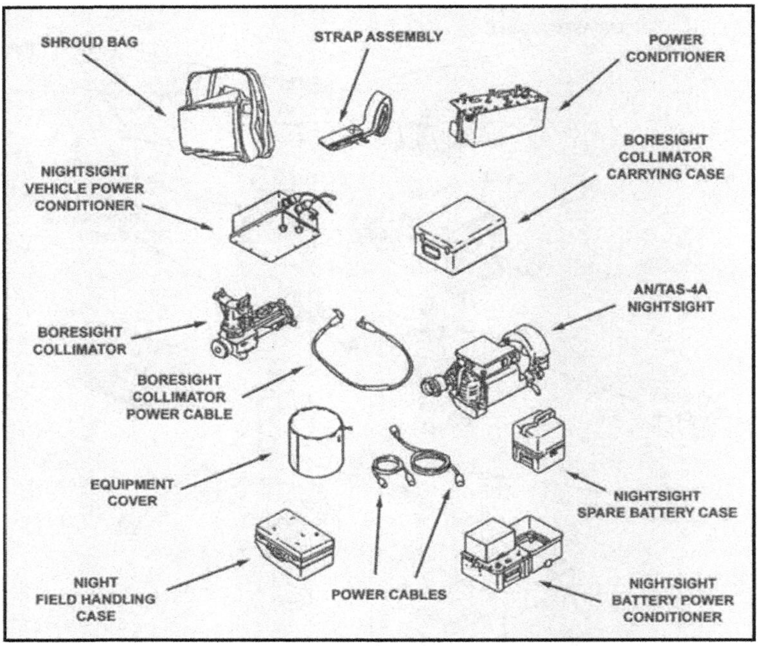

Figure 1-2. M220A2 TOW components (continued).

1-3. MISSILE CONFIGURATIONS AND TYPES

The TOW missile comes in four configurations with numerous types. These types mainly consist of minor modified work orders (MWOs) that are transparent to the operator and are continually added to. This FM will not attempt to deal with the various types but will confine itself to the four major configurations. All configurations use the same basic airframe, aerodynamic control system, command-link wire, and missile electronics designs.

a. The first configuration is the improved TOW (ITOW). These missiles have improved 5-inch warheads that include extended probes for greater standoff and penetration.

b. The second configuration is the TOW 2. This missile has a full-caliber 6-inch warhead that includes an extended probe. In addition to the infrared radiator of the ITOW missile, TOW 2 has a second infrared radiator to provide hardened system performance against battlefield obscurants and countermeasures. The second radiator is called the thermal beacon and provides link compatibility with the electrooptical infrared nightsight, which is part of the TOW 2 launcher system.

c. The third configuration is the TOW 2A. The TOW 2A has an added small explosive charge in the tip of the extended probe providing improved performance against reactive armor by premature detonation.

d. The fourth configuration is the TOW 2B. The TOW 2B has an entirely different warhead and kill mechanism than the other TOW missiles. It is a top-attack missile (fly over/shoot down [FOSD]) that defeats enemy armor at its most vulnerable point—the top deck of the turret and hull. The TOW 2B has a tandem warhead that fires two explosively formed projectiles (EFPs) down through the thin upper deck armor of the enemy vehicle. The gunner tracks the target the same as any other TOW missile with the crosshairs on center mass, but the missile automatically flies 2.25 meters above the line of sight (LOS). When the missile senses that it is directly above the target (by means of the target's shape and magnetic field), it automatically fires its warhead.

1-4. TOW MISSILE IDENTIFICATION

With the increased number of TOW missile types and wide variations in their capabilities (especially armor penetration), TOW crews and range safeties must be able to quickly and accurately identify the specific type of missile they have.

a. The configuration of the missile can be readily identified by the various identification decals, tapes, and stencils on the missile case (Figure 1-3 and Table 1-1). (See TM 9-1410-470-34 for current information on TOW missiles.)

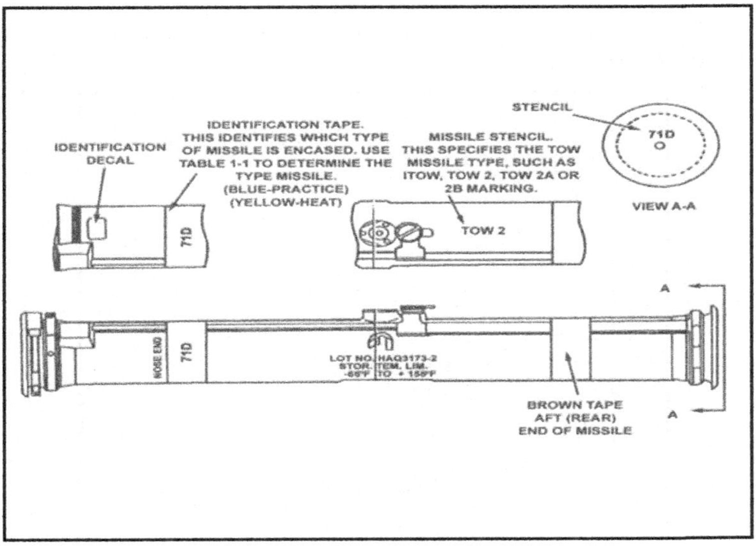

Figure 1-3. TOW missile case markings.

TYPE OF MISSILE	DECAL PLATE DATA
Improved TOW Practice	Guided Missile Practice, STD, BTM-71A1
Improved TOW HE	Guided Missile Surface Attack, ITOW-BGM-71C
TOW 2 Practice	Guided Missile Practice, TOW 2, BTM-71D-1B
TOW 2 HE	Guided Missile Surface Attack, TOW 2, BGM-71D-3B
TOW 2A Practice	Guided Missile Practice, TOW 2A, BGM-71E-3B
TOW 2B Practice	Guided Missile Practice, TOW 2B, BTM-71E-2B
TOW 2B HE	Guided Missile Surface Attack, TOW 2B, BGM-71F
Decals are located on the nose end of the launcher tube. They describe the type of round and model. BLUE—indicates practice round. YELLOW—indicates HE round. WHITE—indicates HE or practice round. BROWN—indicates low explosive (live rocket motor).	

Table 1-1. TOW missile identification.

b. The TOW 2B missile is unique because it is an FOSD missile. It is marked with a decal as shown in Figure 1-4.

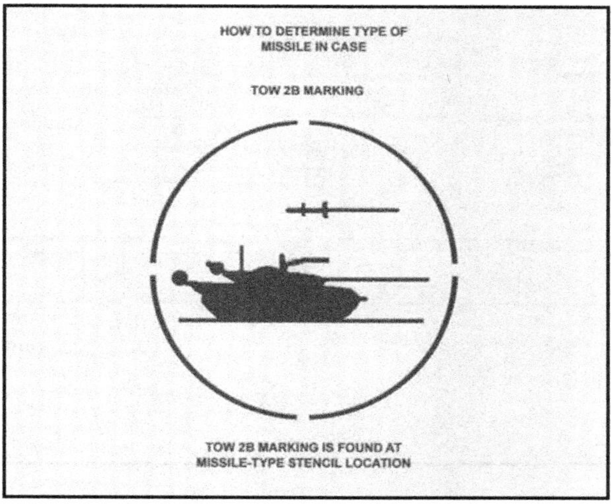

Figure 1-4. TOW 2B marking.

1-5. INTERNAL COMPONENTS AND SEQUENCE OF OPERATIONS

A TOW crewman or leader who thoroughly understands his weapon will be able to employ it to its maximum effectiveness. This paragraph describes the internal missile operations from trigger depression to missile impact, which will increase the soldier's knowledge of how his weapon functions.

a. **Major Internal Sections and Components.** The missile is divided into three major sections and the launch container.

(1) ***Front Section.*** The front section contains the warhead and the electronics.

(a) The warhead contains an extended probe on the ITOW and TOW 2-series (17 inches on TOW 2, 10 inches on ITOW). All missile types have a crush ogive switch to detonate the missile (located in the end of the probe tip on ITOW and TOW 2). The warhead is a conventional shaped charge (except TOW 2B) with a copper cone and liner and a detonator at the rear of the charge. The diameter of the warhead is 5.2 inches on ITOW and 6 inches on TOW 2 and 2A.

(b) All electronic circuitry necessary to perform on-board electronic functions while the missile is in flight is contained in the front section. This circuitry receives steering signals from the MGS and corrects the missile flight path to conform to LOS.

(2) ***Center Section.*** The center section contains the thermal batteries, the gyroscope, the flight motor, and wings.

(a) The missile has three thermal batteries, which are used instead of storage batteries because their shelf life is much longer. The chemical reaction that produces both heat and electricity is started by an electrical charge that is part of the prefire signal sent when the trigger is depressed. Within approximately two tenths of a second the batteries begin producing electricity.

(b) The gyroscope stabilizes the missile in flight, and it provides information about the attitude of the missile to the MGS from sensors in its housing. When the prefire signal is sent to the missile, an explosive squib on the neck of a bottle of compressed nitrogen is ignited. This produces the popping sound and whirring noise that is heard after the gunner depresses the trigger and before the missile launches. The escaping nitrogen causes the gyroscope to begin spinning in its bracket. When the gyroscope reaches 42,000 rpm, it automatically disengages from its bracket and begins to spin freely in its housing. This causes an electrical current that ignites the launch motor launching the missile from the weapon system.

NOTE: The process of the thermal batteries heating up to produce electricity and the gyroscope coming up to speed and igniting the launch motor takes about 1.5 seconds. This causes the delay between the time when the gunner presses the trigger and the missile actually fires.

(c) The flight motor is a solid fuel rocket motor with two nozzles protruding through the body of the missile at 30-degree angles. The flight motor burns for 1.6 seconds. During this time, it provides all the momentum necessary for the full flight of the missile.

(d) For the rest of its flight, the missile coasts on momentum and the lift provided by the four wings. The wings are located just aft of the flight motor, recessed into the body of the missile during storage and spring loaded. When the missile leaves the launch container, the wings open and lock into place. The wings are set at 45-degree angles to the body of the missile.

(3) ***Aft Section.*** The aft section contains the wire spools, the launch motor, the actuator control system and control surface push rods, a bottle of compressed helium, the flight control surfaces, the xenon beacon (on all missiles), and the thermal beacon (on TOW 2-series missiles only).

(a) *Wire Spools.* Two wire spools, each holding 3,750 meters of wire, are located at the rear of the missile. The wire is made of fine steel and is coated with a thin layer of varnish for insulation. There is virtually no tension on the wire, so it droops down on the ground behind the missile as the missile flies downrange.

(b) *Launch Motor.* The launch motor is a solid fuel rocket that burns entirely inside the launch tube. It provides just enough thrust to propel the missile to a point where the flight motor can ignite safely without injuring the gunner.

(c) *Flight Control Surfaces.* Four flight control surfaces are located on the aft end of the TOW missile set at 90-degree angles to the body of the missile. The TOW, unlike the Dragon, does not spin in flight but maintains a stable position. One pair of flight control surfaces controls the "up and down" (pitch) corrections while the other pair controls the "left and right" (yaw) corrections. Like the wings, the flight control surfaces are recessed into the body of the missile while in the launch container. They spring out and lock into place when the missile leaves the launch container.

(d) *Actuator Control System/Push Rods/Helium Bottle.* The TOW missile makes steering corrections by moving the flight control surfaces. It moves them by opening and closing the valve on a bottle of compressed helium. The escaping helium moves a system of push rods back and forth. The push rods are connected to the flight control surfaces and move them in turn. The valve on the helium bottle opens and closes in response to signals sent by the on-board electronics, which are responding to signals sent down the wires by the MGS.

(e) *Xenon Beacon.* The xenon beacon emits an infrared beacon, which is picked up by the infrared receiver on the daysight tracker. The infrared receiver measures the angle at which the infrared beam strikes, thus providing the major source of data on the position of the missile to the MGS. The xenon beacon consists of a bulb filled with xenon gas and two electrodes. When an electric current passes between the two electrodes, it creates a spark, which excites the gas. This excited gas emits infrared light that exits from a window on the rear of the beacon as a narrow beam. All basic TOW infrared beacons operate on the same frequency (the infrared light turns on and off at the same rate of speed). This causes two problems: First, two systems cannot be placed closer than 300 meters because the beacons overlap and the MGS has no means of distinguishing between the missiles, causing it to lose control. Second, jamming the daysight tracker is fairly simple if the enemy knows the correct frequency. The TOW 2 overcomes these problems by having the MGS send a signal to the missile that controls the frequency at which the xenon beacon is operating. The MGS varies this pattern randomly, speeding it up and slowing it down in no apparent pattern. The MGS is always able to distinguish its missile from other missiles because no two missiles will be operating on the same frequency at the same time. For the same reason, the enemy cannot jam the system.

(f) *Thermal Beacon.* Under some battlefield conditions such as heavy smoke, dust, or fog, the xenon beacon, which operates on the lower end of the infrared spectrum, cannot penetrate the obscuration. With the TOW missile and weapon system it is sometimes possible for the gunner to be able to acquire the target, yet the infrared receiver on the daysight tracker cannot acquire the infrared beacon and the system loses the missile. TOW 2-series missiles have a thermal beacon to correct this deficiency. The thermal beacon operates on the upper end of the infrared spectrum and will penetrate any obscurant the nightsight can see through. A postamplifier was added to the AN/TAS-4A nightsight on the TOW 2 weapon system to acquire the thermal beacon and track it in much the same manner as the infrared beacon. The basic AN/TAS-4 is a passive sight, but the AN/TAS-4A is an active sight with the ability to steer the missile to any target that the gunner can see through the sight itself.

(4) ***Launch Container.*** The launch container is the fiberglass tube that holds the missile. It protects the missile from environmental and physical damage while in storage, secures it in the weapon system when about to be fired, and electrically connects the missile to the weapon system during firing.

b. **Sequence of Operations.** The sequence of operations (Table 1-2) starts from the time that the gunner presses the trigger. The time that each event occurs is given in negative or positive numbers, with 0 being the time that the launch motor fires.

TIME (SECONDS)	EVENT
-1.5	Gunner presses trigger; system sends prefire sequence to missile, starts chemical reaction in thermal batteries, blows explosive squib off nitrogen bottle; gyroscope starts spinning.
0	Gyroscope reaches 42,000+ rpm and disengages; sends electrical current to launch motor; launch motor fires.
+0.5	Missile exits launch tube.
+0.8	Wing 2 switch closes.
+.10	Wing 4 switch closes; helium bottle opens; timed part of arming sequence starts; flight motor ignites.
+.145	Delay switch changes state (part of arming sequence).
+.18	Safety and arming unlock occurs. As missile accelerates to 19 Gs, the "G" force causes a rotor in the warhead to unlock.
+.53	S&A clock mechanism rotates and aligns detonator with warhead. Missile is armed.
+1.6	Flight motor burns out. Missile is fully armed and travelling at maximum velocity. It will coast the rest of the way on momentum and the lift from the four wings.

Table 1-2. Sequence of operations.

c. **Tracking and Steering of the Missile While in Flight**. As the missile flies downrange, the infrared beam from the rear of the missile enters the infrared receiver on the daysight tracker, is reflected by a spinning mirror in the rear of the infrared receiver called a nutater, and strikes the infrared sensor in the front of the infrared receiver. The pattern that this reflected beam makes on the sensor measures the angle of the beam and tells the MGS where the missile is. The MGS has two other sources of information on the location and attitude of the missile: the gyroscope and the traversing unit. Both of these have sensors. The gyroscope measures the attitude of the missile while in flight, and the traversing unit measures how quickly and in what direction the gunner moves it. The MGS uses this information to make a very precise determination of where the missile is in relation to the gunner's aiming point. This automatically steers the missile toward that point. This process continues until the missile impacts the target.

1-6. TOW TRAINING STRATEGY

TOW training strategy synchronizes institutional and unit training to produce TOW crews and units that win on the battlefield. (See Appendix A for training tips and Appendix B for a train-the-trainer program.)

a. The training strategy for TOW begins in 11B OSUT where the soldier is taught critical Skill Level 1 common skill tasks and critical 11B Skill Level 1 tasks. He is trained to be a loader, gunner, and driver.

b. Much individual and most collective, tactical, and gunnery training takes place in the unit. Primary sources for tactical and leader training are FM 7-91, ARTEP 7-91-MTP, and ARTEP 7-91-Drill.

c. The Skill Level 1 and 2 soldiers, squads, and crews are trained in the unit. Gunners take part in an individual gunner's qualification program that includes TOW Tables 1 and 3 and then qualify on Tables 2 and 4. The squads and crews practice on TOW Tables 5 and 7 and then qualify on Tables 6 and 8. (See Chapter 4 for more detailed information on TOW training. See DA Pam 350-381 for strategies and resource requirements.)

d. The section is the basic TOW fighting organization; therefore, section training is critical. Sections practice on TOW Table 9 and qualify on Table 10.

e. Platoon gunnery events consist mainly of Table 11 for practice and Table 12 for qualification. Also, the platoon must conduct a semiannual force-on-force exercise to demonstrate its proficiency with a maneuver element

f. TOW training in units is supported by MOS-specific tracks in BNCOC and ANCOC, and the Antiarmor Leader Course for NCOs (SGT[P] through SFC).

g. Army Correspondence Course Program (ACCP) courses also support TOW training. (See ACCP catalog for specific courses).

CHAPTER 2
TRIPOD-MOUNTED TOW/TOW 2

Regardless of the type of carrier the TOW weapon system is mounted on, at times it will be employed as a ground-mounted system. This may be due to tactical requirements, such as helicopter insertion for an antiarmor ambush, or it may be due to mechanical failure or battle damage to the carrier. The following information and procedures apply to all TOW units. During combat or field training, TOW crews will sometimes be required to dismount the TOW system. Each TOW crew dismounts the TOW IAW dismount procedures for the appropriate vehicle as outlined in this manual.

Section I. OPERATION PROCEDURES

Operation procedures for the TOW weapon system include performing a system checkout, conducting preventive maintenance checks and services (PMCS), selecting firing positions, and preparing TOW missiles for operation.

2-1. SYSTEM CHECKOUT

The TOW weapon system is set up on the tripod to conduct a system checkout to ensure that the system is working properly before engaging targets. System checkout procedures are conducted before and after installation of an M220-series launcher on its organic carrier; daily, after the system is in use; or anytime the system has been in storage for one week. (See TM 9-1425-472-12 [M220A1] or TM 9-1425-450-12 [M220A2] for complete details.) After the system checkout procedures are completed, the weapon is mounted or installed on its organic carrier IAW the appropriate TM.

2-2. PREVENTIVE MAINTENANCE CHECKS AND SERVICES

A preventive maintenance checks and services (PMCS) should be performed before, during, and after operation of the TOW weapon system. PMCS are performed daily if the system is in continuous use. (See TM 9-1425-472-12 [M220A1] or TM 9-1425-450-12 [M220A2] for complete details.)

2-3. SELECTION OF FIRING POSITIONS

When the TOW is assembled for ground emplacement, the tripod must be level for effective operation of the weapon. If the tripod is not level, lateral (azimuth) movement of the traversing unit produces a change in the elevation angle of the launch tube; therefore, acquiring and tracking the target may be difficult. An unleveled tripod can cause an electronic control error during the final tracking operation (while the command-link wire is guiding the missile to its target). For ground emplacement, proper adjustment of the tripod legs compensates for uneven ground up to a slope of 30 degrees. Therefore, the firing position must not slope more than 30 degrees. When the tripod legs are properly positioned, the bubble in each level will be between the marks on the level indicator. The TOW is not fired from a position where the traversing unit has a lateral cant of more than 10 degrees. The ideal firing position is as level as possible. However, the missile can be fired from a position where the tripod is canted as much as 30 degrees directly downhill or uphill. (See Appendix C for more information on fighting positions.) After the site is selected and the weapon emplaced, the squad leader and gunner make a range card for the position. (For detailed information on making a range card, see Appendix D.)

2-4. PREPARATION OF MISSILES

The TOW missile requires complete attention to detail when preparing it for firing missions. Missile identification, inspection, and correct handling procedures eliminate safety hazards to personnel and equipment. (See TM 9-1425-450-12 or TM 9-1425-472-12 for complete details.)

Section II. PERFORMANCE PROBLEMS AND DESTRUCTION PROCEDURES

TOW crew members must recognize and overcome performance problems and limitations to survive in battle. Each unit develops and maintains an SOP for the proper destruction of ammunition and weapon systems.

2-5. MALFUNCTIONS

A malfunction is the failure of ammunition to function correctly when fired, or the unexpected functioning of explosive components during a test. Premature functioning during handling, maintenance, storage, transportation, or deployment constitutes a malfunction. Any abnormal functioning such as failure of the missile to leave the launch container on launch motor ignition also constitutes a malfunction. During operation of the TOW, a malfunction occurs if the missile fails to launch after the normal 1.5-second delay after the firing trigger has been pressed or if the missile fails to function properly in flight. Two major types of malfunctions are *hangfire* and *misfire*. In-flight malfunctions are classified as two basic types—duds and erratic flight. Defective missiles (either fired or encased) must be disposed of by an explosive ordnance disposal (EOD) team only.

NOTE: When appropriate, weapon and vehicle TM checklists should be used when correcting malfunctions. (See TM 9-1425-450-12.)

a. **Hangfire**. A hangfire is an unexpected delay in the functioning of a series of ammunition components. It may occur anywhere in the series of events, from pressing the firing trigger to launch motor ignition. After the trigger is pressed, the distinctive impulse of the squibs firing (to activate the missile batteries and gyro) and the gyro spin noise can be heard.

(1) If the missile fails to fire, alert the crew by announcing "Hangfire" and continue to track the target for one minute.

(2) If the missile fails to launch after one minute, close the trigger cover, lower the arming lever, place the elevation lock in the 8-degree down position, and place the azimuth lock in the locked position (do not lock the azimuth if it points the missile out of the left and right limits). Open the bridge clamp and remove the missile from the launch tube. Carry it to a dud pit or a safe distance from friendly troops.

(3) Point the missile downrange at all times and notify EOD personnel.

WARNING
The encased missile will become hot to the touch if the TOW thermal batteries, which generate temperatures higher than 800 degrees F, are activated. The encased missile will produce smoke if the thermal beacon has been ignited.

b. **Misfire.** A misfire (complete failure to fire) occurs if the launch motor fails to ignite after the firing trigger has been pressed. This type of failure may be caused by a faulty firing circuit, a failure of electrical power, poor electrical connections, short circuits, or faulty components in the ignition or propellant systems. When the missile fails to launch, the prefire functions are not heard and the condition is a misfire. The target is tracked for one minute and the trigger is pressed again.

(1) Alert the crew by announcing, "Misfire," and check the battery (see notes 2 and 3). Then, check the MGS to ensure the coil cable connector is properly attached to connector J1. Turn the coil cable connector locknut clockwise until it is tight to make sure the coil cable connector is locked in place. Ensure the bridge clamp locking handle is properly seated. Lower the arming lever and check the backblast area to ensure no personnel are in danger; then, raise the arming lever. Press the trigger again.

(2) If the missile still fails to launch, continue tracking the target for one minute or at the gunner's discretion (in combat). Close the trigger cover, lower the arming lever, place the elevation lock in the 8-degree down position, and place the azimuth lock in the locked position. Open the bridge clamp and remove the missile from the launch tube.

(3) Move the missile a safe distance away (minimum 100 meters) and place it on the ground, preferably in a dud pit. Point the missile downrange at all times and notify EOD personnel.

c. **Dud**. A dud is any missile that has been launched (fired from the launch tube) but failed to fly to the target or failed to detonate on target impact. The warhead of all duds must be considered to be armed, and therefore, extremely dangerous. The location of all duds will be noted and referred to EOD personnel for disposal.

d. **Erratic Flight.** An erratic or uncontrolled missile flight may result due to any of the following conditions:

- Shorted or broken command-link wires.
- Defective missile wing or control surface.
- Defective rocket motor.
- Excessively high gunner tracking errors.

NOTES:
1. In a training situation, wait 30 minutes after announcing, "Hangfire" before proceeding.
2. For the M220A1, (MGS self-test position 1) check for proper band reading. If out of band reading, replace the battery.

3. For the M220A2, do MGS self-test for battery. If battery fails, replace the battery.
4. In a training situation, lower the arming lever, close trigger cover, place elevation lock in locked position, place azimuth lock in locked position, and wait 30 minutes.
5. Report all misfires, hangfires, and other missile malfunctions IAW AR 75-1.

2-6. FIRING LIMITATIONS

Some conditions may limit the firing and engagement capabilities of the TOW. The following information should be considered before engaging targets. (See TM 9-1425-450-12 for updated firing limitations.)

a. **Firing Over Water.** Firing across bodies of water wider than 1,100 meters can reduce the range of the TOW. Signals being sent through the command-link wires are shorted out when a large amount of wire is submerged in water. Maximum and limited range firing over water varies according to missile type. If the range is less than 1,100 meters, the missile's range is not affected. A TOW position should be as high above and as far back from the water as the tactical situation allows. The squad or section leader should analyze his sector as soon as the position is occupied to determine if water will affect the employment of the TOW.

(1) The various types of TOWs listed may be fired over water to a maximum range as shown in Figure 2-1 (TOW launcher and target at water surface level).

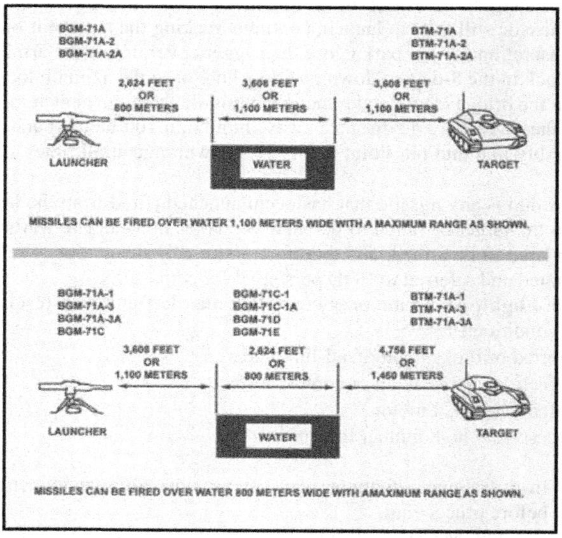

Figure 2-1. Maximum TOW missile range.

(2) The various types of TOWs listed may be fired over water to a limited range as shown in Figure 2-2 (TOW launcher and target at water surface level).

(3) To determine the extended range of missiles when firing over water (when TOW launcher or target is above surface level of water), use Table 2-1 (pages 2-6 and 2-7).

(a) Determine the type of missile being used.

(b) Determine the height above the water surface of the TOW launcher and place a tick mark at the corresponding height in the left-hand column of the table.

(c) Determine the height above the water surface of the target. Place a tick mark at the corresponding height in the right-hand column of the table.

(d) Connect the two tick marks with a straight line. The point where the lines intersect the center column is the maximum range the missile can travel without getting too close to the water.

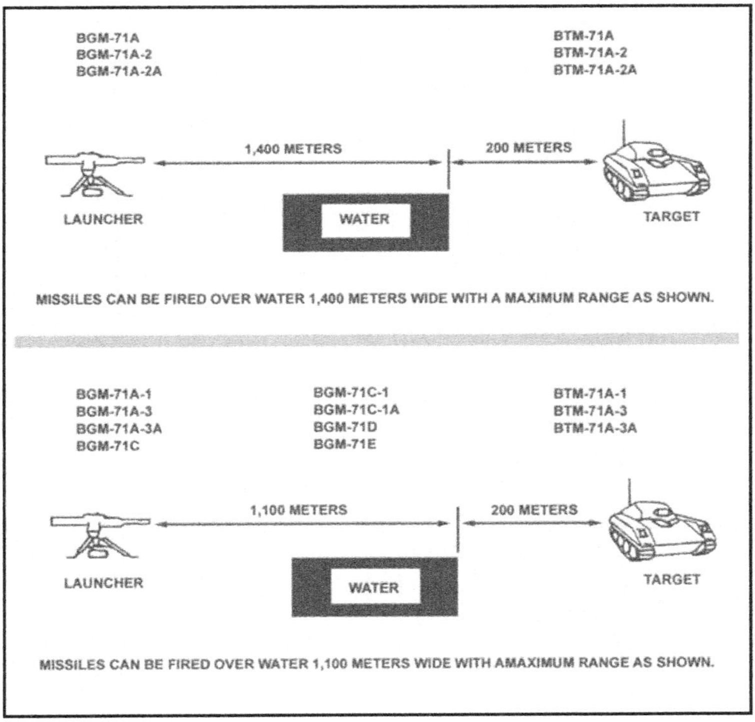

Figure 2-2. Limited TOW missile range.

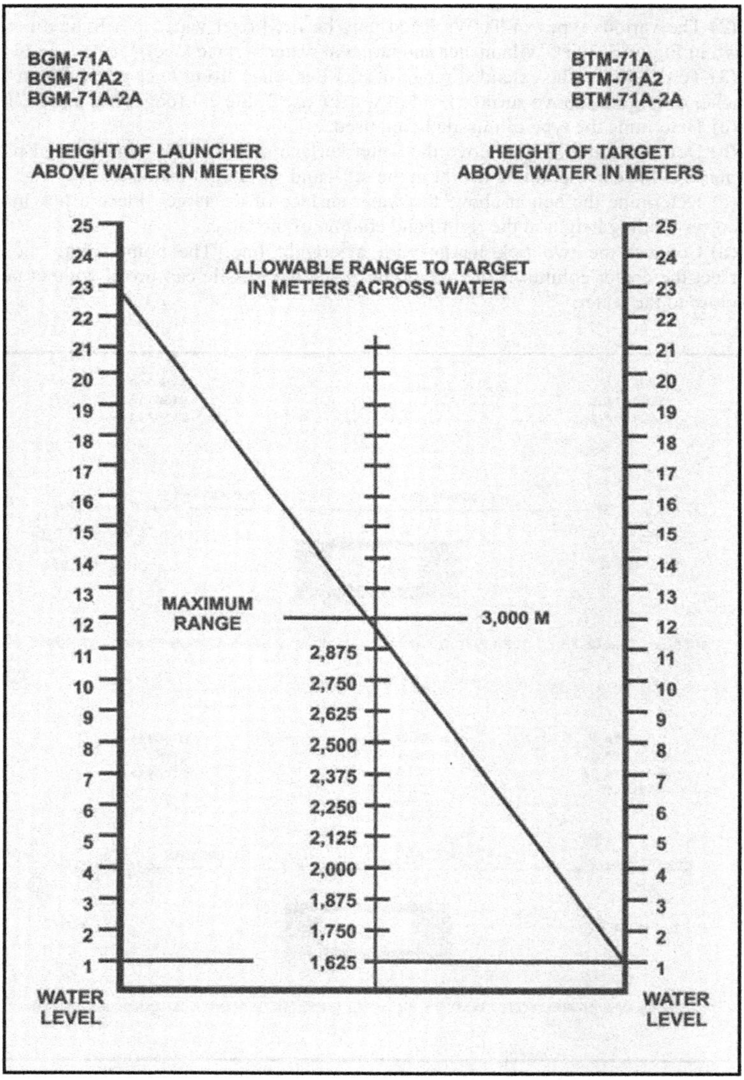

Table 2-1. Extended range over water.

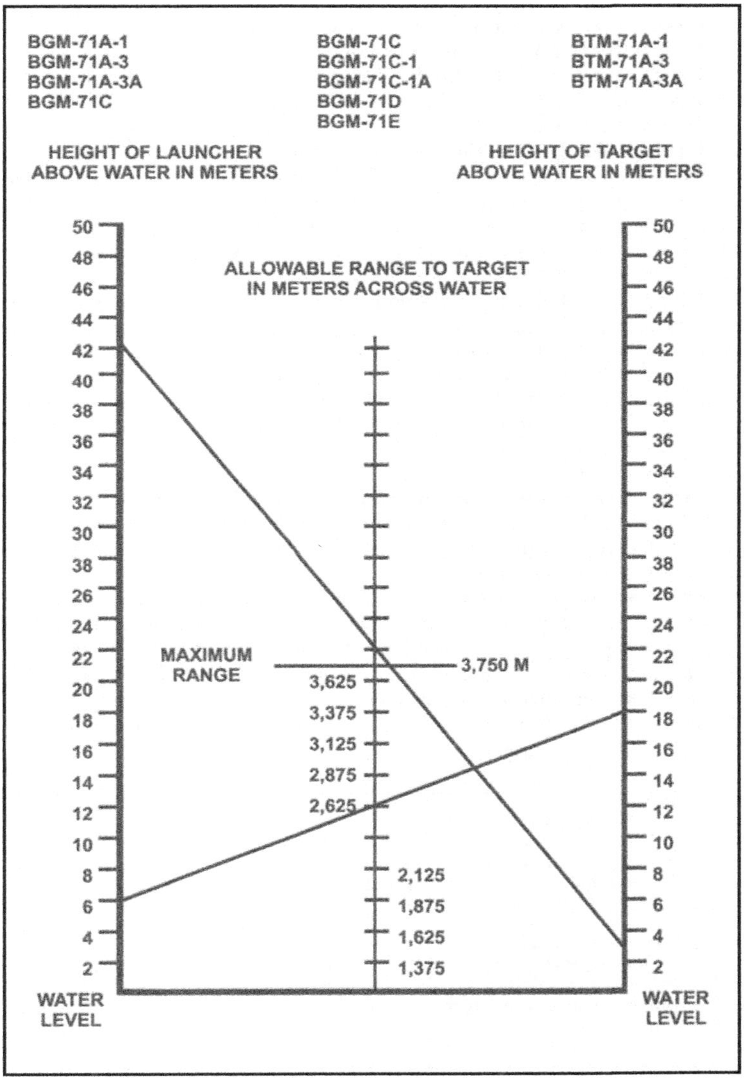

Table 2-1. Extended range over water (continued).

b. **Firing Over Electrical Lines.** If the command-link wires contact a live high-voltage power line, personnel can be injured or control of the missile can be lost. The launcher electronics may also be damaged. In addition to power lines, other high-voltage sources include street cars, electric train ways, and some moving target trolleys on training ranges.

c. **Firing in Windy Conditions.** Gusty, flanking, or quartering winds can cause the launch tube to vibrate and spoil the tracking performance. The effect is similar to driving in a strong crosswind. Erecting a windscreen next to the launcher helps to reduce this problem. Strong winds can move the missile around during flight, but as long as the crosshairs are kept on the center mass of the target, the weapon system itself can compensate for wind effects.

d. **Firing Through Smoke and Area Fires.** Smoke can obscure the line of sight and hide the target when using the daysight tracker. A smooth tracking rate should be maintained as the target disappears into a smoke cloud so that the missile will still be on target or very close as the vehicle goes out the other side of the smoke cloud. (This technique should be practiced during field tracking exercises.) A fire can burn through the command-link wire, causing loss of control of the missile. The gunner should avoid firing through fire and over fires if there is a possibility that the wires will contact the fire before missile impact.

e. **Firing from Bunkers and Buildings.** In accordance with DA Pam 385-63, TOWs will not be fired from buildings, bunkers, or within 100 meters of a vertical or nearly vertical backstop without the approval of the commanding general. (See Appendix E for further safety precautions.)

f. **Clearance Requirements.** The TOW muzzle must have at least nine inches of clearance at the end of the launch tube so the wings and control surfaces of the missile will not be damaged when they extend after clearing the launch tube. The muzzle of the launch tube must extend beyond any enclosure, window sill, or aperture, and at least 30 inches of clearance must be between the line of sight and any obstruction from 500 to 900 meters downrange. A 30-inch line-of-sight clearance ensures a high probability the missile will not strike the ground on the way to the target (Figure 2-3).

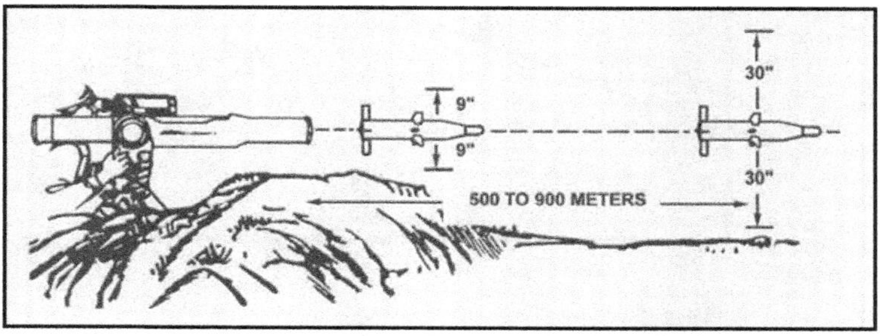

Figure 2-3. Clearance requirements.

2-7. DESTRUCTION OF TOW WEAPON SYSTEMS

When subject to capture in the combat zone, the encased missile and weapon system and its components are destroyed by the TOW squad. This action is taken when, in the judgment of the unit commander, the tactical situation warrants it, and when it is necessary IAW orders or policy established by the commander.

NOTE: The following information is for guidance only. Certain procedures outlined require the use of explosives and incendiary grenades that may not be authorized items for the TOW squad. The issue of these and related materials and the conditions for destruction are command decisions.

Destruction of TOW components by mechanical means, explosives, gunfire, or burning renders them useless to the enemy. The best method to destroy everything is to set fire to the carrier. To prevent the enemy from obtaining useful equipment, the encased missiles and certain weapon system parts should be destroyed. When lack of time or personnel prevents destruction of all parts, the encased missiles and the priority parts of each weapon system must be destroyed. Destroying the same parts of each weapon system prevents the enemy from assembling a complete weapon system. Priority is used to select the parts for destruction. If destruction is directed, the TOW squad selects a point of destruction that will cause the greatest obstruction to enemy movement and not create a hazard to friendly troops.

Each organization and installation that uses, maintains, or stores the encased missiles or weapon system parts should have an SOP for TOW destruction. The procedure should contain priorities of destruction, methods of destruction, quantities of explosives required, and instructions for destruction. The destruction plan should be flexible to cover any situation.

a. **Destruction of Encased Missiles.** The encased missile is a priority item and should be destroyed by any of the following methods. The expended launch containers should also be destroyed or removed from the battlefield so that enemy troops may not use them to make improvised munitions and booby traps.

WARNING

Do not attempt to destroy encased missiles by mechanical methods.

(1) ***Launch the Missiles.*** If the system is operational and time permits, launch the missiles into enemy territory before destroying the vehicle. Missile detonation should hinder the enemy in every way possible without endangering friendly units or equipment.

(2) ***Use Explosives.*** To destroy the TOW using explosives, become familiar with the provisions of FM 5-250.

(a) Using three pounds of TNT or equivalent, prepare and place a 1-pound explosive charge 6 to 8 inches from the nose end, a 1-pound explosive charge 4 to 6 inches behind the electrical connector, and a 1-pound explosive charge 6 to 8 inches from the motor end of the missile (Figure 2-4, page 2-10).

NOTE: Because the missile contains high explosive, a smaller explosive charge may be used to destroy it.

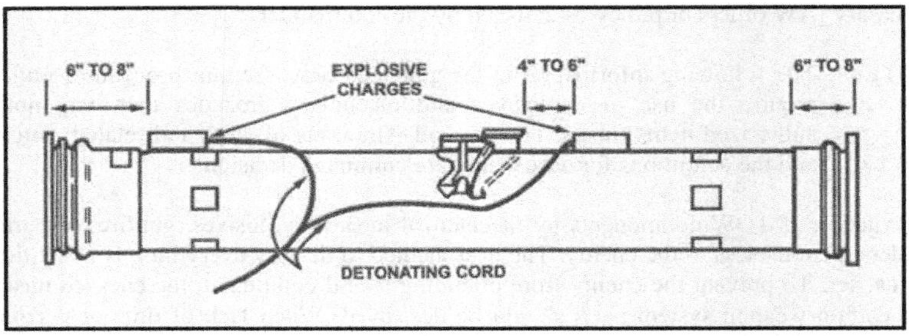

Figure 2-4. Explosive charges on an encased missile.

(b) Determine if electrical blasting caps and wire or nonelectric blasting caps and safety fuses will be used for priming and detonating the explosive charges. If nonelectric caps are used, crimp them to a safety fuse at least 2 meters long.

WARNING

Safety fuse burns at the rate of 1 foot in 30 to 40 seconds. Safety fuse, which contains black powder, and blasting caps must be protected from moisture at all times.

(c) Connect the charges with detonating cord to produce a simultaneous detonation. Dual prime the charges to reduce the possibility of a misfire.

WARNING

The blasting caps, detonating cord, and safety fuses must be kept separate from the charges until needed.

(d) If the charges are primed with nonelectric blasting caps, ignite the safety fuses and take cover at once. If the charges are primed with electric blasting caps, take cover before firing.

(3) ***Use Gunfire.*** Use well-aimed shots from artillery, rockets, rifle grenades, or tank guns to destroy the encased missile by gunfire.

> **WARNING**
> **The live rocket motors and high-explosive antitank warhead are a hazard to soldiers firing at the encased missile. Therefore, small arms should not be used. Firing artillery from ranges of 1,000 meters or less must be done from cover. Cover must also be used when rifle grenades or rockets are fired.**

b. **Destruction of Weapon Systems.** Destruction of TOW parts should be done so that the use or restoration to serviceable condition is impossible. In addition, certain weapon system parts could provide important information to the enemy.

(1) ***Priority of Destruction.*** If time does not permit the destruction of the complete weapon system, destroy as many parts as possible in the following priority:

- Daysight tracker.
- Nightsight.
- Missile guidance set.
- Traversing unit.
- Battery assembly.
- Launch tube.
- Tripod.
- Battery power conditioner (TOW 2).

(2) ***Destruction Methods.*** Consider the location of friendly units and the availability of personnel, materials, and time required for destruction when selecting the method of destroying TOW parts. The following list of destruction methods is in priority sequence; however, the considerations stated above could change the priority or method of destruction used.

(a) *Explosives.* Use at least one pound of explosive on each part. Place the explosive inside the part, if possible, to cause the greatest destruction. For example, remove the coil cable and place the charge in the traversing unit, or remove the battery assembly from the MGS, place the charge in the battery well, and replace the battery. After placing the charge, cover the parts with filled sandbags or other heavy materials to increase the explosive effect.

(b) *Gunfire.* Use well-aimed shots from artillery, rockets, tank guns, rifle grenades, rifle fire, or other small-arms fire to make the weapon system useless to the enemy. When using small-arms fire, aim for critical parts. The vehicle can be destroyed by artillery or tank guns from a range of 1,000 meters.

(c) *Mechanical.* Smash the TOW parts with axes, picks, crowbars, or rocks, or drive over each part with a tracked vehicle (make sure the parts are on a hard surface or they must be dug up to ensure they have been made useless). Destroy the priority parts if enough time and personnel are available.

WARNING
Do not try to destroy the battery assembly by mechanical methods. An electrical safety hazard may result if the power output terminals are short-circuited. Discard the battery assembly if no other means of destruction is available.

(d) *Burning*. After maximum damage has been done to the weapon system and vehicle parts by other destruction methods, and if time permits, set fire to the resulting debris. Use vehicle fuels and lubricants to aid burning. Place one or more incendiary grenades on each part.

CHAPTER 3

M966-MOUNTED TOW/TOW 2

The M966-mounted TOW is a one-vehicle (1 1/4-ton truck) combat system that is air-transportable, versatile, maintainable, and survivable (Figure 3-1). Its 16-inch ground clearance, four-wheel independent suspension, steep approach and departure angles of 60-percent gradation, side slope of 40 percent, and 30-inch (without kit, 60-inch with kit) water-fording capability provides the off-road mobility and speed needed in combat. The vehicle carries one complete launcher system, six encased missiles, and a three-man crew. This chapter discusses the duties of the crew and the detailed operations of the weapon system.

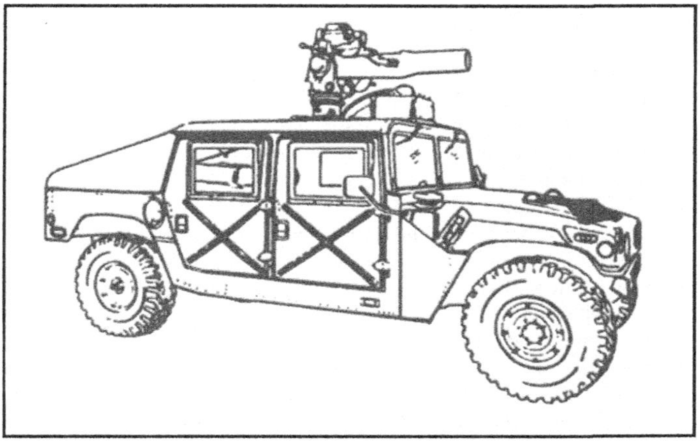

Figure 3-1. M966-mounted TOW.

Section I. INDIVIDUAL DUTIES AND RESPONSIBILITIES

The three-man crew consists of the squad leader, the gunner, and the driver-loader. The duties and responsibilities of each crew member are discussed in the following paragraphs.

3-1. SQUAD LEADER

The squad leader is responsible for the combat readiness and tactical employment of his squad. This includes discipline, health and welfare, training, and maintenance of equipment and weapons. Other specific duties include the following:

- Directing the driver over the exact route.
- Detecting and identifying targets.
- Issuing fire commands.

- Controlling fire and movement.
- Supervising and assisting in dismounting and assembly of the M220-series TOW launcher.
- Employing the squad according to the orders of the section leader.
- Selecting primary, alternate, and supplementary firing positions.
- Requesting squad resupply.
- Informing the section leader of events that affect the tactical situation and of the status of his crew.
- Carrying the tripod, launch tube, coolant cartridges, battery case or BPCs, and binoculars when the TOW is used in the ground mode.
- Acting as the gunner when needed.
- Assisting the gunner in system check-out procedures.
- Supervising construction of the TOW firing position.
- Determining the direction of fire.

3-2. GUNNER

Specific duties of the gunner include the following:

- Conducting system check-out procedures.
- Acquiring, recognizing, and identifying armored vehicles.
- Determining if a target can be engaged.
- Engaging targets.
- Recognizing and eliminating firing-angle limitations.
- Adjusting the gunner's platform before stowing the MGS.
- Stowing the TVPC in the MGS and stowing the MGS, nightsight, and traversing unit.
- Preparing range cards for the TOW.
- Removing the nightsight, daysight tracker, launch tube, traversing unit, and MGS when the TOW is used in the ground mode.
- Carrying the MGS and daysight tracker to the firing position when the TOW is used in the ground mode.

3-3. DRIVER-LOADER

The driver-loader is a dual-purpose position in the M966 TOW squad. Specific duties of the driver-loader include the following:

- Driving the M966.
- Maintaining the vehicle.
- Camouflaging the vehicle.
- Carrying the traversing unit, encased missile, nightsight, and collimator to the firing position when the TOW is used in the ground mode.
- Securing the area.
- Stowing battery assembly, collimator, encased missiles, daysight tracker, spare battery case, coolant cartridges and batteries (or BPCs), and tripod on the M966 with the assistance of the squad leader.
- Maintaining radio equipment.
- Removing misfired missiles.

Section II. OPERATION PROCEDURES

Before using the M966-mounted TOW/TOW 2, the squad must install the M220-series TOW launcher and encased missiles and must prepare the M966 for TOW firing. (To install an M220A1 TOW launcher and encased missiles on an M966, see TM 9-1425-472-12. To install an M220A2 TOW launcher and encased missiles on an M966, see TM 9-1425-450-12.)

3-4. EQUIPMENT STORAGE

The TOW system components are stored in the interior of the vehicle (Figure 3-2 [M220A1], page 3-4 and Figure 3-3 [M220A2], page 3-5).

a. Six encased missiles are stowed in the missile racks on the left rear of the cargo area.

b. The traversing unit is stowed on the traversing unit adapter behind the right front passenger seat.

c. The MGS is stowed on the gunner's platform between the left and right rear passenger seat.

d. The nightsight and collimator are stowed on the right cargo shelf.

e. The launch tube is stowed on the right side of the cargo shell door.

f. The BPC and spare battery pack, or the nightsight coolant cartridges and the 4.8-volt nickel cadmium battery, are stowed in the floor bracket next to the right wall of the cargo shelf.

g. The daysight tracker is stowed in the floor bracket next to the missile rack.

h. The battery assemblies are stowed in the battery rack above the missile rack.

i. The tripod is stowed retracted, lying across the rear of the cargo area.

j. The TVPC is stowed in the MGS.

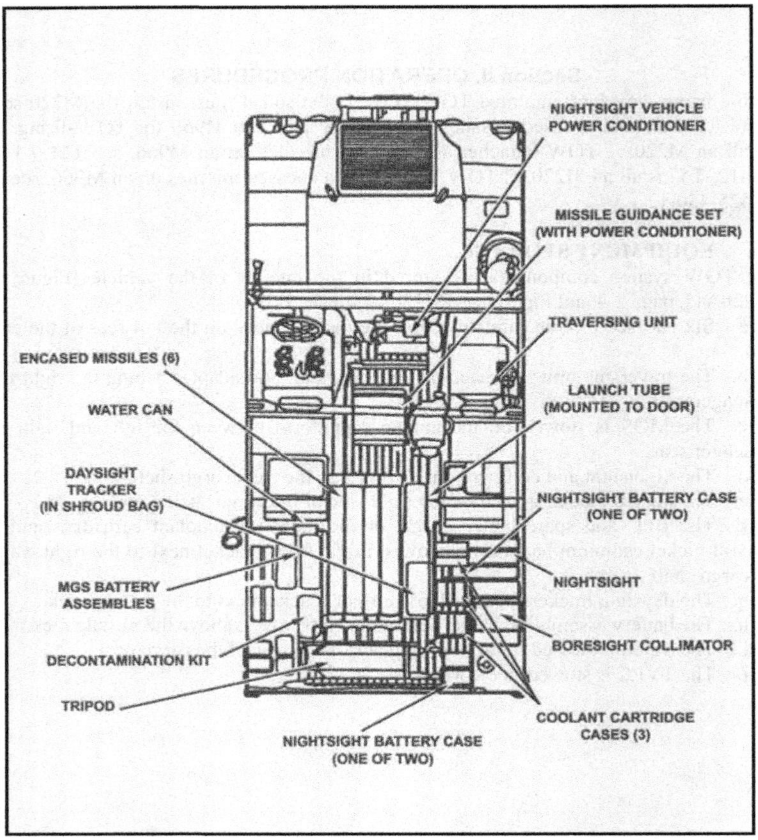

Figure 3-2. Stowing M220A1 on M966.

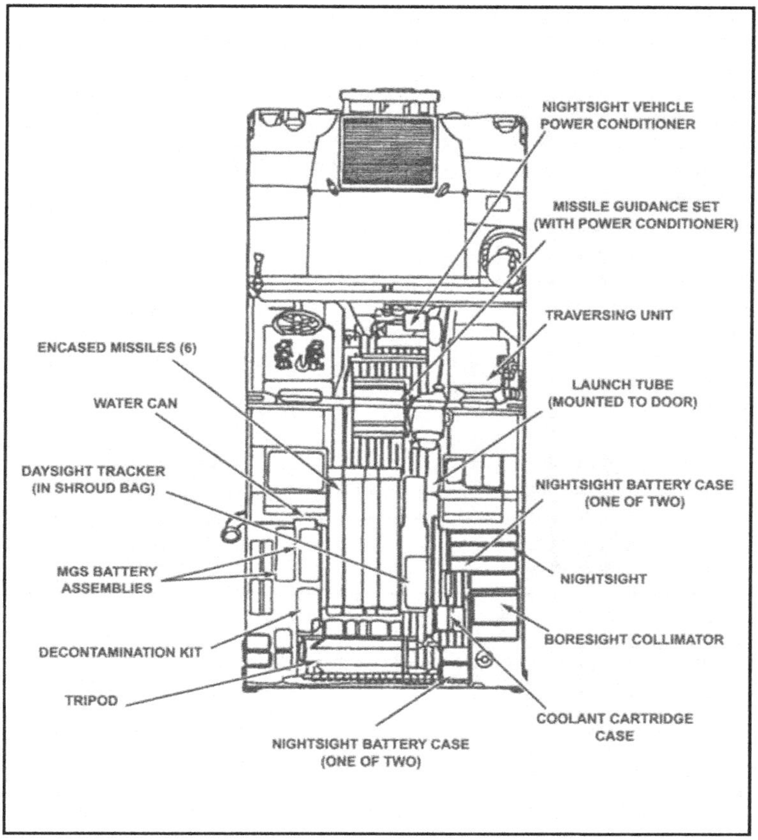

Figure 3-3. Stowing M220A2 on M966.

3-5. HMMWV INTERCHANGEABLE MOUNT SYSTEM

The HMMWV Interchangeable Mount System (HIMS) is used to convert an M966 TOW HMMWV carrier into an automatic weapons platform (Figure 3-4, page 3-6). The HIMS provides units equipped with M966 TOW HMMWV carriers with low cost, quick, and effective options for tailoring forces for contingency missions. The HIMS contains two parts. (The HIMS technical data package contains the information needed to operate the HIMS and where to find the materials needed. TM 9-2320-280-10 contains the combat loading plans for the vehicle.)

a. Part one is an armament mount panel assembly with pintle that mounts on the M1025 armament HMMWV carrier. This part is interchangeable with the missile guidance tray on the M966 TOW HMMWV carrier. The common turret ring of the two vehicles will accommodate the armament mount panel assembly using existing holes. No modifications are required.

b. Part two consists of a locally fabricated internal floor stowage plate that has automatic weapons stowage brackets, ammunition stowage trays, and footman tie-down straps already mounted for quicker emplacement. This plate is made of sheet aluminum and is mounted in the floor of the M966 TOW HMMWV carrier after removal of selected items of the TOW bracketry.

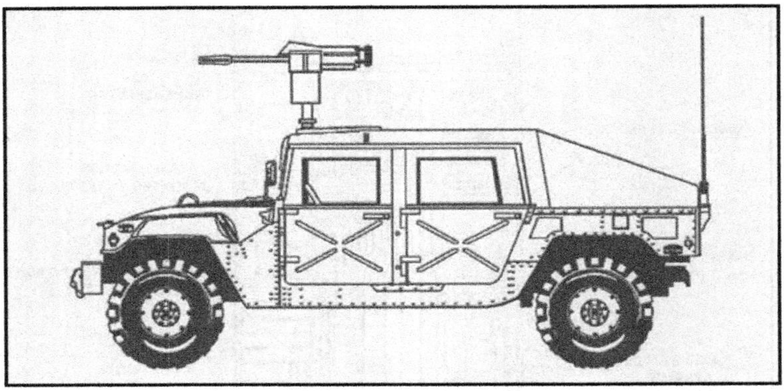

Figure 3-4. M966 with HIMS.

3-6. PREPARATION OF AN M966 FOR TOW FIRING

The three-man crew prepares the M966 for TOW firing by performing the following:

a. The gunner, driver-loader, and squad leader release the latches securing the hatch cover (Figure 3-5). The driver-loader enters the right rear passenger door.

b. The gunner removes the MGS from its stowed position.

(1) The gunner unlatches the retaining strap securing the MGS.

(2) The gunner releases the latches to remove the MGS cover and hands the cover to the squad leader.

(3) The squad leader stows the MGS cover in the front passenger's footwell of the vehicle.

(4) The gunner tilts the MGS to the vertical position on the gunner's platform and releases the lower coupling clamp on the traversing unit adapter.

c. The gunner positions the weapon station. The gunner opens the hatch and enters the weapon station, secures the hatch with the retaining latch, and rotates the weapon station by pulling up on the brake handle so that the backrest is toward the rear of the vehicle. To lock the weapon station, the gunner pushes down on the brake handle.

d. The gunner lifts the MGS through the weapon station, places it on the MGS tray, and secures it with the straps (Figure 3-6, page 3-8).

e. The gunner installs the traversing unit.

(1) The gunner opens the pedestal mount cover and secures it (Figure 3-7, page 3-8).

(2) The gunner lifts the traversing unit with adapter attached from the stowage base and secures the traversing unit to the pedestal mount with the coupling clamp ensuring that the traversing unit azimuth lock is toward the rear of the vehicle.

(3) The squad leader pulls the coil cable out of its retaining cup and hands it to the gunner (Figure 3-8, page 3-9).

f. The gunner connects the coil cable connector by aligning the yellow indexing line on the coil cable connector with the yellow indexing line on the J1 connector on the MGS (Figure 3-9, page 3-9). He turns the nut on the coil cable connector in the direction of the arrow until the red mark (band) on the M1 connector on the MGS cannot be seen. (Do not apply downward pressure on the coil cable connector. This can cause misalignment and bent pins.)

CAUTION
Ensure that no dirt is in the end of the coil cable connector. Dirt can cause a bad connection with the MGS, resulting in poor system operations.

NOTE: On the MGS for the basic TOW, push down to seat the coil cable connector on the M1 connector and tighten the locking nut.

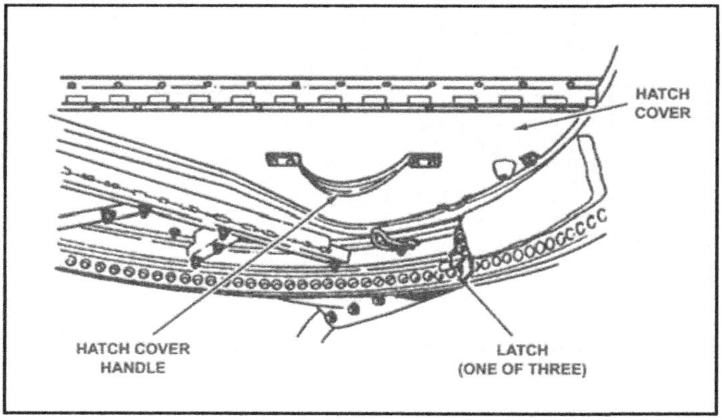

Figure 3-5. Hatch cover and latch.

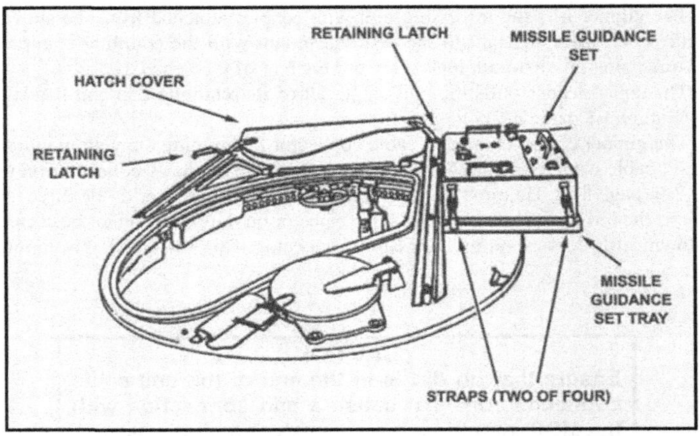

Figure 3-6. Hatch cover retaining latches and MIGS tray.

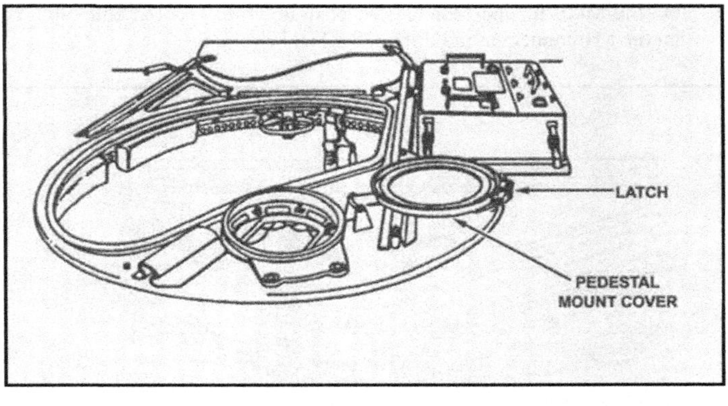

Figure 3-7. Pedestal mount cover and latch.

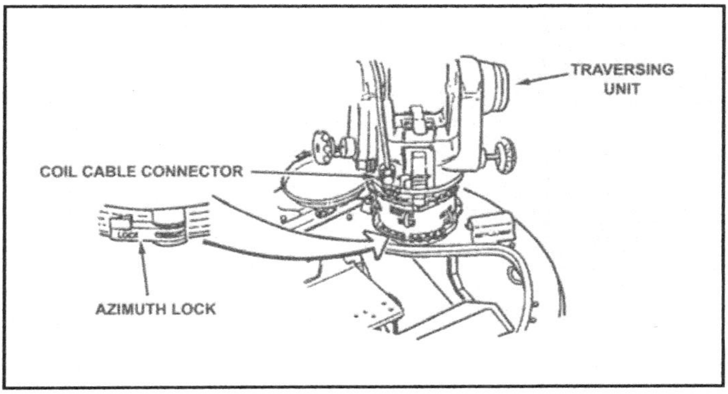

Figure 3-8. Coil cable connector.

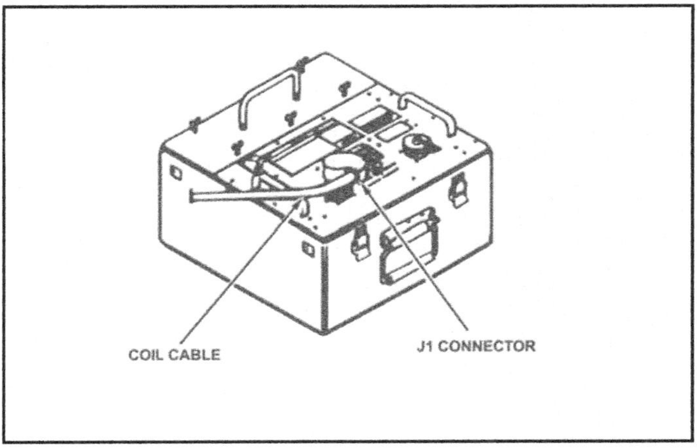

Figure 3-9. Coil cable connector and J1 connector.

g. The gunner installs the launch tube.

(1) The driver-loader uses the forward latch to open the cargo shell door and hands the launch tube to the gunner.

(2) The gunner installs the launch tube on the traversing unit and secures it with the launch tube latch (Figure 3-10, page 3-10).

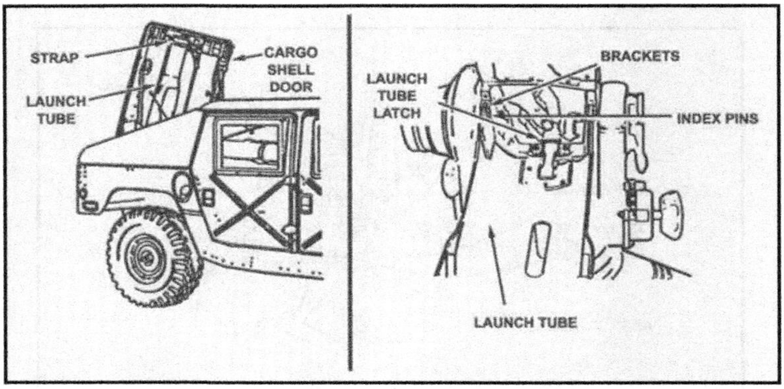

Figure 3-10. Launch tube latch.

h. The gunner installs the daysight tracker.

(1) The driver-loader hands the daysight tracker to the gunner.

(2) The gunner mounts the daysight tracker on the traversing unit boresight plate. He secures the daysight tracker by locking the latch handle (Figure 3-11).

(3) The driver-loader stows the nightsight case in its bracket.

(4) The squad leader connects connector P2 of the TVPC power cable to the power conditioner cable.

(5) The squad leader unwinds the nightsight power conditioner cable and TVPC cable and hands both to the gunner.

(6) The gunner connects the nightsight power cable 2W2 to the nightsight input power conditioner J1 (Figure 3-12, page 3-12), then connects the TVPC power cable P1 to connector J1.

(7) The squad leader turns on the TVPC.

(8) The gunner turns on the power conditioner.

(9) The gunner checks steps 10 through 15 of system self-test on TOW 2 (positions 1 and 7 on basic TOW), then collimates the nightsight.

NOTE: The limiter assembly limits elevation to 20 degrees and depression to 10 degrees. If the limiter is used, the gunner makes sure both elevation and depression limiters are in the up position and pinned (Figure 3-13, page 3-12).

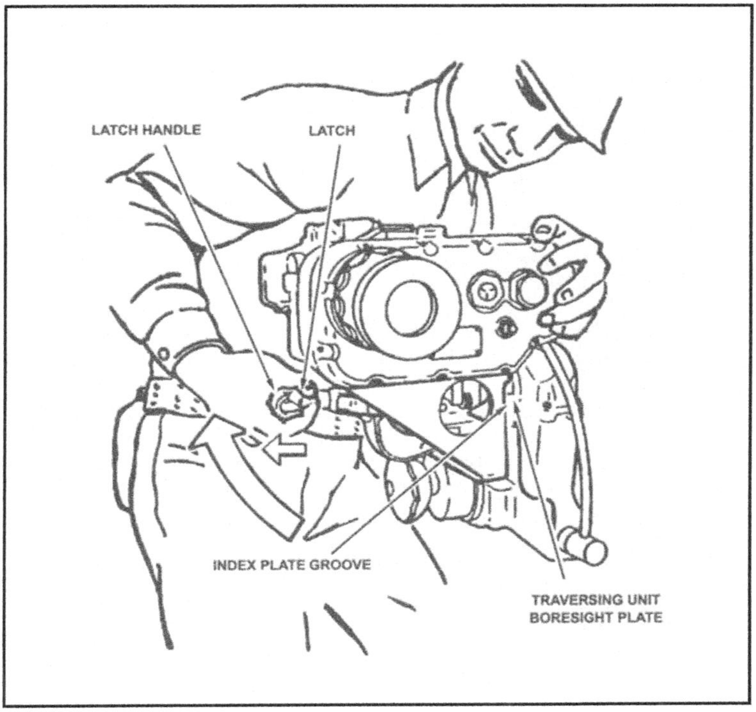

Figure 3-11. Daysight tracker on traversing unit boresight plate.

> **WARNING**
>
> **Injury to the gunner and damage to the equipment can occur if the TOW system is fired at angles greater than 20 degrees or less than -20 degrees for all firing positions except over the cargo shell door. If firing over the cargo shell door, firing angles should not be greater than 20 degrees or less than -13 degrees. Pressure waves and flying debris can cause injury to personnel and damage to equipment.**

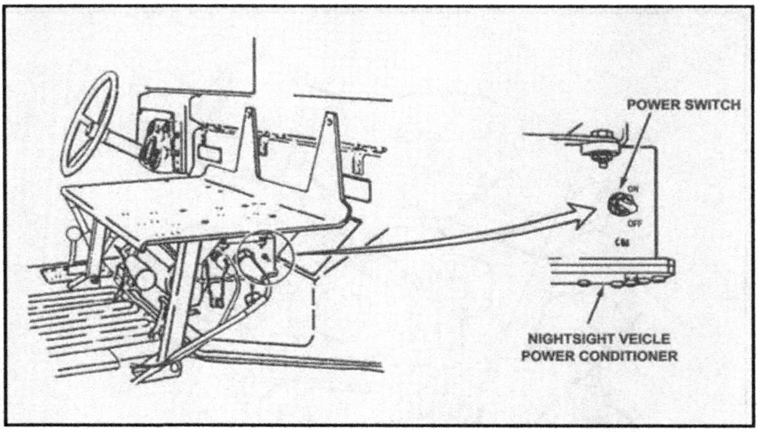

Figure 3-12. Nightsight power cable 2W2, J1 connector, and vehicle power conditioner.

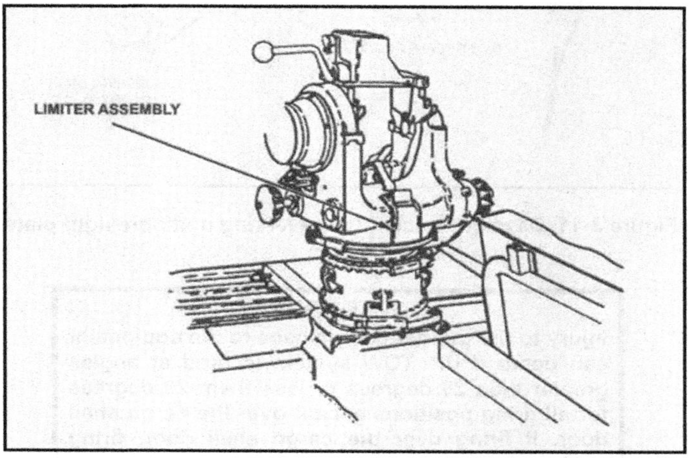

Figure 3-13. Limiter assembly location.

3-7. LOADING, ARMING, AND UNLOADING

The procedures for loading, arming, and unloading the M966-mounted TOW/TOW 2 are as follows:

a. **Load.** Missiles must be unloaded from the missile racks in numerical order. The forward handling ring, preformed packing, and quick-release clamp are saved in case the missile is not fired. If missile diaphragms get damaged while loading, the missile can still be fired. The following procedures are performed to load the M966-mounted TOW/TOW 2.

> **WARNING**
>
> **Never open one end of the cargo shell door until the opposite end is securely closed. Personnel injury, equipment damage, or mission abort will occur if both ends open at the same time.**

(1) On the traversing unit, the gunner ensures that the launch tube is locked in the 8-degree down position. He pushes the locking handle forward, then up, and opens the bridge clamp.

> **CAUTIONS**
>
> 1. Do not drop an encased missile. Handling rings or launch container may be bent or damaged.
> 2. Do not break the diaphragm on the end of an encased missile. If water gets into the launcher container, the missile can be damaged.
> 3. Do not dispose of bad missiles. If an encased missile is damaged or handling rings are badly bent, return the encased missile to the ammunition unit for inspection.
> 4. Be careful not to punch a hole in an encased missile diaphragm when the forward handling ring is removed.

(2) The driver-loader unlatches and raises the cargo shell door using the forward latch. He removes the quick-release clamp and forward handling ring from the encased missile.

> **WARNING**
>
> **Clear all personnel from the firing danger zone before proceeding with missile loading procedure.**

(a) The driver-loader turns the inside portion of the protective cover two complete turns counterclockwise, and he removes it from the electrical connector. (The protective cover is kept for future use.)

(b) He turns the encased missile so that the electrical connector is facing up and hands it to the gunner.

(3) The gunner takes the encased missile from the driver-loader. He raises the back end of the missile and slides it forward and down into the launch tube until the missile indexing lugs are firmly in place. The gunner lowers the back end of the encased missile ensuring that the electrical connector joins with the bridge clamp. He lowers the bridge clamp and pushes down on the top of it. He pulls the bridge clamp locking handle downward and backward to lock the encased missile in the launch tube.

DANGER

CARGO SHELL DOOR MUST BE CLOSED AND SECURELY LATCHED AT BOTH ENDS BEFORE FIRING A MISSILE. FIRING A MISSILE WITH THE DOOR NOT SECURE WILL CAUSE INJURY TO PERSONNEL, DAMAGE TO EQUIPMENT, OR MISSION ABORT.

b. **Arm.** The following procedures are performed to arm the M966-mounted TOW/TOW 2.

(1) The driver-loader closes the cargo shell door.

(2) The gunner makes sure the cargo shell door is securely latched at the forward end.

WARNING

Do not raise arming lever until ready to fire at selected target.

(3) The gunner checks the backblast area before raising the arming lever.

CAUTION

Do not press the trigger with an empty missile case in the launch tube. Possible shorts can result, damaging the prefire, fire, and wire-cutter circuits.

c. **Unload.** The gunner lowers the trigger cover over the trigger after the missile has impacted and locks the traversing unit in the 8-degree down position. If a missile is prepared for firing and then not fired, it must be immediately tagged to show the time, date, and using unit. If the missile is not used within 90 days of initial preparation, it must be turned in to the ammunition supply point.

CAUTION

If the missile is not fired, lower the arming lever before opening the bridge clamp. The command-link wires will be cut if the arming lever is not lowered.

3-8. TARGET ENGAGEMENT

The crew members perform the following target engagement procedures.

a. When the squad leader issues the fire command, the gunner releases the azimuth and elevation locks on the traversing unit and turns on the nightsight if necessary. If crew members other than the gunner are in or leaning on the M966, they must be very still while the gunner is tracking. Even small movements can cause the vehicle to move and adversely affect the gunner's aim.

b. The gunner acquires and begins to track the target.

(1) If the target cannot be acquired in the sight, he announces, "Lost."

(2) When the target has been acquired, he identifies it as friendly or enemy.

(3) If the target cannot be identified, he announces, "Cannot identify."

(4) If the target is identified as enemy, he places the crosshairs on the center mass of the exposed portion of the target, presses the trigger, and smoothly tracks the target until missile impact.

(5) If the target is sighted out of the sector of fire, the gunner must rotate the weapon station to engage the target after the squad leader verifies the target's engageability.

3-9. IMMEDIATE ACTION

The crew members perform the following immediate action procedures in the event of a hangfire or misfire. If the gunner presses the trigger and nothing happens after 1.5 seconds, a misfire has occurred. If he presses the trigger and the pop of the explosive squib on the gyroscope is heard but the missile does not launch after 1.5 seconds, a hangfire has occurred.

DANGER

THE CREW MUST USE EXTREME CARE TO AVOID INJURY FROM A DELAYED MISSILE LAUNCH. A DELAYED LAUNCH USUALLY OCCURS WITHIN 1 MINUTE. HOWEVER, IT COULD OCCUR AS LONG AS 30 MINUTES AFTER PRESSING THE TRIGGER. KEEP ALL PERSONNEL OUT OF FIRING DANGER ZONES.

DANGER
DO NOT OPEN THE CARGO SHELL DOOR DURING FAILURE TO LAUNCH PROCEDURES. SHOULD THE MISSILE FIRE WITH THE CARGO SHELL DOOR OPEN, SEVERE INJURY TO PERSONNEL OR DAMAGE TO EQUIPMENT WILL RESULT.

a. For a *hangfire*, the gunner alerts the crew by announcing, "Hangfire" and continues to track the target for one minute.

(1) If the missile fails to launch after one minute, he closes the trigger cover, lowers the arming lever and locks the elevation lock and the azimuth lock, ensuring that the launch tube is locked in the 8-degree down position (Figure 3-14).

NOTE: In a training situation, the crew dismounts the vehicle, closes the doors, calls EOD, and waits 30 minutes. The gunner may then remount the vehicle.

DANGER
IF THE VEHICLE IS NOT POSITIONED WITH ITS FRONT OR REAR IN LINE WITH THE TARGET, IT MUST BE REPOSITIONED SO THAT EITHER THE FRONT OR REAR OF THE VEHICLE IS IN LINE WITH THE TARGET. THE WEAPON STATION MUST BE ROTATED DURING VEHICLE REPOSITIONING SO THE FORWARD END OF THE MISSILE IS POINTING TOWARD THE TARGET AT ALL TIMES.

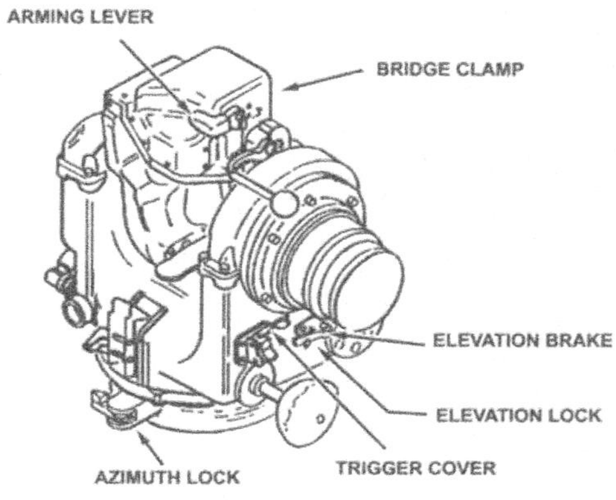

Figure 3-14. Trigger cover, arming lever, elevation lock, and azimuth lock.

(2) He opens the bridge clamp (Figure 3-15, page 3-18), removes the missile from the launch tube (keeping the missile pointed downrange at all times), and hands the missile over either side of the vehicle to the driver-loader (Figure 3-16, page 3-18).

WARNING

Do not hand off a malfunctioned missile to the front or rear of the vehicle. A malfunctioned missile must be handed off from the vehicle to either the left or right side.

(3) The driver-loader, keeping the missile pointed downrange, places the missile on the ground a safe distance from the vehicle. (In training, he places the missile on the ground at least 200 meters from the vehicle and marks the location.)

b. For a *misfire,* the gunner continues to track the target for one minute and presses the trigger again. He alerts the crew by announcing, "Misfire" and continues to track the target.

(1) If the missile still fails to launch, he disarms the missile by closing the trigger cover, lowers the arming lever, and locks the azimuth and elevation locks ensuring that the launch tube is locked in the 8-degree down position.

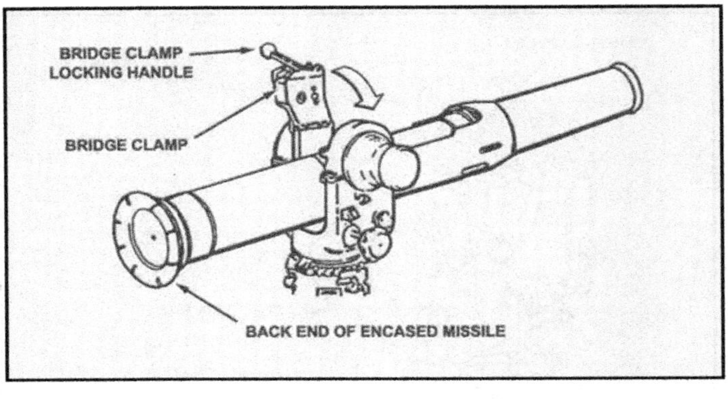

Figure 3-15. Bridge clamp, missile, and launch tube.

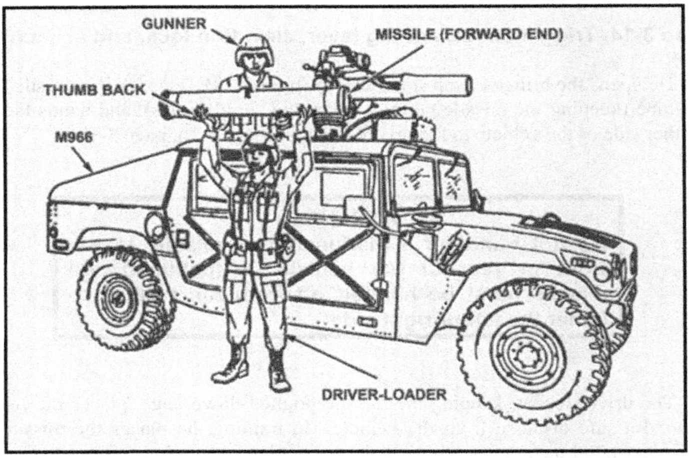

Figure 3-16. Driver-loader's position for receiving the missile from the gunner.

(a) The gunner presses the TEST-OPERATE switch to check the battery power. If he receives an OUT-OF-BAND/BATTERY FAIL, he checks to ensure that the TVPC switch is in the ON position and that the P1 cable is securely mated to the container. If all cables are connected securely, he replaces the power conditioner with a battery and checks the battery (Figure 3-17).

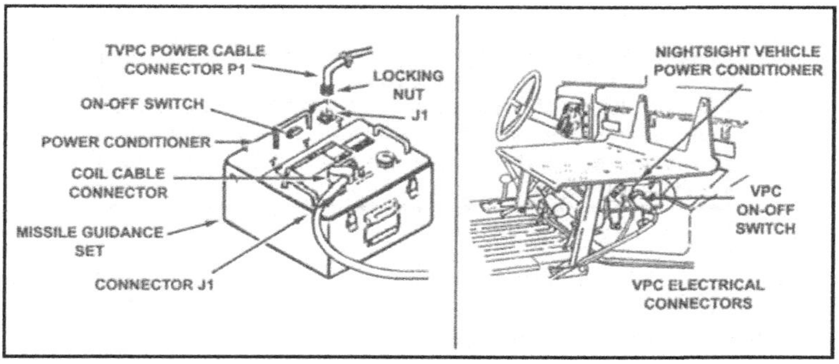

Figure 3-17. MGS, TVPC cable connector P1, power conditioner.

(b) He checks the coil cable connection to J1 connector. If it is not connected properly, he corrects it.

(c) He ensures the bridge clamp locking handle is properly seated, unlocks the azimuth and elevation locks, and raises the arming lever.

(d) He ensures the electrical connector is properly mated, raises the trigger cover, and presses the trigger.

(2) If the missile still fails to launch, the gunner may continue tracking the target for one minute (at the gunner's discretion in combat).

(a) He closes the trigger cover, lowers the arming lever, and locks the azimuth and elevation locks, ensuring the launch tube is locked in the 8-degree down position.

NOTE: In a training situation, the crew dismounts the vehicle, closes the doors, calls EOD, and waits 30 minutes. The gunner may then remount the vehicle.

DANGER

IF THE VEHICLE IS NOT POSITIONED WITH ITS FRONT OR REAR IN LINE WITH THE TARGET, REPOSITION IT SO THAT EITHER THE FRONT OR REAR OF THE VEHICLE IS IN LINE WITH THE TARGET. ROTATE THE WEAPONS STATION DURING VEHICLE REPOSITIONING SO THE FORWARD END OF THE MISSILE IS POINTING TOWARD THE TARGET AT ALL TIMES.

(b) The gunner opens the bridge clamp, removes the missile from the launch tube (keeping the missile pointed downrange at all times), and hands the missile over either side of the vehicle to the driver-loader.

WARNING
Do not hand off malfunctioned missile to the front or rear of the vehicle. A malfunctioned missile must be handed off from the vehicle to either the left or right side.

(3) The driver-loader, keeping the missile pointed downrange, places the missile on the ground a safe distance from launcher. (In training, he places the missile on the ground at least 200 meters from the launcher and marks the missile location.)

3-10. M220A1 DISMOUNT

Specific actions must be followed when dismounting the M220A1 and moving it to the firing site. All actions happen simultaneously.

a. **Dismount Actions.** Each crew member begins executing his assigned responsibilities as soon as the squad leader commands, DISMOUNT THE TOW.

(1) ***Squad Leader's Actions***. The squad leader—

(a) Turns the VPC to the OFF position, places the MGS cover on the driver-loader's seat, and places the coil cable in its retaining cup on the traversing unit.

(b) Dismounts and goes to the rear of the vehicle; opens the cargo shell door using the rear latch; lowers the tailgate; removes the tripod, battery case, and coolant cartridge case and places them on the ground beside the launch tube (on left front side of vehicle in front of the MGS and the daysight tracker); closes the tailgate and cargo shell door.

(c) Enters the right rear passenger door, gets the collimator case and the nightsight case, dismounts the vehicle, and places the collimator case and the nightsight case on the ground beside the vehicle.

(d) Carries the tripod, the coolant cartridge case, the battery case, and the launch tube to the firing site.

(2) ***Gunner's Actions*** (assisted by driver-loader when needed). The gunner—

(a) Disconnects the coil cable from the MGS; disconnects the VPC and TVPC cables.

(b) Removes the nightsight and hands it to the driver-loader who places it on the ground. Removes the daysight tracker and hands it to the driver-loader who places it on the ground in front of the left side of the vehicle. (The nightsight should be put into its case and placed behind daysight tracker.)

(c) Removes the launch tube and hands it to the driver-loader who places it on the ground in front of the daysight tracker.

(d) Unlocks the top coupling clamp on the traversing unit adapter; removes the traversing unit (minus adapter); hands the traversing unit to the driver who places it on the ground behind the daysight tracker.

(e) Removes the TVPC from the MGS and places it to his rear on top of the vehicle within reach of the cargo hatch (see next step); unstraps the MGS and hands it to the driver-loader who places it on the ground beside the daysight tracker.

(f) Opens the cargo shell door by using the forward latch; switches the TVPC for the TOW battery assembly, and hands the battery assembly to the driver-loader to install in the MGS.

(g) Unstraps a missile and hands it to the driver-loader who places the missile on the ground behind the traversing unit. The gunner then closes the cargo shell door and dismounts the vehicle.

(h) Carries the daysight tracker and the MGS to the firing site.

(3) ***Driver-Loader's Actions***. The driver-loader—

(a) Dismounts the vehicle and waits to receive equipment from the gunner.

(b) Aligns equipment as described in paragraphs 2a through 2g; installs the battery assembly in the MGS and then replaces the MGS cover on the MGS.

(c) Carries the traversing unit on the first trip, the missile on the second trip, and the nightsight and boresight collimator on the third trip to the firing site.

b. **Firing Position Actions.** Upon reaching the firing position, the crew proceeds with the following actions.

(1) ***Squad Leader's Actions.*** The squad leader—

(a) Places the coolant cartridge case, battery case, and launch tube on the ground. Sets up the tripod with the coupling clamp handle facing the direction of fire and unlocked.

(b) Pulls the coil cable out of the traversing unit retaining cup, installs the launch tube, and observes downrange for enemy targets.

(c) Issues fire commands.

(2) ***Gunner's Actions***. The gunner—

(a) Places the MGS on the ground next to the tripod and installs the daysight tracker on the traversing unit.

(b) Removes the MGS cover and connects the coil cable to the MGS.

(c) Installs the nightsight and turns the actuator switch to the ON position.

(d) Conducts the system self-test, boresight, and collimation.

(e) Engages the target upon receipt of the fire command.

(3) ***Driver-Loader's Actions***. The driver-loader—

(a) Installs the traversing unit on the tripod.

(b) Returns to the vehicle for the missile; places the missile on the ground at the firing position to the right side of the weapon system.

(c) Returns to the vehicle to get the nightsight case and the boresight collimator case; returns to the firing position and hands the nightsight to the gunner.

(d) Places the collimator on the ground, removes it from its case, and installs it on the nightsight. Attaches the battery to the collimator.

(e) Removes the collimator from the nightsight; removes the battery from the collimator; places both the collimator and the battery back in their cases (if time permits).

(f) Loads the missile and arms it if commanded.

3-11. M220A1 REMOUNT

The following actions occur at the same time, when the squad leader commands, CEASE TRACKING, OUT OF ACTION.

a. **Firing Position Actions.** The following actions take place at the firing position before returning to the vehicle.

(1) ***Squad Leader's Actions***. The squad leader—

(a) Removes the launch tube and places it on the ground.

(b) Folds the tripod legs and commands MOVE OUT. (The squad leader remains at the firing position until the driver-loader returns and all equipment is en route to the vehicle.)

(c) Carries the coolant cartridge case, tripod, battery case, and launch tube back to the vehicle. Places the launch tube on the ground on the driver's side of the vehicle and the rest of the equipment on the ground behind the vehicle.

(2) ***Gunner's Actions***. The gunner—

(a) Disconnects the coil cable.

(b) Replaces the nightsight lens cover, removes the nightsight, and places it on the ground near the driver-loader.

(c) Removes the daysight tracker and picks up the MGS; carries them to the vehicle.

(d) Places the daysight tracker and the MGS on top of the vehicle and enters the gunner's station.

(3) ***Driver-Loader's Actions***. The driver-loader—

(a) Removes the encased missile (or empty launch container if the missile was fired).

(b) Removes the traversing unit from the tripod, carries it to the vehicle, and hands it to the gunner when the gunner reenters the gunner's station.

(c) Returns to the firing position to get the nightsight and collimator; returns to the vehicle and hands the nightsight to the gunner; sets the collimator and the nightsight case on the ground near the rear of the vehicle.

(d) Returns to the firing position to get the encased missile (if a missile was not fired) and carries it to the vehicle. Hands the missile to the gunner.

b. **Remount Actions.** The following actions are performed to remount the vehicle.

(1) ***Squad Leader's Actions***. The squad leader—

(a) Stows the coolant cartridge, battery case, tripod, nightsight case, and collimator on the vehicle.

(b) Hands the TVPC and VPC cables to the gunner. Enters the squad leader's seat in the vehicle and turns the VPC to the ON position.

(c) Places the MGS cover on the floorboard.

(d) Assists other crew members as needed.

(2) ***Gunner's Actions***. The gunner—

(a) Receives the traversing unit from the driver-loader, installs it on the traversing unit adapter, and secures it with the top coupling clamp.

(b) Installs the MGS in the MGS tray and secures it with the straps; hands the MGS cover to the squad leader; connects the coil cable.

(c) Installs the daysight tracker and nightsight, and connects the VPC.

(d) Moves into the cargo area and places the unfired missile in the missile rack (if necessary).

(e) If time permits, conducts the system self-test, boresight, and collimation; exchanges the battery assembly for the TVPC; connects the TVPC cable.

(3) ***Driver-Loader's Actions.*** The driver-loader assists the squad leader and gunner as needed. He closes the tailgate and cargo shell door at the end of the drill.

NOTE: If a missile was fired, the empty launch container is destroyed.

3-12. M220A2 DISMOUNT (READY-TO-FIRE CONFIGURATION)

Specific actions must be followed when dismounting the M220A2 and moving it to the firing site. All actions happen simultaneously.

NOTE: Infantry units using the M966 conduct many types of missions. Therefore, during dismount operations, all equipment must be properly stored before leaving the vehicle.

a. **Dismount Actions.** The following actions occur at the same time when the squad leader commands, DISMOUNT THE TOW.

(1) ***Squad Leader's Actions***. The squad leader—

(a) Turns the VPC to the OFF position, places the MGS cover on the driver-loader's seat, and places the coil cable in its retaining cup on the traversing unit.

(b) Dismounts and goes to the rear of the vehicle; opens the cargo shell door using the rear latch; lowers the tailgate; removes the tripod and the BPC and places them on the ground beside the launch tube (on the left front side of the vehicle in front of the MGS and the daysight tracker); closes the tailgate and cargo shell door.

(c) Enters the right rear passenger door, gets the collimator case and the nightsight case, dismounts the vehicle, and places the collimator case and the nightsight case on the ground beside the vehicle.

(d) Carries the tripod, BPC, and launch tube to the firing site.

(2) ***Gunner's Actions*** (assisted by driver-loader when needed). The gunner—

(a) Disconnects the coil cable from the MGS and disconnects the VPC and TVPC cables.

(b) Removes the nightsight and hands it to the driver-loader who places it on the ground. Removes the daysight tracker and hands it to driver-loader who places it on the ground in front of the left side of the vehicle. (The nightsight should be put into its case and placed behind the daysight tracker.)

(c) Removes the launch tube and hands it to driver-loader who places it on the ground in front of the daysight tracker.

(d) Unlocks the top coupling clamp on the traversing unit adapter, removes the traversing unit (minus adapter), and hands it to the driver-loader who places it on the ground behind the daysight tracker.

(e) Removes the TVPC from the MGS and places it to his rear on top of the vehicle within reach of the cargo hatch (see next step); unstraps the MGS and hands it to the driver-loader who places it on the ground beside the daysight tracker.

(f) Opens the cargo shell door by using the forward latch; exchanges the TVPC for the TOW battery assembly; hands the battery assembly to the driver-loader to install in the MGS.

(g) Unstraps a missile and hands it to the driver-loader who then places the missile on the ground behind the traversing unit. The gunner then closes the cargo shell door and dismounts the vehicle.

(h) Carries the daysight tracker and the MGS to the firing site.

(3) ***Driver-Loader's Actions***. The driver-loader—

(a) Dismounts the vehicle and waits to receive equipment from the gunner.

(b) Aligns equipment as described in paragraphs 2a through 2g; installs the battery assembly in the MGS and then installs the MGS cover on the MGS.

(c) Carries the traversing unit on the first trip, the missile on the second trip, and the nightsight and boresight collimator on the third trip to the firing site.

b. **Firing Position Actions.** Upon reaching the firing position, the crew proceeds with the following actions:

(1) ***Squad Leader's Actions***. The squad leader—

(a) Places the BPC and launch tube on the ground; sets up the tripod with the coupling clamp handle facing the direction of fire and unlocked; connects the BPC cable and turns the BPC to the ON position; places the BPC cable within reach of the gunner.

(b) Pulls the coil cable out of the traversing unit retaining cup; installs the launch tube; observes downrange for enemy targets.

(c) Issues fire commands.

(2) ***Gunner's Actions***. The gunner—

(a) Places the MGS on the ground next to the tripod and installs the daysight tracker on the traversing unit.

(b) Removes the MGS cover and connects the coil cable to the MGS.

(c) Installs the nightsight and turns the actuator switch to the ON position.

(d) Conducts the system checkout, boresight, and collimation.

(e) Engages the target upon receipt of the fire command.

(3) ***Driver-Loader's Actions***. The driver-loader—

(a) Installs the traversing unit on the tripod.

(b) Returns to the vehicle for the missile; places it on the ground at the firing position to the right side of the weapon system.

(c) Returns to the vehicle and gets the nightsight case and collimator case; returns to the firing position, takes the nightsight out of the case, and hands it to the gunner.

(d) Places the collimator case on the ground; removes the collimator from its case and installs it on the nightsight. Attaches the boresight cable to the collimator.

(e) Disconnects the boresight cable and removes the collimator from the nightsight; places the collimator in its case (time permitting).

(f) Loads the missile and arms it upon receipt of the fire command.

3-13. M220A2 REMOUNT (READY-TO-FIRE CONFIGURATION)

The following actions occur at the same time, when the squad leader commands, CEASE TRACKING, OUT OF ACTION.

a. **Firing Position Actions**. The following actions are performed at the firing position.

(1) ***Squad Leader's Actions***. The squad leader—

(a) Removes the launch tube and places it on the ground.

(b) Places the coil cable in the retaining cup in the bottom of the traversing unit.

(c) Folds the tripod legs and commands MOVE OUT. (The squad leader remains at the firing position until the driver-loader returns and all equipment is enroute to the vehicle.

(d) Carries the BPC, tripod, and launch tube back to the vehicle. Places the launch tube on the ground on the driver's side of the vehicle and the rest of the equipment on the ground behind the vehicle.

(2) ***Gunner's Actions***. The gunner—

(a) Disconnects the coil cable.

(b) Replaces the nightsight lens cover; removes the nightsight and sets it on the ground near the driver-loader.

(c) Removes the daysight tracker, picks up the MGS, and carries them to the vehicle.

(d) Places the daysight tracker and MGS on top of the vehicle and enters the gunner's station.

(3) ***Driver-Loader's Actions***. The driver-loader—

(a) Removes the encased missile (or empty launch container if the missile was fired).

(b) Removes the traversing unit from the tripod, carries it to the vehicle, and hands it to the gunner when the gunner reenters the gunner's station.

(c) Returns to the firing position to get the nightsight and collimator; returns to the vehicle and hands the nightsight to the gunner; sets the collimator and the nightsight case on the ground near the rear of the vehicle.

(d) Returns to the firing position, gets the encased missile (if the missile was not fired), and carries it to the vehicle. Hands the missile to the gunner.

b. **Remount Actions.** The following actions are performed to remount the vehicle.

(1) ***Squad Leader's Actions.*** The squad leader—

(a) Stows the BPC, tripod, collimator, and nightsight case on the vehicle.

(b) Hands the TVPC and VPC cables to the gunner. Enters the squad leader's seat in the vehicle and turns the VPC to the ON position.

(c) Places the MGS cover on the floorboard.

(d) Assists the other crew members as needed.

(2) ***Gunner's Actions.*** The gunner—

(a) Receives the traversing unit from the driver-loader, installs it on the traversing unit adapter, and secures it with the top coupling clamp.

(b) Installs the MGS in the MGS tray and secures it with the straps; hands the MGS cover to the squad leader; connects the coil cable.

(c) Installs the daysight tracker and the nightsight, and connects the VPC.

(d) Moves into the cargo area and places the unfired missile in the missile rack (if necessary).

(e) Conducts the system checkout, boresight, and collimation (if time permits); exchanges the battery assembly for the TVPC and connects the TVPC cable.

(3) ***Driver-Loader's Actions.*** The driver-loader assists the squad leader and gunner as needed.

NOTE: If a missile was fired, the empty launch container is destroyed.

3-14. FIRING LIMITATIONS

Firing limitations for the M966-mounted TOW/TOW 2 are the same as for the tripod-mounted TOW/TOW 2 with an additional limitation: Restricted firing angles are those greater than +20 degrees or less than -20 for all firing positions except firing over the cargo shell door. The firing angle over the cargo shell door should not exceed +20 degrees or -13 degrees.

CHAPTER 4

TOW TRAINING PROGRAM

The TOW Training Program is a comprehensive program of required training and testing that begins with individual Skill Level 1 tasks and progresses through individual gunner qualification to squad-, section-, and platoon-level testing and verification. This process has been condensed and unified into a program that consists of 12 TOW gunnery tables plus a quarterly Gunner's Skill Test. All mandatory TOW training and testing are included in this program. Additional types of TOW training, such as STXs and CO EXEVALs, will be performed with the frequency prescribed in STRAC and in the manner described in ARTEP 7-91. The unit commander must assess the training status of the individual unit to determine what additional TOW training is necessary to support the 12 mandatory training tables and must prepare his crews to pass their required testing.

Section I. TRAINING ASSESSMENT AND PLANNING

Differences between resources required and resources available will affect both the time required to conduct and sustain training and the unit's ability to meet the required standards. Required resources and maintenance assistance should be planned for and requested well in advance. Nothing should be left to chance, because resources will be wasted and training opportunities lost. A good program must have provisions for routine inspection of the TOW weapon systems by direct support units. Leaders must adjust the training program to meet the changing needs of the unit. (See Table 4-1, page 4-2, for ammunition requirements according to DA Pam 350-38).

4-1. COMMANDER'S RESPONSIBILITY

All mandatory TOW gunnery training and testing are included in the 12 TOW Gunnery Tables plus the quarterly Gunner's Skill Test. However, the commander must still conduct an accurate assessment of the state of TOW training in his unit to determine what additional training is necessary to prepare his unit for the quarterly and semiannual tables. *He must remember that his units must meet the standards outlined in DA PAM 350-38 to be certified in the training status C1, fully trained.* A good assessment of training requirements and thorough planning produce an effective unit training program. To accomplish this, the commander should start early, be thorough, and be flexible and creative. All aspects of the training program should be coordinated to produce the desired results.

a. The commander should refer to FM 25-101 when developing his unit training program. This FM applies the doctrine established in FM 25-100 and assists leaders in the development and execution of training programs.

b. The heart of an effective training program is the development of a METL and the soldier tasks, leader tasks, and collective tasks that support the METL. Battle focus drives the METL development process; the METL is based on the wartime mission, and the unit must train as it plans to fight. Chapter 2 of FM 25-101 illustrates this process in clear and practical terms.

EVENT/TABLE	TRC A* (M80 TOW blast simulator)	TRC B (M80 TOW blast simulator))	TRC C (M80 TOW blast simulator)	TRC D (M80 TOW blast simulator)
GST	0	0	0	
TABLES 1 & 2	0	0	0	
TABLES 3 & 4	0	0	0	
TABLES 5 & 6	240	60	0	
TABLES 7 & 8	40	20	0	
TABLES 9 & 10	48	24	0	
TABLES 11 & 12	48	24	0	
*The STRAC determines a unit's ammunition requirements based on its training readiness condition (TRC). The four major categories are: TRC A, active Army; TRC B, National Guard enhanced brigades; TRC C, USAR and other National Guard units; and TRC D, Reserve training divisions.				

Table 4-1. Required TOW gunnery tables and ammunition needed per system per year (IAW DA Pam 350-38 STRAC).

4-2. SUSTAINMENT TRAINING AND EVALUATION

As individual TOW crewmen, squads, sections and platoons become qualified, the commander should maintain that status by sustainment training and evaluation, and by crew stabilization. Personnel changes are inevitable in units. Before TOW squads lose their crew integrity and combat readiness, personnel changes should be examined in detail and alternative solutions sought where possible. Figure 4-1 is an example of a crew stability management chart used to record training milestones and qualifications.

4-3. TOW TRAINING GUIDELINES

A soldier must track all targets and perform all individual and collective tasks and battle drills discussed in this manual to become a fully trained heavy antiarmor weapon system crewman. Soldier training publications and ARTEP battle drills for TOW provide tasks, conditions, and standards for combat-critical skills.

a. When a crew member becomes skilled in his job, he should be trained to do another crew member's job (cross-trained). Therefore, the loss of one crew member will not make the squad combat-ineffective.

b. Training should be conducted under realistic conditions. To properly develop combat skills, training must be conducted in a tactical environment with emphasis on the type of threat the crew can expect to face in combat.

c. Multiechelon training should be conducted to save time and resources. For example, while squads are training in crew tasks, leaders can train to execute their tactical command and control responsibilities.

d. Training should be conducted with supported units. When possible, tactical training should be conducted with the type of units to be supported in combat—infantry, mechanized infantry, and armor. TOW units must be able to interface with the units they support.

VEH NO.	TOW DUTY POSITION	JAN 93	FEB 93	MAR 93	APR 93	MAY 93	JUN 93
A-31	SECTION LEADER	8278 (NOTE 2)	8278	8632 (NOTE 2)	8632	8632	8632
A-31	GUNNER	7257	7257	7257 GST (NOTE 3)	7257	7257	7257
A-31	LOADER	4410	4410	4410 GST	4410	4410	4410
A-31	DRIVER	3981	3981 GST	1850 SQD QUAL	1850	1850	1850
A-32	SQUAD LEADER	8273	8273 GST	1711 SQD QUAL	1711	1711	1711
A-32	GUNNER	1984	1984 GST	1984 SQD QUAL	1984	1984	1984
A-32	LOADER	2029	2029 GST	2029 SQD QUAL	2029	2029	2029
A-32	DRIVER	9699	9699 GST	9699 SQD QUAL	9699	9699	9699

NOTES:

1. This number represents the last four of the individual's social security number. It is written in black in this particular block and remains black until the individual leaves the duty position.
2. This number is written in red, signifying the change in personnel. April 93 number goes to black.
3. Training data is colored in green to indicate successful completion; red indicates failure.

Figure 4-1. Example of a crew stability management chart.

e. Training should be concentrated in areas where it is most needed. Time should not be wasted by training skills the unit and soldiers have already mastered. The trainer must know where the weak spots are and train to correct them.

f. Sections and squads will not always be at full strength. The mission in combat and in training can still be accomplished if under-strength units are organized with the following rules in mind:

(1) Key leadership positions should always be filled. For example, the gunner may have to fill the position of the squad leader.

(2) The primary weapon system should always be manned. If the loader becomes a casualty, the squad leader may have to load the TOW and control the squad at the same time. If the gunner becomes a casualty, the squad leader must operate and fire the TOW.

4-4. TOW GUNNERY TABLES

The TOW Training Program consists of 12 gunnery tables plus the Gunner's Skill Test. The STRAC establishes the frequency of a unit's training requirements based on its TRC. A TRC A (active Army) unit, for example, is required to conduct Tables 1 through 6 and the Gunner's Skill Test and Tables 7 through 12 semiannually. (See Table 4-2 for frequency of events as required by DA Pam 350-38 STRAC.)

NOTE: The second quarterly GST should be conducted within seven days before conducting Tables 7 through 12.

EVENT/TABLE	TRC A*	TRC B	TRC C	TRC D
Weapons Training	12	4	1	1
Gunner's Sustainment	12	4	1	1
GST	4	1	1	1
TABLES 1 & 2*	4	2	1	1
TABLES 3 & 4	4	2	1	1
TABLE 5	4	1	1	1
TABLE 6	4	1	1	0
TABLES 7 & 8	2	1	0	0
TABLES 9 & 10	2	1	0	0
TABLES 11 & 12	2	1	0	0
*See STRAC Table 5-11 for details on personnel required to conduct training and any training devices used.				

Table 4-2. Frequency of required TOW gunnery events.

Tables 1 and 2 will be performed by all of the crew with the TOW Gunnery Trainer (TGT). Only gunners will perform tables 3 and 4. Tables 5 through 12 will be performed with the TOW Field Tactical Trainer (TFTT). The TFTT replicates the flight and performance characteristics of the TOW. (See Appendix F for more information on TOW training devices.)

4-5. CONDUCT OF UNIT TOW TRAINING

TOW training involves individual and collective training divided into four phases. Each unit must complete these phases as required.

a. **Phase I—Individual Training.**

- Conduct refresher training on STP tasks as deemed necessary to prepare for the Gunner's Skill Test.
- Conduct sustainment training with the TGT as deemed necessary to prepare for TOW Gunnery Tables 1 through 4.
- Conduct Gunner's Skill Test.
- Conduct TOW Gunnery Table 1, Individual Gunnery Practice.

- Conduct TOW Gunnery Table 2, Individual Gunnery Qualification.
- Conduct TOW Gunnery Table 3, Advanced Gunnery Practice.
- Conduct TOW Gunnery Table 4, Advanced Gunnery Qualification.

b. **Phase II—Squad Training.**

- Conduct refresher training on the installation and operation of TFTT or MILES equipment as deemed necessary in preparation for TOW Gunnery Tables 5 and 6.
- Conduct TOW Gunnery Table 5, Baseline Gunnery Practice.
- Conduct TOW Gunnery Table 6, Baseline Gunnery Qualification.
- Conduct squad-level sustainment training as deemed necessary to prepare for TOW Gunnery Tables 7 and 8.
- Conduct TOW Gunnery Table 7, Squad Gunnery Practice, in preparation for Table 8.
- Conduct TOW Gunnery Table 8, Squad Gunnery Qualification.

c. **Phase III—Section Training.**

- Conduct section-level sustainment training as deemed necessary to prepare for TOW Gunnery Tables 9 and 10.
- Conduct TOW Gunnery Table 9, Section Gunnery Practice, in preparation for Table 10.
- Conduct TOW Gunnery Table 10, Section Gunnery Qualification.

d. **Phase IV—Platoon Training.**

- Conduct platoon-level sustainment training as deemed necessary to prepare for TOW Gunnery Tables 11 and 12.
- Conduct TOW Gunnery Table 11, Platoon Gunnery Practice, in preparation for Table 12.
- Conduct TOW Gunnery Table 12, Platoon Gunnery Qualification.

4-6. COLLECTIVE TRAINING

Collective training is the coordinated performance of individual skills and tasks as a team. Collective tasks are normally performed by more than one soldier.

a. Common crew task training is the first level of collective training. Common crew tasks provide the basic technical collective tasks performed by the crew to accomplish the wartime mission. These tasks tell "how to" and allow each soldier in the crew to perform his tasks. As a result, the squad leader, gunner, driver, and loader combine their learned individual technical skills to perform as one crew on their assigned vehicle.

b. The critical squad and platoon collective tasks are called battle drills. Battle drills are mostly independent of METT-TC and require minimal leader actions to execute. They are standardized throughout the U.S. Army. Battle drills are in the applicable drill book for each echelon and equipment-type unit. The battle drills for antiarmor platoons are in ARTEP 7-91-Drill.

c. Common crew tasks are tasks performed by a TOW crew that are not battle drills but are performed in the same manner. An example of a common crew task is Dismount/Remount the TOW. Crew tasks are crucial to applying tactics to enhance survival and to accomplish the TOW crew's mission of providing antiarmor direct fire support to destroy the enemy.

d. ARTEP training exercises are used to practice, evaluate, and sustain collective tasks and mission proficiency. They have specific goals and are modified based on METT-T factors. Exercises are structured to the availability of resources, time, and training areas. (See appropriate ARTEP for details on conducting ARTEP exercises.)

SECTION II. TOW TRAINING PHASES

The TOW training program is divided into four phases: individual training, squad training, section training, and platoon training. Phase I (individual training) consists of the Gunner's Skill Test and TOW Gunnery Tables 1 through 4. Phase II (squad training) consists of TOW Gunnery Tables 5 through 8. Phase III (section training) consists of TOW Gunnery Tables 9 and 10. Phase IV (platoon training) consists of TOW Gunnery Tables 11 and 12. (All scoring forms are located in the back of this manual for local reproduction.)

4-7. PHASE I: INDIVIDUAL TRAINING

Individual training covers all tasks that are performed by a single TOW crewman. These tasks include Skill Level 1 11B soldier's manual tasks and individual gunnery. They are the building blocks that form the foundation of TOW training.

a. **Gunner's Skill Test.** The Gunner's Skill Test comprises the Skill Level 1 tasks for the TOW weapon system listed in STP 7-11B1-SM-TG plus some additional tasks (Figure 4-2, page 4-7 and Figure 4-3, page 4-8). All soldiers assigned to TOW-equipped units must take this portion of the test and receive a GO on at least 70 percent of the tasks. Soldiers must be tested on the group of tasks that applies to the vehicle his unit is equipped with and receive a GO on at least 70 percent of those tasks. Some of the tasks are combined individual/crew tasks. Since every member of the crew must be cross trained, the combined individual/crew tasks will be performed in rotation and every member of the crew will be tested on each part.

TASK: Perform the TOW gunner's tasks listed in (1) below.
CONDITIONS: Testing will be hands on, with the appropriate equipment. Specific conditions will be found in the reference cited for each task.
STANDARDS: Specific standards for each task listed will be found in the appropriate reference. The overall minimum standard to receive a GO on the Gunners' Skill Test is for each individual to receive a GO on at least 70 percent of the tasks he is required to test on.

---

(1) The GST consists of the following tasks.

(a) 071-056-0004 Maintain an M220-series launcher system.

(b) 071-056-0007 Load an M220-series launcher system.

(c) 071-056-0008 Unload an M220-series launcher system.

(d) 071-056-0009 Engage targets with an M220-series launcher system.

(e) 071-056-0010 Perform immediate action for an M220-series launcher system malfunction.

(f) 071-056-0005 Conduct a system check-out on an M220A1 launcher system. (Tested only if unit is equipped with M220A1 systems.)

(g) 071-056-0013 Conduct a system check-out on an M220A2 launcher system. (Tested only if unit is equipped with M220A2 systems.)

(h) 071-056-0030 Determine TOW-series launcher system firing limitations.

(i) No Task Number Identify combat vehicles. (See Appendix G.)

(j) No Task Number Determine if a target can be engaged by an M220-series launcher system. (See Chapter 5.)

(k) 071-317-0000 Prepare an antiarmor range card.

(l) No Task Number Assemble an M220-series launcher system. (See TM 9-1425-472-12 [M220A1] or TM 9-1425-450-12 [M220A2].)

GUNNER'S SKILL TEST - PART 1

For use of this form, see FM 3-22.34. The proponent agency is TRADOC.

NAME	RANK	UNIT	DATE
THOMPSON, DAVID B.	SGT	C 2/2 IN	13 AUG 03

TEST#	TASK NUMBER	TASK TITLE	GO/NO GO
1	071-056-0004	MAINTAIN AN M220-SERIES LAUNCHER SYSTEM	GO H.H.R.
2	071-056-0007	LOAD AN M220-SERIES LAUNCHER SYSTEM	GO K.L.C.
3	071-035-0008	UNLOAD AN M220-SERIES LAUNCHER SYSTEM	GO K.L.C.
4	071-056-0009	ENGAGE TARGETS WITH AN M220-SERIES LAUNCHER SYSTEM	GO K.L.C
5	071-056-0010	PERFORM IMMEDIATE ACTION FOR AN M220-SERIES LAUNCHER SYSTEM	1. NO GO B.L.C. 2. GO B.L.C
6	071-056-0005	CONDUCT A SYSTEM CHECK OUT ON AN M220A1 LAUNCHER SYSTEM	
7	071-056-0013	CONDUCT A SYSTEM CHECK OUT ON AN M220A2 LAUNCHER SYSTEM	GO K.M.
8	071-056-0030	DETERMINE M220-SERIES LAUNCHER SYSTEM FIRING LIMITATIONS	GO R.H.
9	NO TASK NUMBER	IDENTIFY COMBAT VEHICLES	GO A.Q.
10	NO TASK NUMBER	DETERMINE IF A TARGET CAN BE ENGAGED BY AN M220-SERIES LAUNCHER SYSTEM (SEE FM 3-22.34, CH5)	GO R.P.
11	071-317-0000	PREPARE AN ANTIARMOR RANGE CARD	GO R.P.
12	NO TASK NUMBER	ASSEMBLE AN M220-SERIES LAUNCHER SYSTEM	GO W.E.

DA FORM 7331-R, AUG 2003

APD V1.00

Figure 4-2. Example of completed DA Form 7331-R, Gunner's Skill Test—Part 1.

GUNNER'S SKILL TEST - PART 2

For use of this form, see FM 3-22.34. The proponent agency is TRADOC.

NAME	RANK	UNIT	DATE
WILLIAMS, RAY	SPC	CSC 2/1 IN	13 AUG 03

TEST#	TASK NUMBER	TASK TITLE	GO/NO GO
1	NO TASK NUMBER	PLACE THE M966 VEHICLE IN THE READY-TO-FIRE CONFIGURATION (SEE FM 3-22.34 CH 3)	GO K.L.C.
2	071-056-9932	CONDUCT DISMOUNTING AND REMOUNTING OF AN M220A1 LAUNCHER SYSTEM ON AN M966 VEHICLE (M220A1- EQUIPPED UNITS ONLY)	
3	071-056-0034	CONDUCT DISMOUNTING AND REMOUNTING OF AN M220A2 LAUNCHER SYSTEM ON AN M966 VEHICLE (M220A2 - EQUIPPED UNITS ONLY)	GO K.L.C.
4	071-034-0001	LOAD AN M243 OR M259 SMOKE GRENADE LAUNCHER	
5	071-034-0002	UNLOAD AN M243 OR M259 SMOKE GRENADE LAUNCHER	

DA FORM 7332-R, AUG 2003 APD V1.00

Figure 4-3. Example of completed DA Form 7332-R, for Gunner's Skill Test—Part 2.

(4) The following are the tasks, conditions, and standards for the unnumbered tasks in the Gunner's Skill Test.

TASK: Identify combat vehicles.
CONDITIONS: Given a classroom with appropriate audiovisual equipment and assorted vehicle GTAs. (See the local Training Support Center for the combat vehicle slides.)
STANDARDS: Given a 12-second exposure time for each slide, the soldier must correctly identify 36 out of 40 vehicles by nomenclature as described in Appendix G.

TASK: Determine if a target can be engaged by an M220-series TOW launcher system.
CONDITIONS: In the field, given an operational, assembled M220-series TOW launcher system and tactical vehicles placed in realistic situations; or indoors, given a series of graphic presentations of target engageability situations or using the TGT.
STANDARDS: Correctly determine target engageability for the M220-series TOW launcher system in terms of range and obstructions using both the daysight tracker and nightsight as described in Chapter 5, paragraph 5-1.

TASK: Assemble an M220-series launcher system.
CONDITIONS: Given a complete but unassembled M220-series launcher.
STANDARDS: Correctly assemble an M220-series launcher system as described in TM 9-1425-472-12 (M220A1) or TM 9-1425-450-12 (M220A2) within 3 minutes.

TASK: Place the M966 HMMWV in the ready-to-fire configuration.
CONDITIONS: Given an M966 HMMWV, an M220-series TOW launcher complete, and an MSR.
STANDARDS: Correctly place the M966 in the ready-to-fire configuration as described in TM 9-1425-475-12 (M220A1) or TM 9-1425-450-12 (M220A2) within 5 minutes.

---

b. **TOW Gunnery Table 1, Individual Gunnery Practice.** TOW Gunnery Table 1 consists of firing a minimum of two practice events (instructional firing) in preparation for actual gunner qualification on TOW Gunnery Table 2. Units will use the TGT to conduct this table.

TASK: Verify on TOW Gunnery Table 1, Individual Gunnery Practice using the TGT.
CONDITIONS: Indoors, using the TGT. The gunner will fire a minimum of two 10-shot events using the TGT. They will be selected from the events listed in Table 4-3 (page 4-10).
STANDARDS: To verify on TOW Gunnery Table 1 and progress to TOW Gunnery Table 2, the gunner must fire at least one 10-shot event with a cumulative score of at least 625 points and at least 6 hits.

---

(1) ***Preparation of the TGT for Gunner Practice.*** The TGT is not designed for use as an outdoor trainer. Therefore, it should always be set up and used indoors. It may be used in tents or under shelter as long as it is not exposed to the elements, particularly dust or blowing sand. It may be used in either the tripod-mounted mode or in the HMMWV-mounted mode. The TGT should *never* be mounted on a moving HMMWV. It should be assembled and operated according to the instructions in TM 9-6920-452-10.

(2) ***Personnel Duties.*** Normally, only two people at a time, the gunner and the trainer, will be involved in training with the TGT. The trainer will usually be either the squad or section leader. Other members of the squad should be occupied with concurrent training. Other gunners should not be allowed to watch during conduct of the event. Once the TGT has been assembled and warmed up, the trainer must assemble a plan group that conforms to one of the 10-shot events listed in Table 4-3 (page 4-15). The trainer should have the TGT inspected and warmed up, and the plan group constructed before the gunner is brought in to begin engaging targets. When constructing the plan group, the trainer ensures battle sounds are ON, target size is set at 100 percent, and obscuration is set to last for 1.5 seconds for the thermal sight and for the daysight tracker. When everything is ready, the trainer should bring the gunner in and brief him before beginning the event. The briefing should include the following:

- Safety considerations.
- Breath control—remind the student of proper procedures.
- Obscuration—tell the student the obscuration setting used.
- Sight—tell the student which sight he will use.
- Special conditions—tell the student any special conditions (multiple targets and so forth) that apply.

(3) ***Conduct of Training.*** When the TGT is ready for operation, the plan group has been constructed, and the gunner has been briefed and is in place, the trainer begins conducting the table. Before the trainer starts each mission, he gives the gunner a fire command and tells him to fire when ready. During the mission, the trainer does not coach the gunner in any way. The gunner must determine when to fire and at which target. After each mission, the trainer gives the gunner the results and then proceeds to the next mission.

NOTE: The trainer has the option of thoroughly debriefing the gunner on his performance at the end of each mission or of saving all 10 missions and reviewing them after the table is complete. The trainer must remember that the TGT will only save a maximum of 20 missions at one time.

TABLE NO.	1	2	3
DISK NO.	DFT 10	DFT 11	DFT 12
MISSION NO.	2	1	22
	3	2	24
	6	4	26
	8	3	28
	9	18	29
	18 (THERMAL)	26 (THERMAL)	2 (THERMAL)
	14 (THERMAL)	27 (THERMAL)	3 (THERMAL)
	12 (THERMAL)	28 (THERMAL)	4 (THERMAL)
	15 (THERMAL)	29 (THERMAL)	9 (THERMAL)
	22 (THERMAL)	30 (THERMAL)	10 (THERMAL)

Table 4-3. 10-shot events for the TGT.

(4) ***Scoring.*** The two scoring criteria for TOW Gunnery Table 1 are total point score and number of hits. The point score for the table consists of the cumulative point score for all 10 missions. The minimum number of hits is 6 out of 10. The trainer should use DA Form 7333-R to record results (Figure 4-4).

(5) ***Sustainment Training.*** Units are required to conduct monthly sustainment training. This training should not be confused with the quarterly tables. Missions for monthly sustainment training may be taken from any of the videodiscs and are chosen at the trainer's discretion. The missions in Table 4-3 should not be used for sustainment training, if possible.

TOW GUNNERY TABLES 1 AND 2: INDIVIDUAL GUNNERY

☒ PRACTICE ☐ QUALIFICATION

For use of this form, see FM 3-22.34. The proponent agency is TRADOC.

DATA REQUIRED BY PRIVACY ACT OF 1974

AUTHORITY: 10USC 3012Bg. Executive Order 9397. PRINCIPAL PURPOSE(S): Records individual's performance on record fire range. ROUTINE USE(S): Evaluation of individual's proficiency and basis for determination of award of proficiency badge; SSN is used for positive identification purposes only. MANDATORY OR VOLUNTARY DISCLOSURE AND EFFECT ON INDIVIDUAL NOT PROVIDING INFORMATION: Voluntary; individuals not providing information cannot be rated/scored on a mass basis.

NAME	RANK	SSN	UNIT
ERVIN, D.	SGT	000-00-0011	C 2/2 IN
DATE		DISK#	

MISSION#	INDIVIDUAL GUNNERY TASK TESTED	SIGHT	SCORE	SUCCESSFUL	
# 22		OS	90	(YES)	NO
# 24		OS	90	(YES)	NO
# 26		OS	90	(YES)	NO
# 28		OS	80	(YES)	NO
# 29		OS	85	(YES)	NO
# 2		TS	88	YES	(NO)
# 3		TS	81	(YES)	NO
# 4		TS	78	(YES)	NO
# 9		TS	66	(YES)	NO
# 10		TS	75	(YES)	NO
	TOTAL TRACKING SCORE		823	SUCCESSFUL MISSIONS: 9	

GUNNER CLASSIFICATION

EXPERT / 875 - 1,000 1ST CLASS / 750 - 874 2ND CLASS / 625 - 749

SCORING INDIVIDUAL GUNNERY TABLES

IN THE INDIVIDUAL GUNNERY TABLES, THE GUNNER'S RATING IS DETERMINED BY HIS TOTAL POINT SCORE FOR 10 MISSIONS WITH AT LEAST 6 SUCCESSFUL MISSIONS. A SUCCESSFUL MISSION IS ONE WHERE THE GUNNER CORRECTLY FOLLOWS THE TRAINERS FIRE COMMANDS AND ENGAGES AND DESTROYS THE CORRECT TARGET.

DA FORM 7333-R, AUG 2003 APD V1.00

Figure 4-4. Example of completed DA Form 7333-R, TOW Gunnery Tables 1 and 2: Individual Gunnery Practice and Qualification.

c. **TOW Gunnery Table 2, Individual Gunnery Qualification.** TOW Gunnery Table 2 will be set up and run the same as TOW Gunnery Table 1. It will be performed with the TGT.

TASK: Qualify on TOW Gunnery Table 2, Individual Gunnery Qualification using the TGT.
CONDITIONS: Indoors using the TGT. The gunner will fire a 10-shot event selected from Table 4-3.
STANDARDS: To qualify on TOW Gunnery Table 2 and progress to TOW Gunnery Table 3, the gunner must fire one 10-shot event with a cumulative score of at least 625 points and at least 6 hits. The gunner's classification (expert, first class, second class) will be determined by the cumulative point score using the point scale as shown in Figure 4-4.

NOTE: Preparation of the TGT, personnel duties, and conduct of training are the same as for TOW Gunnery Table 1.

d. **TOW Gunnery Table 3, Advanced Gunnery Practice.** TOW Gunnery Table 3 (Figure 4-5, page 4-14) consists of a series of missions performed with the TGT that train and test advanced gunnery skills. These skills include determining a target's engageability, identifying targets (to include friend or foe), prioritizing targets, deciding which fire control methods to use, and engaging evasive targets that are hard to track such as helicopters, obscured targets, or extreme-range targets. Only the assigned gunners in each squad are required to verify on TOW Gunnery Table 3. (Other members of the squad may perform the table if time permits and the trainer thinks it useful.) TOW Gunnery Table 3 can only be performed with the TGT. Gunners must verify on TOW Gunnery Table 3 to progress to TOW Gunnery Table 4, Advanced Gunnery Qualification.

(1) ***Gunner Verification on TOW Gunnery Table 3, Advanced Gunnery Practice.***

TASK: Verify on Table 3, Advanced Gunnery Practice.
CONDITIONS: Indoors, using the TGT. The gunner will fire a minimum of two 10-shot events using the TGT. The 10-shot events will be selected from Table 4-4 (page 4-15).
STANDARDS: To verify on TOW Gunnery Table 3 and progress to TOW Gunnery Table 4, the gunner must score at least 625 points. The gunner must also receive a GO on at least 6 out of 10 missions.

---

(2) ***Personnel Duties.*** The trainer (usually the squad or section leader) sets up the TGT IAW TM 9-6920-452-10, creates the planning group for each 10-shot event, briefs the gunner, gives the fire commands for each mission, and completes DA Form 7334-R (TOW Gunnery Tables 3 and 4, Advanced Gunnery Practice and Qualification Scorecard, Figure 4-5, page 4-14). He also debriefs and critiques the gunner at the end of each 10-shot event.

(3) ***Conduct of Training.*** Conduct of advanced gunnery training with the TGT is more complex than individual gunnery. The trainer must not only observe and record the gunner's tracking performance but also brief, observe, and score the particular advanced gunnery task that is presented in each separate mission. The trainer selects the 10-shot event from Table 4-4 (page 4-15). The sequence of missions in each 10-shot event should be changed continuously to prevent the gunner from becoming familiar with them.

(a) Before each mission, the trainer must present a fire command to the gunner that clearly states what the gunner must do. For example, if a particular mission involves prioritizing targets (such as tanks are higher priority targets than APCs), the trainer gives the command "GUNNER! TARGETS TO YOUR FRONT! ENGAGE TANKS FIRST! FIRE WHEN READY!"

(b) In some missions, the task involved may not require specific fire commands. These tasks include determining a target's engageability, engaging evasive/obscured targets, and identifying targets (to include friend or foe). When giving fire commands for these missions, the trainer need only give general instructions, such as "GUNNER! TARGETS TO YOUR FRONT! FIRE WHEN READY!"

(4) ***Scoring.*** The trainer has two criteria for scoring each of these missions: tracking and success of the mission. The trainer will record the gunner's performance on the DA Form 7334-R (Figure 4-5, page 4-14).

(a) *Tracking Score.* TOW Gunnery Table 3 is scored in the same manner as TOW Gunnery Table 1 except for target identification (to include friend or foe) missions. In these missions, if the gunner correctly identifies a target as friendly and does not fire on it, he receives 100 points. If he fires on a friendly target, he receives 0 points whether he hits it or not. The total tracking score from the 10 missions (Table 4-4, page 4-15) determines the gunner's ranking.

(b) *Successful Mission Score.* The gunner must achieve a GO on at least 6 out of 10 missions. The requirements for a mission to be successful vary according to the advanced gunnery task that is presented in each mission. A successful mission is defined as follows:

- The gunner successfully engages (destroys) the correct target as determined by the trainer's instructions for that particular mission. In a mission to determine a target's engageability, a hit on any target presented is scored as a success. In scenarios to prioritize targets or determine the method of fire control, the gunner must engage the correct target according to the instructions given to him in the fire command.
- In a scenario to identify targets (including friend or foe), the gunner must not fire on friendly targets.
- Engaging evasive and or obscured targets does not require special skills or actions from the gunner. These targets are simply hard to hit. A hit on any target is scored as a successful mission.

TOW GUNNERY TABLES 3 AND 4: ADVANCED GUNNERY

☐ PRACTICE ☒ QUALIFICATION

For use of this form, see FM 3-22.34. The proponent agency is TRADOC.

DATA REQUIRED BY PRIVACY ACT OF 1974

AUTHORITY: 10USC 30129g. Executive Order 9397. **PRINCIPAL PURPOSE(S):** Records individual's performance on record fire range. **ROUTINE USE(S):** Evaluation of individual's proficiency and basis for determination of award of proficiency badge; SSN is used for positive identification purposes only. **MANDATORY OR VOLUNTARY DISCLOSURE AND EFFECT ON INDIVIDUAL NOT PROVIDING INFORMATION:** Voluntary; individuals not providing information cannot be rated/scored on a mass basis.

NAME	RANK	SSN	UNIT
HEIDER, R.	SPC	000-01-0000	C 2/2 IN

DATE	DISK#
30 AUG 03	10

MISSION#	INDIVIDUAL GUNNERY TASK TESTED	SIGHT	SCORE	SUCCESSFUL	
# 1	EVASIVE TARGET	OS	81	(YES)	NO
# 7	TARGET ENGAGEABILITY	OS	78	(YES)	NO
# 12	FIRE CONTROL METHODS	TS	73	(YES)	NO
# 14	EVASIVE/OBSCURE TARGET	TS	66	(YES)	NO
# 22	TARGET PRIORITY	TS	81	YES	(NO)
# 24	TARGET ENGAGEABILITY	OS	83	(YES)	NO
# 28	EVASIVE TARGET	OS	91	(YES)	NO
# 29	TARGET PRIORITY	OS	78	(YES)	NO
# 26	TARGET PRIORITY	OS	80	(YES)	NO
# 13	FRATRICIDE	OS	100	(YES)	NO
	TOTAL TRACKING SCORE		811	SUCCESSFUL MISSIONS: 9	

GUNNER CLASSIFICATION

EXPERT / 875 - 1,000 1ST CLASS / 750 - 874 2ND CLASS / 625 - 749

SCORING ADVANCE GUNNERY TABLES

IN THE INDIVIDUAL GUNNERY TABLES, THE GUNNER RATING IS DETERMINED BY HIS TOTAL POINT SCORE FOR 10 MISSIONS, PLUS HE MUST HAVE AT LEAST 6 SUCCESSFUL MISSIONS. A SUCCESSFUL MISSION IS ONE WHERE THE GUNNER CORRECTLY FOLLOWS THE TRAINERS FIRE COMMANDS AND ENGAGES AND DESTROYS THE CORRECT TARGET. IF THE MISSION IS A FRATRICIDE MISSION, THE GUNNER SHOULD **NOT** ENGAGE THE TARGET. IF HE **DOES NOT** ENGAGE THE FRIENDLY VEHICLE, HE SCORES 100 POINTS AND A SUCCESSFUL MISSION. IF HE **DOES** ENGAGE THE FRIENDLY VEHICLE, HE SCORES 0 POINTS AND AN UNSUCCESSFUL MISSION. IN TARGET ENGAGEABILITY AND EVASIVE TARGET SCENARIOS IT DOES NOT MATTER WHICH TARGET IS ENGAGED.

DA FORM 7334-R, AUG 2003 APD V1.00

Figure 4-5. Example of completed DA Form 7334-R, TOW Gunnery Tables 3 and 4: Advanced Gunnery Practice and Qualification.

DISK NO.	MISSION NO.
DFT 10	1 Evasive Target/Optical Sight 7 Target Engageability/Optical Sight 12 Fire Control Methods/Thermal Sight 14 Evasive/Obscure Target/Thermal Sight 22 Target Priority/Thermal Sight 24 Target Engageability/Optical 28 Evasive Target/Optical 29 Target Priority/Optical 26 Target Priority/Optical 13 Fratricide/Optical
DFT 12	7 Fratricide/Thermal 9 Fratricide/Thermal 22 Target Priority/Optical 23 Target Priority/Thermal 5 Target Priority/Thermal 6 Target Priority/Thermal 10 Target Priority/Thermal 12 Target Priority/Thermal 15 Evasive Target/Thermal 11 Evasive Target/Thermal
DFT 13	30 Target Engageability/Daysight Tracker 7 Target Priority/Daysight Tracker 26 Target Engageability/Daysight Tracker 21 Fratricide/Daysight Tracker 6 Target Priority/Thermal 10 Target Priority/Thermal 8 Target Priority/Thermal 15 Target Priority/Thermal 12 Target Priority/Thermal
DFT 15	2 Target Engageability/Dual 5 Target Priority/Dual 6 Target Priority/Dual 7 Target Priority/Dual 9 Target Priority/Dual 11 Target Engageability/Dual 12 Evasive Target/Dual 8 15 Target Engageability/Dual 3 15 Target Engageability/Dual 15 Target Engageability/Dual
Note: As reviewer discs become available and are verified, they will be added to the manual as a change.	

Table 4-4. 10-shot events for advanced gunnery tables.

NOTES: 1. In the mission descriptions in Table 4-4, the term "evasive target" refers to evasive, obscured, or difficult targets and helicopters.
2. The term "target priority" refers to missions in which different types of targets are presented (such as tanks and APCs) or in which some are moving and some are stationary. (Using these differences as criteria, the trainer chooses the high priority target for the gunner.)
3. In the missions where the type of sight is listed as dual, both sights are active and the gunner may choose whichever sight he prefers for the engagement or may change sights during the engagement.
4. In the fire control missions, the trainer gives the gunner a choice of fire control methods (frontal, depth, or crossfire), and the gunner must engage the appropriate target out of the group presented.

(5) ***Sustainment Training.*** Each month every soldier assigned to a TOW-equipped unit is required to perform sustainment training on the TGT. The trainer may choose from all missions available for this training. He must remember that he is preparing his soldiers for the quarterly gunnery tables and should choose scenarios that train his soldiers in the skills and engagements in which they are weakest. Since only assigned gunners are *required* to perform and pass TOW Gunnery Tables 3 and 4, the trainer must be sure they are trained on the difficult advanced scenarios. Other members of his TOW crews should be trained as thoroughly as time permits and their training weaknesses require.

e. **TOW Gunnery Table 4, Advanced Gunnery Qualification.** TOW Gunnery Table 4 is similar to TOW Gunnery Table 3. It consists of a series of missions performed with the TGT that train and test advanced gunnery skills. Gunners will be evaluated on their ability to track and hit targets, and on their ability to perform advanced gunnery skills. These skills include determining a target's engageability, identifying targets (to include friend or foe), prioritizing target's, choosing a method of fire control, and engaging targets that are difficult to track, such as helicopters, obscured targets, and extreme-range targets. Only assigned gunners in each squad are required to qualify on TOW Gunnery Table 4. (Other members of the squad may perform the table if time permits and the trainer thinks it useful.) TOW Gunnery Table 4 can only be performed with the TGT. Therefore, units that do not have access to the TGT are not required to perform this table. Gunners must qualify on TOW Gunnery Table 4 to progress to TOW Gunnery Table 5, Baseline TFTT Gunnery Practice.

(1) ***Gunners' Qualification on TOW Gunnery Table 4, Advanced Gunnery Qualification.***

TASK: Qualify on TOW Gunnery Table 4, Advanced Gunnery Qualification.
CONDITIONS: Indoors, using the TGT. The gunner will fire one 10-shot event using the TGT. The 10-shot event will be selected from Table 4-4.
STANDARDS: To qualify on TOW Gunnery Table 4 and progress to TOW Gunnery Table 5, the gunner must score a total of at least 625 points. The precise ranking of the gunner (Expert, 1st class, or 2d class) is determined according to the scale shown on the scorecard. In addition, the gunner must receive a GO on at least 6 out of 10 missions.

---

(2) ***Personnel Duties.*** Personnel duties, conduct of training, and sustainment training are the same as for TOW Gunnery Table 3.

(3) ***Scoring.*** TOW Gunnery Table 4 is scored the same as TOW Gunnery Table 3. The form shown in Figure 4-5, TOW Gunnery Tables 3 and 4, Advanced Gunnery Practice and Qualification Scorecard, is used to score the table.

4-8. PHASE II: SQUAD TRAINING

Phase II begins squad-level TOW training. TOW Gunnery Tables 5 and 6 are static gunnery exercises from a baseline. TOW Gunnery Tables 7 and 8 are moving gunnery exercises where the vehicle moves from point to point on a course and engages a series of targets. These gunnery exercises are used to train and evaluate the TOW unit's combat capability. Each element has two tables: a practice table to train and a qualification table for evaluation. Each table is performed once during the day and once during the night under realistic (simulated) battlefield conditions. Point scores for target engagements are based on a combat model that predicts success based on the time from target detection to target engagement, type of target, and range to target. These scores are taken from the live-fire point calculation sheet (Figure 4-16, 4-35). Point scores are also affected by crew cuts (crew cuts are crew errors, either major or minor, that reduce the point score for a target engagement). The actual rating is calculated by adding the scores for the day and night phases and dividing them by two to get an average score. During move-out scenarios, the senior trainer is normally to the rear of the leader's vehicle (75 meters by 90 degrees) for evaluation and training purposes. An assistant evaluator of the appropriate rank and experience is normally assigned to each additional vehicle that is evaluated as part of the same exercise. Evaluators must monitor crew conversations and fire commands.

a. **TOW Gunnery Table 5, Baseline TFTT Gunnery.** TOW Gunnery Table 5 reacquaints crews with TFTT and allows them to troubleshoot the equipment. It refreshes crews on TFTT gunnery and target engagement techniques to prepare the crews for the practice and qualification tables that follow.

TASK: Conduct baseline TFTT gunnery practice (day and night).

CONDITIONS: On an MPRC or in the field, given a TOW weapon system, its organic carrier, TFTT system, M80 blast simulators, and either pop-up moving targetry or tactical vehicles equipped with TFTT retroreflector assembly, as appropriate. (Use of MILES harness and LTIDs is optional.) The table will be conducted once during daylight and once at night.

STANDARDS: To complete TOW Gunnery Table 5 and progress to TOW Gunnery Table 6, the crew must successfully install, troubleshoot, and operate the TFTT system once during daylight and again at night IAW the appropriate TM. The crew must also successfully acquire, engage, and destroy 6 out of 10 targets presented in Table 5 (once during daylight and once at night).

---

(1) ***Conduct***. In TOW Gunnery Table 5, the system and or carriers are set up in static positions on a baseline. A range with pop-up or moving targets equipped with TFTT retroreflectors should be used if available. (Use of MILES harness or LTIDs is optional.) If such a range is not available, tactical vehicles equipped with TFTT retroreflectors may be used. (Use of tactical vehicles equipped with MILES harness or LTIDs is optional.) The TOW systems will not move but will engage a series of moving and or pop-up targets from the baseline. TOW Gunnery Table 5 will be fired twice; once during daylight and once at night. Some of the engagements will be designated as NBC engagements and will be performed with the crew in MOPP 4.

(2) ***Scoring.*** To count as a successful engagement, the target must be killed within 30 seconds or less after exposure. Use DA Form 7335-R (TOW Gunnery Tables 5 and 6, Baseline TFTT Gunnery Practice and Qualification Scorecard, Figure 4-6) to record the crew's score (day or night).

(3) ***Suggested Support Requirements.*** The following support requirements are recommended when conducting TOW Gunnery Tables 5 through 8.

- Minimum evaluators (evaluators will not assist the gunner in finding or identifying targets):
  - One evaluator per squad.
  - One evaluator in the range tower to record scores on engagements.
- Opposing forces: None.
- Support troops: None.
- Vehicles/communications:
  - One AN/PRC-119 radio per squad evaluator.
  - Two AN/PRC-119 radios for the tower evaluator (one primary, one spare).
  - Two frequencies (one for the tower and squad evaluator and one for the antiarmor squad).
- Maneuver area:
  - Multipurpose range complex.
  - A route for the element to move from the assembly area to the first firing point without damaging the underground wiring.
- Firing area:
  - The TOW backblast area (75 meters by 90 degrees) will be clear.
  - Smoke may be used.
- Training aids, devices, and special equipment:
  - TFTT system.
  - Sufficient TFTT retroreflectors to cover targets. (Use of MILES harness and LTIDs is optional.)
- Ammunition: Ten M80 blast simulators per TOW, per iteration.
- Key references: FM 3-21.91(FM 7-91), FM 3-22.34(FM 23-34), and live-fire point calculation sheet.

(4) ***Trainer/Evaluator Guidelines.*** Because this table is conducted on an MPRC, marking the firing positions may be necessary. At night, the roads must be marked with luminous tape.

b. **TOW Gunnery Table 6, Baseline TFTT Gunnery Qualification.** TOW Gunnery Table 6 is conducted according to the same conditions and standards as TOW

Gunnery Table 5. DA Form 7335-R (TOW Gunnery Tables 5 and 6, Baseline TFTT Gunnery Practice and Qualification Scorecard, Figure 4-6) is used to score the table.

TOW GUNNERY TABLES 5 AND 6: BASELINE TFTT GUNNERY

☐ PRACTICE ☒ QUALIFICATION

For use of this form, see FM 3-22.34. The proponent agency is TRADOC.

SQUAD	SECTION	PLATOON	COMPANY
1ST	1ST	1ST	C 3/60 IN

DATE	DAY	NIGHT
3 SEP 03	X	

ENGAGEMENT	TARGET DESCRIPTION	HIT/MISS/TIME	GO/NG GO
TARGET 1	STATIONARY T72/FLANK/ 1,000 - 1,500 METERS	(HIT) MISS TIME: 21 SEC	GO
TARGET 2	STATIONARY T72/FLANK/ 2,100 - 3,750 METERS	(HIT) MISS TIME: 25 SEC	GO
TARGET 3	STATIONARY T72/FRONTAL/ 2,100 - 3,750 METERS	(HIT) MISS TIME: 23 SEC	GO
TARGET 4	STATIONARY T72/OBLIQUE/ 2,100 - 3,750 METERS	(HIT) MISS TIME: 24 SEC	GO
TARGET 5	STATIONARY T72/DEFILADE/ 800 - 1,500 METERS/NBC	HIT (MISS) TIME: 21 SEC	NG
TARGET 6	MOVING T72/FLANK/ L-R, 30 KPH 1,600 - 2,000 METERS/NBC	(HIT) MISS TIME: 28 SEC	GO
TARGET 7	MOVING T72/FLANK/ L-R, 30 KPH 1,600 - 2,000 METERS/NBC	(HIT) MISS TIME: 26 SEC	GO
TARGET 8	MOVING T72/FLANK/ L-R, 5 KPH 1,000 - 1,500 METERS	(HIT) MISS TIME: 27 SEC	GO
TARGET 9	MOVING T72/FLANK/ L-R, 10 KPH 2,100 - 3,750 METERS	(HIT) MISS TIME: 26 SEC	GO
TARGET 10	MOVING T72/FLANK/ L-R, 40 KPH 1,000 - 1,500 METERS	(HIT) MISS TIME: 23 SEC	GO
		OVERALL GO/NO GO:	9

TO RECEIVE A "GO" ON ANY TARGET ENGAGEMENT, THE GUNNER MUST HIT THE TARGET WITHIN 30 SECONDS AFTER THE TARGET IS EXPOSED. TO RECEIVE A "GO" ON TABLE 5, THE CREW MUST RECEIVE A "GO" ON "INSTALL TFTT" AND A "GO" ON AT LEAST 6 OUT OF 10 TARGET ENGAGEMENTS, BOTH DAY AND NIGHT.

DA FORM 7335-R, AUG 2003 APD V1.00

Figure 4-6. Example of completed DA Form 7335-R, TOW Gunnery Tables 5 and 6: Baseline TFTT Gunnery Practice and Qualification.

c. **TOW Gunnery Table 7, Squad Gunnery Practice.** This table trains the TOW squad to engage stationary and moving targets in a simulated battlefield scenario (Figure 4-7, page 4-22). It prepares the TOW squad for TOW Gunnery Table 8, Squad Gunnery Qualification, by presenting engagement scenarios and task requirements similar to those for evaluation. TOW Gunnery Table 7 may be conducted by a TOW HMMWV on an MPRC light.

(1) ***Conduct.*** All firings will be from the hull down position. Reload procedures will be executed from a defilade position. All stationary targets should be mounted on pop-up mechanisms to facilitate target acquisition and scoring. If pop-up mechanisms are not available, exposed stationary targets will be placed in an appropriate array. If exposed targets must be used, the exercise should be designed so that targets are presented only to the position from which they should be engaged. Moving targets should be presented at speeds from 5 to 40 kilometers per hour.

(2) ***Scoring.*** The squad is rated based on the total point score. For scoring gunnery, time will start for the first target when the target is raised and will stop when the target is hit. All reloads will be made from a hull defilade position. Time for the second target begins when the vehicle reoccupies the firing position and ends when the target is hit. DA Form 7336-R (TOW Gunnery Table 7, Squad Gunnery Practice Scoresheet, Figure 4-8, page 4-23) will be used to record scores.

(a) The table consists of five tasks. Each task consists of one or more target engagements from one firing position. Each target engagement is worth a maximum of 100 points. The point score is determined by the time taken to kill the target using the live-fire point calculation sheet (Figure 4-16, page 4-35). Any crew or leader cuts are deducted from this score (Table 4-5, page 4-21 and Table 4-6, page 4-26).

(b) Each task is also worth a maximum of 100 possible points. This score is calculated by adding the scores for all engagements made during the task and dividing by the number of targets engaged. For example, three targets were engaged and the scores were 75, 82, and 93 for a total of 250 points divided by 3 (the number of targets presented) for a task score of 81 points.

(c) Each table is performed once during the day and once at night. Scoring for day engagements will be made by the evaluator in the range tower. Scoring for night engagements will be made by both evaluators. The scores of the two events are added together for a maximum possible score of 1,000 points. This is the final score for the gunnery table.

NOTE: This scoring process is the same for Tables 7 through 12.

CREW CUTS	
MINOR (Deduct 5 points)	**MAJOR (Deduct 30 points)**
Improper fire commands. Firing before receiving command to fire. Incorrect engagement techniques (such as engaging the least dangerous target before the most dangerous target).	Failure to conduct system self-test/boresight before engaging first target. Failure to collimate day sight and or nightsight before engaging first target. Failure to be in MOPP4 and close all hatches during NBC engagement. Engagement of friendly targets (if friendly targets are presented).

Table 4-5. Crew cuts.

d. **TOW Gunnery Table 8, Squad Gunnery Qualification**. TOW Gunnery Table 8 evaluates the TOW squad on its ability to engage stationary and moving targets in a simulated battlefield scenario (Figure 4-7, page 4-22). The conduct, scoring, suggested support requirements, and trainer/evaluator guidelines are the same as for TOW Gunnery Table 7. The squad's scores will be recorded on DA Form 7337-R, (TOW Gunnery Table 8, Squad Qualification Scoresheet, Figure 4-9, page 4-24).

OPERATION ORDER

SITUATION:

Enemy. The enemy is trying to gain control of key military bases, landing strips, and ports. The 125th and 45th Motorized Rifle Regiments have pushed north towards the objectives of the airfields at Ft. Campbell and Duck AFB. They have been held along the 00 East-West gridline in the vicinity of McKenna. The 125th MPR has swung northeast and is advancing on objectives at EH 42250770 (Watts airstrip) and EH 45101235 (Duffell railhead). Their intent is to secure these areas to facilitae resupply for themselves and for future reinforcements. The 125th is operating at 75 per cent strength and morale is high with the anticipation that they will meet their objectives with little resistance from American forces in the area. The 125th has T72s and BRMs. They are well trained in chemical warfare and have used nonpersistant chemical agents in recent engagements.

Friendly. The section mission is to deny enemy reconnaissances the ability to gain information on the defense of Watts airstrip and Duffell railhead. To our left is __________ (whatever section is appropriate). A Company of the 15th Engineer Battalion has constructed some firing positions and barriers. Our platoon will have priority of fires from one section of our battalion 120mm mortars.

MISSION:

The mission of your squad will be to defend in sector north of Marshall Creek to destroy enemy reconnaissance to prevent them from gaining intelligence concerning Watts airstrip.

EXECUTION:

Concept of the operation. Your squad will occupy on order five battle positions: BP 1 at EH 42260840; BP 2 at EH 42560855; BP 3 at EH 42000858; BP 4 at EH 41810860; and BP 5 at EH 41460872. On order, you will depart from the release point at EH 42340866 and move to BP 1.

Figure 4-7. Example squad gunnery practice and qualification scenario.

TOW GUNNERY TABLE 7: SQUAD GUNNERY PRACTICE

For use of this form, see FM 3-22.34. The proponent agency is TRADOC.

TC: SSG MARK LDR: ________ GNR: SPC ERICKS DVR: SPC MAXWELL | DATE: 5 AUG 03 TIME: 1645

PLATOON: 2D COMPANY: B BATTALION: 2/2 IN. LAB

TASK	CONDITIONS/ TARGETS/ SITUATION	TOTAL TARGET EXPOSURE TIME	HITS	ENGAGEMENT TIMES	POINTS	CREW/ LDR CUTS	TOTAL POINTS
1. ENGAGE A SINGLE TARGET	FIRED FROM BASELINE. 3,000 - 3,750 METERS 1 STATIONARY TANK	TFTT 25 SECS ALL VEHICLES	①	ST TK-17	92		92
2. ENGAGE MULTIPLE TARGETS	FIRED FROM PHASE LINE 2. 2,000 - 2,500 METERS 1 - ST TANK, 1 - MV TANK, 1 - MV BMP	TFTT HMMWV/2:30	① ② ③	ST TK-18 MV TK-40 MV BMP-35	86 83 24		64
3. ENGAGE MULTIPLE TARGETS	FIRED FROM PHASE LINE 3. 2,500 - 3,000 METERS 1 - MV TANK, 1 - ST BMP	TFTT HMMWV/1:40	1 ②	ST TK-0 ST BMP-18	0 86	-5 IMPROPER FIRE COMMAND	38
4. ENGAGE MULTIPLE TARGETS	FIRED FROM PHASE LINE 2. 2,500 - 3,000 METERS 1 ST TANK, 1 - MV TANK (NBC)	TFTT HMMWV/1:40	① ②	ST TK-18 MV TK-34	86 94		90
5. ENGAGE MULTIPLE TARGETS	FIRED FROM BASELINE. 2,500 - 3,000 METERS 1 - MV TANK, 1 - MV BMP (NBC)	TFTT HMMWV/1:40	① ②	MV TK-34 MV BMP-18	94 86		60
					TOTAL SCORE		344

NOTES:

1. ENGAGEMENT TIMES ARE CALCULATED FROM THE TIME THE TARGET IS EXPOSED TO THE TIME THE TARGET IS HIT. USE THE LIVE-FIRE POINT CALCULATION SHEET FOR SECOND AND THIRD ENGAGEMENTS FROM THE SAME POSITION. THE TIME BEGINS WHEN THE VEHICLE REOCCUPIES ITS FIRING POSITION (STOPS MOVING) AND ENDS WHEN THE TARGET IS HIT.
2. EACH TASK IS WORTH 100 POINTS. EACH TARGET IS ALSO WORTH 100 POINTS FOR TASKS WITH MULTIPLE ENGAGEMENTS. TOTAL THE NUMBER OF POINTS AND DIVIDE BY THE NUMBER OF TARGETS PRESENTED.
3. CREW/LEADER CUTS: SEE FM 3-22.34, CH 4.
4. SCORING TABLE 7: TABLE 7 IS PERFORMED TWICE, ONCE DURING THE DAY AND ONCE AT NIGHT. EACH EVENT IS WORTH A MAX OF 500 POINTS. ADD THE TWO SCORES TOGETHER FOR A POSSIBLE TOTAL OF 1,000 POINTS. SQUADS MUST ACHIEVE A MINIMUM TOTAL OF 700 POINTS TO VERIFY ON TABLE 7 AND PROGRESS TO TABLE 8.

DA FORM 7336-R, AUG 2003 APD V1.00

Figure 4-8. Example of completed DA Form 7336-R, TOW Gunnery Table 7: Squad Gunnery Practice.

TOW GUNNERY TABLE 8: SQUAD GUNNERY QUALIFICATION

For use of this form, see FM 3-22.34. The proponent agency is TRADOC.

TC: SSG MAXWELL LDR: ________ GNR: SPC COOKS DVR: PFC ASHLEY | DATE: 6 AUG 03 TIME: 1945

PLATOON: 1 ST COMPANY: D BATTALION: 1/502 IN

TASK	CONDITIONS/ TARGETS/ SITUATION	TOTAL TARGET EXPOSURE TIME	HITS	ENGAGEMENT TIMES	POINTS	CREW/ LDR CUTS	TOTAL POINTS
1. ENGAGE A SINGLE TARGET	FIRED FROM BASELINE. 1,500 - 2,000 METERS 1 STATIONARY TANK	TFTT 25 SECS ALL VEHICLES	1	ST TK-20	74		74
2. ENGAGE MULTIPLE TARGETS	FIRED FROM PHASE LINE 1. 2,000 - 2,500 METERS 1 - MV TANK, 1 - ST BMP	TFTT HMMWV/1:40	1 2	MV TK-34 ST BMP-18	94 86		90
3. ENGAGE MULTIPLE TARGETS	FIRED FROM PHASE LINE 2. 3,000 - 3,750 METERS 2 - MV TANK, 1 - ST BMP	TFTT HMMWV/2:45	1 2 3	MV TK-30 MV TK-25 ST BMP-0	94 98 0	-5 IMPROPER FIRE COMMAND	60
4. ENGAGE MULTIPLE TARGETS	FIRED FROM PHASE LINE 1. 1,500 - 2,000 METERS 1 - ST TANK, 1 - MV TANK (NBC)	TFTT HMMWV/1:40	1 2	ST TK-20 MV BMP-19	74 84		79
5. ENGAGE MULTIPLE TARGETS	FIRED FROM BASELINE. 2,500 - 3,000 METERS 1 - MV TANK, 1 - MV BMP (NBC)	TFTT HMMWV/1:40	1 2	MV TK-34 MV BMP-18	94 86	-5 IMPROPER FIRE COMMAND	85
					TOTAL SCORE		388

NOTES:

1. TABLE 8 IS CONDUCTED AND TIMED IN EXACTLY THE SAME MANNER AS TABLE 7.
2. TABLE 8, UNLIKE TABLE 7, IS FOR QUALIFICATION. THE SQUAD IS RATED ON ITS FINAL SCORE, RATHER THAN SIMPLY HAVING TO MEET A MINIMUM STANDARD.
3. CREW/LEADER CUTS: SEE FM 3-22.34, CH 4.
4. SCORING TABLE 8: TABLE 8 IS PERFORMED TWICE, ONCE DURING THE DAY AND ONCE AT NIGHT. EACH EVENT IS WORTH A MAX OF 500 POINTS. ADD THE TWO SCORES TOGETHER FOR A POSSIBLE TOTAL OF 1,000 POINTS. SQUADS ARE RATED ON THE FOLLOWING BASIS:

DISTINGUISHED: 900 - 1,000 POINTS SUPERIOR: 800 - 899 POINTS QUALIFIED: 700 - 799 POINTS UNQUALIFIED: 0 - 699 POINTS

DA FORM 7337-R, AUG 2003 APD V1.00

Figure 4-9. Example of completed DA Form 7337-R, TOW Gunnery Table 8: Squad Gunnery Qualification.

4-9. PHASE III: SECTION TRAINING

Phase III begins section-level TOW training. TOW Gunnery Tables 9 and 10 are moving gunnery exercises where the vehicle moves from point to point on a course and engages a series of targets. These gunnery exercises are used to train and evaluate the TOW unit's combat capability. Each element has two tables: a practice table to train and a qualification table for evaluation. Each table is performed once during the day and once during the night under realistic (simulated) battlefield conditions. Point scores for target engagements are based on a combat model that predicts success based on the time from target detection to target engagement, type of target, and range to target. These scores are taken from the live-fire point calculation sheet (Figure 4-16, page 4-35). Point scores are also affected by crew cuts (crew cuts are crew errors, either major or minor, that reduce the point score for a target engagement). The actual rating is calculated by adding the scores for the day and night phases and dividing them by two to get an average score. During move-out scenarios, the senior trainer is normally to the rear of the leader's vehicle for evaluation and training purposes. An assistant evaluator of the appropriate rank and experience is normally assigned to each additional vehicle that is evaluated as part of the same exercise. Evaluators must monitor crew conversations and fire commands.

a. **TOW Gunnery Table 9, Section Gunnery Practice.** This table trains the TOW section to control fire and distribution. It contains offensive and defensive engagements in a simulated battlefield scenario (Figure 4-10, page 4-27). TOW Gunnery Table 9 prepares the TOW section for the qualification table by presenting engagement scenarios and task requirements similar to those for evaluation. It may be conducted by a TOW HMMWV on an MPRC light.

(1) ***Conduct.*** All firings will be from the hull down position. Reload procedures will be executed from a defilade position. All stationary targets should be mounted on pop-up mechanisms to facilitate target acquisition and scoring. If pop-up mechanisms are not available, exposed stationary targets will be placed in an appropriate array. If exposed targets must be used, the exercise should be designed so that targets are presented only to the position from which they should be engaged. Moving targets should be presented at speeds from 5 to 40 kilometers per hour.

(2) ***Scoring.*** The section is rated based on the total point score accumulated. For scoring gunnery, times for engagements will be scored separately for each vehicle. Start time for the section leader's vehicle will be the same as in TOW Gunnery Table 7. Start time for the squad leader's vehicle will be upon completion of the fire directive by the section leader, which will include the method of engagement. Stop time will be when the target is hit. Points will be awarded IAW the live-fire point calculation sheet (Figure 4-16, page 4-35). Deductions will be made from this point score when a major or minor crew cut error (Table 4-5, page 4-21) or a section leader cut error (Table 4-6, page 4-26) is made. The evaluator in the range tower will score day engagements. Both evaluators will score night engagements. DA Form 7338-R (TOW Gunnery Table 9, Section Gunnery Practice Scoresheet, Figure 4-11, page 4-28) will be used to record the scores.

LEADER CUTS	
MINOR (Deduct 10 points)	MAJOR (Deduct 30 points)
Give incorrect fire commands. Fail to give SITREP to tower.	— Fail to properly distribute and control section fires.

Table 4-6. Section leader cuts.

(3) ***Suggested support requirements.*** The following support requirements are recommended when conducting TOW Gunnery Tables 9 and 10.

- Minimum evaluators (evaluators will not assist the gunner in finding or identifying targets):
  - — One evaluator per section and squad.
  - — Two evaluators in the range tower to record scores on engagements.
- Opposing forces: None.
- Support troops: None.
- Vehicles/communications:
  - — One AN/PRC-119 radio per section and squad evaluator.
  - — Two AN/PRC-119 radios for the tower evaluator (one primary, one spare).
  - — Two frequencies (one for the tower and section evaluator and one for the antiarmor squad).
- Maneuver area:
  - — Multipurpose range complex.
  - — A route for the element to move from the assembly area to the first firing point without damaging the underground wiring.
- Firing area:
  - — The TOW backblast area (75 meters by 90 degrees) will be clear.
  - — Smoke may be used.
- Training aids, devices, and special equipment:
  - — TFTT system.
  - — Sufficient TFTT retroreflectors to cover the targets. (Use of MILES harness and LTIDs is optional.)
- Ammunition: Six M80 blast simulators per TOW, per iteration.
- Key references: FM 3-21.91(FM 7-91), FM 3-22.34(FM 23-34), and live-fire point calculation sheet.

(4) ***Trainer/Evaluator Guidelines***. Because this table is conducted on an MPRC, marking the firing positions may be necessary. At night, the roads must be marked with luminous tape.

b. **TOW Gunnery Table 10, Section Gunnery Qualification.** TOW Gunnery Table 10 evaluates the TOW section on controlling fire and distribution. It contains offensive and defensive engagements in a simulated battlefield scenario (Figure 4-10, page 4-27). The conduct, scoring, suggested support requirements, and trainer/evaluator guidelines are the same as for TOW Gunnery Table 9. The section's scores will be recorded on

DA Form 7339-R (TOW Gunnery Table 10, Section Gunnery Qualification Scoresheet, Figure 4-12, page 4-29).

OPERATION ORDER

SITUATION:

Enemy. **The enemy is trying to gain control of key military bases, landing strips, and ports. The 125th and 45th Motorized Rifle Regiments have pushed north towards the objectives of the airfields at Ft. Campbell and Duck AFB. They have been held along the 00 East-West gridline in the vicinity of McKenna. The 125th MRR has swung northeast and is advancing on objectives at EH 42250770 (Watts airstrip) and EH 45101235 (Duffell railhead). Their intent is to secure these areas to facilitate resupply for themselves and for future reinforcements. The 125th is operating at 75 per cent strength and morale is high with the anticipation that they will meet their objectives with little resistance from American forces in the area. The 125th has T72s and BRMs. They are well trained in chemical warfare and have used nonpersistant chemical agents in recent engagements.**

Friendly. **The platoon mission is to deny enemy reconnaissances the ability to gain information on the defense of Watts airstrip and Duffell railhead. To our left is __________ (whatever platoon is appropriate). A Company of the 15th Engineer Battalion hs constructed some firing positions and barriers. Our platoon will have priority of fires from one section of our battalion 120mm mortars.**

MISSION:

The mission of your section will be to defend in sector north of Marshall Creek to destroy enemy reconnaissance to prevent them from gaining intelligence concerning Watts airstrip.

EXECUTION:

Concept of the operation. **Your section will occupy on order five battle positions: BP 1 at EH 42260840; BP 2 at EH 42560855; BP 3 at EH 42000858; BP 4 at EH 41810860; and BP 5 at EH 41460872. On order, you will depart from the release point at EH 42340866 and move to BP 1.**

Figure 4-10. Example section gunnery practice and qualification scenario.

TOW GUNNERY TABLE 9: SECTION GUNNERY PRACTICE

For use of this form, see FM 3-22.34. The proponent agency is TRADOC.

SECTION	TC: SSG HEIDER	LDR:	GNR: SPC CONNOR	DVR: PFC BROWN	DATE: 15 AUG 03
SQUAD	TC: SGT QUEEN	LDR:	GNR: PFC MARKUS	DVR: PFC WILSON	TIME: 0900

PLATOON: ______ COMPANY: ______ BATTALION: ______

TASK	CONDITIONS/ TARGETS/ SITUATION	TOTAL TARGET EXPOSURE TIME	HITS	ENGAGEMENT TIMES	POINTS	CREW/ LDR CUTS	TOTAL POINTS
1. ENGAGE A SINGLE TARGET	FIRED FROM BASELINE. 3,000 - 3,750 METERS 1 - ST TANK, 1 - ST BMP	TFTT 40 SECS ALL VEHICLES	(1) (2)	ST - 20 SB - 30	74 35		55
2. ENGAGE MULTIPLE TARGETS	FIRED FROM PHASE LINE 2. 2,000 - 2,750 METERS 1 - ST BMP	TFTT 25 SECS ALL VEHICLES	(1)	SB - 20	74		74
3. ENGAGE MULTIPLE TARGETS	FIRED FROM PHASE LINE 3. 1,000 - 1,500 METERS 2 - MV BMPS, 1 - ST TANK	TFTT HMMWV/1:40	(1) (2) (3)	MB - 16 MB - 18 ST - 13	91 87 100		93
4. ENGAGE MULTIPLE TARGETS	FIRED FROM PHASE LINE 2. 2,500 - 3,000 METERS 2 - MV TANKS, 2 - ST TANKS (NBC)	TFTT HMMWV/2:38	(1) (2) (3) (4)	MT - 19 MT - 21 ST - 20 ST - 22	100 99 74 64	-10 INCORRECT FIRE COMMAND	74
5. ENGAGE MULTIPLE TARGETS	FIRED FROM BASELINE. 1,000 - 1,500 METERS 1 - ST TANK, 1 - ST BMP (NBC)	TFTT 25 SECS ALL VEHICLES	(1) (2)	ST - 16 SB - 17	85 92		89
					TOTAL SCORE		385

NOTES:

1. TABLE 9 IS CONDUCTED AND TIMED IN EXACTLY THE SAME MANNER AS TABLES 7 AND 8.
2. SCORING TABLE 9: TABLE 9 IS PERFORMED TWICE, ONCE DURING THE DAY AND ONCE AT NIGHT. EACH EVENT IS WORTH A MAX OF 500 POINTS. ADD THE TWO SCORES TOGETHER FOR A POSSIBLE TOTAL OF 1,000 POINTS. TABLE 9 IS A PRACTICE TABLE. TO VERIFY ON TABLE 9 AND PROGRESS TO TABLE 10, THE SECTION MUST ACHIEVE A MINIMUM COMBINED SCORE OF AT LEAST 700 POINTS, AFTER CREW AND LEADER CUTS.

DA FORM 7338-R, AUG 2003 APD V1.00

Figure 4-11. Example of completed DA Form 7338-R, TOW Gunnery Table 9: Section Gunnery Practice.

TOW GUNNERY TABLE 10: SECTION GUNNERY QUALIFICATION

For use of this form, see FM 3-22.34. The proponent agency is TRADOC.

SECTION	TC: SGT BERRY	LDR: PFC KIM	GNR: SPC MARKS	DVR: PFC SMITH	DATE: 28 AUG 03
SQUAD	TC: SGT WATT	LDR: PFC LUCAS	GNR: SPC COLEMAN	DVR: PFC RICHARDS	TIME: 1600

PLATOON: 2 D COMPANY: E BATTALION: 2/16 IN

TASK	CONDITIONS/ TARGETS/ SITUATION	TOTAL TARGET EXPOSURE TIME	HITS	ENGAGEMENT TIMES	POINTS	CREW/ LDR CUTS	TOTAL POINTS
1. ENGAGE A SINGLE TARGET	FIRED FROM BASELINE. 3,000 - 3,750 METERS 2 - MV TANK, 2 - ST BMPS	TFTT HMMWV/2:38	① ② ③ ④	MT-30 MT-37 SB-17 SB-16	96 93 92 100	-10 INCORRECT ENGAGEMENT TECH.	85
2. ENGAGE MULTIPLE TARGETS	FIRED FROM PHASE LINE 2. 2,000 - 2,500 METERS 1 - MV TANK	TFTT 40 SECS ALL VEHICLES	①	MT-40	83		83
3. ENGAGE MULTIPLE TARGETS	FIRED FROM PHASE LINE 3. 1,000 - 1,500 METERS 1 - MV BMP, 1 - ST TANK	TFTT 25 SECS ALL VEHICLES	① ②	ST-64 MB-59	64 59		62
4. ENGAGE MULTIPLE TARGETS	FIRED FROM PHASE LINE 2. 2,500 - 3,000 METERS 1 - ST BMP, 2 - MV TANKS (NBC)	TFTT HMMWV/2:40	① ② ③	SB-18 MT-34 MT-24	86 94 98		93
5. ENGAGE MULTIPLE TARGETS	FIRED FROM BASELINE. 3,000 - 3,750 METERS 1 - ST TANK, 1 - ST BMP (NBC)	TFTT 25 SECS ALL VEHICLES	① 2	ST-20 MB-0	74 0		37
					TOTAL SCORE		360

NOTES:

1. TABLE 10 IS CONDUCTED AND TIMED IN EXACTLY THE SAME MANNER AS TABLE 8 AND 9.
2. SCORING TABLE 10: TABLE 10 IS PERFORMED TWICE, ONCE DURING THE DAY AND ONCE AT NIGHT. EACH EVENT IS WORTH A MAX OF 500 POINTS. ADD THE TWO SCORES TOGETHER FOR A POSSIBLE TOTAL OF 1,000 POINTS. TABLE 10 IS A QUALIFICATION TABLE. TO PROGRESS TO TABLE 11, THE SECTION MUST ACHIEVE A MINIMUM COMBINED SCORE OF AT LEAST 700 POINTS.

DA FORM 7339-R, AUG 2003 APD V1.00

Figure 4-12. Example of completed DA Form 7339-R, TOW Gunnery Table 10: TOW Section Gunnery Qualification.

4-10. PHASE IV: PLATOON TRAINING

Phase IV begins platoon-level TOW training. TOW Gunnery Tables 11 and 12 are moving gunnery exercises where the vehicle moves from point to point on a course and engages a series of targets. These gunnery exercises are used to train and evaluate the TOW unit's combat capability. Each element has two tables: a practice table to train and a qualification table for evaluation. Each table is performed once during the day and once during the night under realistic (simulated) battlefield conditions. Point scores for target engagements are based on a combat model that predicts success based on the time from target detection to target engagement, type of target, and range to target. These scores are taken from the live-fire point calculation sheet (Figure 4-16, page 4-35). Point scores are also affected by crew cuts (crew cuts are crew errors, either major or minor, that reduce the point score for a target engagement). The actual rating is calculated by adding the scores for the day and night phases and dividing them by two to get an average score. During move-out scenarios, the senior trainer is normally to the rear of the leader's vehicle for evaluation and training purposes. An assistant evaluator of the appropriate rank and experience is normally assigned to each additional vehicle that is evaluated as part of the same exercise. Evaluators must monitor crew conversations and fire commands.

a. **TOW Gunnery Table 11, Platoon Gunnery Practice.** This table trains the TOW platoon to control fire and distribution. It contains offensive and defensive engagements in a simulated battlefield scenario (Figure 4-13, page 4-32). TOW Gunnery Table 11 prepares the TOW platoon for the qualification table by presenting engagement scenarios and task requirements similar to those for evaluation. It may be conducted by a TOW HMMWV on an MPRC light.

(1) ***Conduct.*** All firings will be from the hull down position. Reload procedures will be executed from a defilade position. All stationary targets should be mounted on pop-up mechanisms to facilitate target acquisition and scoring. If pop-up mechanisms are not available, exposed stationary targets will be placed in an appropriate array. If exposed targets must be used, the exercise should be designed so that targets are presented only to the position from which they should be engaged. Moving targets should be presented at speeds from 5 to 40 kilometers per hour.

(2) ***Scoring.*** The platoon is rated based on the total point score accumulated. For scoring gunnery, times for engagements will be scored separately for each vehicle. Start time will begin upon completion of the directive given to the vehicle commander. Stop time will be when the target is hit. Points will be awarded IAW the live-fire point calculation sheet (Figure 4-16, page 4-35). Deductions will be made from this point score when a major or minor crew cut error (Table 4-5, page 4-21) or a platoon leader cut error (Table 4-7) is made. The evaluator in the range tower will score day engagements. Both evaluators will score night engagements. DA Form 7340R (TOW Gunnery Table 11, Platoon Gunnery Practice Scoresheet, Figure 4-14, page 4-33) will be used to record the scores.

PLATOON LEADER CUTS	
MINOR (Deduct 10 points)	MAJOR (Deduct 30 points)
Give incorrect fire commands. Fail to give SITREP to tower.	Fail to properly distribute and control section fires.

Table 4-7. Platoon leader cuts.

(3) ***Suggested Support Requirements.*** The following support requirements are recommended when conducting TOW Gunnery Tables 11 and 12.

- Minimum evaluators (evaluators will not assist the gunner in finding or identifying targets):
  - — One evaluator per platoon, section, and squad.
  - — Two evaluators in the range tower to record scores on engagements.
- Opposing forces: None.
- Support troops: None.
- Vehicles/communications:
  - — One AN/PRC-119 radio per platoon, section, and squad evaluator.
  - — Two AN/PRC-119 radios for the tower evaluator (one primary, one spare).
  - — Two frequencies (one for the tower and platoon evaluator and one for the antiarmor platoon).
- Maneuver area:
  - — Multipurpose range complex.
  - — A route for the element to move from the assembly area to the first firing point without damaging the underground wiring.
- Firing area:
  - — The TOW backblast area (75 meters by 90 degrees) will be clear.
  - — Smoke may be used.
- Training aids, devices, and special equipment:
  - — TFTT system.
  - — Sufficient TFTT retroreflectors to cover the targets. (Use of MILES harness and LTIDs is optional.)
- Ammunition: Six M80 blast simulators per TOW, per iteration.
- Key references: FM 3-21.91(FM 7-91), FM 3-22.34(FM 23-34), and live-fire point calculation sheet.

(4) ***Trainer/Evaluator Guidelines.*** Because this table is conducted on an MPRC, marking the firing positions may be necessary. At night, the roads must be marked with luminous tape.

b. **TOW Gunnery Table 12, Platoon Gunnery Qualification.** TOW Gunnery Table 12 evaluates the TOW platoon on controlling fire and distribution. It contains offensive and defensive engagements in a simulated battlefield scenario (Figure 4-13, page 4-32). The conduct, scoring, suggested support requirements, and trainer/evaluator guidelines are the same as for TOW Gunnery Table 11. The platoon's scores will be recorded on DA

Form 7341-R (TOW Gunnery Table 12, Platoon Gunnery Qualification Scoresheet, Figure 4-15, page 4-34).

SITUATION:

Enemy. The enemy is trying to gain control of key military bases, landing strips, and ports. The 125th and 45th Motorized Rifle Regiments have pushed north towards the objectives of the airfields at Ft. Campbell and Duck AFB. They have been held along the 00 East-West gridline in the vicinity of McKenna. The 125th MRR has swung northeast and is advancing on objectives at EH 42250770 (Watts airstrip) and EH 45101235 (Duffell railhead). Their intent is to secure these areas to facilitate resupply for themselves and for future reinforcements. The 125th is operating at 75 per cent strength and morale is high with the anticipation that they will meet their objectives with little resistance from American forces in the area. The 125th has T72s and BRMs. They are well trained in chemical warfare and have used nonpersistant chemical agents in recent engagements.

Friendly. The company mission is to deny enemy reconnaissances the ability to gain information on the defense of Watts airstrip and Duffell railhead. To our left is __________ (whatever company is appropriate). A Company of the 15th Engineer Battalion hs constructed some firing positions and barriers. Our platoon will have priority of fires from one section of our battalion 120mm mortars.

MISSION:

The mission of your platoon will be to defend in sector north of Marshall Creek to destroy enemy reconnaissance to prevent them from gaining intelligence concerning Watts airstrip.

EXECUTION:

Concept of the operation. Your platoon will occupy on order five battle positions: BP 1 at EH 42260840; BP 2 at EH 42560855; BP 3 at EH 42000858; BP 4 at EH 41810860; and BP 5 at EH 41460872. On order, you will depart from the release point at EH 42340866 and move to BP 1.

Figure 4-13. Example platoon gunnery practice and qualification scenario.

TOW GUNNERY TABLE 11: PLATOON GUNNERY PRACTICE

For use of this form, see FM 3-22.34. The proponent agency is TRADOC.

SECTION 1:	TOW CREW 1:	TC: SSG CAMPBELL	LDR:	GNR: SPC COOK	DVR: PFC BROWN	DATE: 15 AUG 93
	TOW CREW 2:	TC: SGT HAMMON	LDR:	GNR: SPC ADAMS	DVR: PFC GONZALES	
SECTION 2:	TOW CREW 1:	TC: SSG QUEEN	LDR:	GNR: SPC McCOY	DVR: PFC ZALAS	
	TOW CREW 2:	TC: SGT WEBB	LDR:	GNR: SPC COOKS	DVR: PFC SMITH	TIME: 1100
SECTION 3:	TOW CREW 1:	TC: SSG PAINE	LDR:	GNR: SPC ERVIC	DVR: PFC FOWLER	
	TOW CREW 2:	TC: SGT MILLER	LDR:	GNR: SPC SHUTT	DVR: PFC ANDERSON	

PLATOON: 3RD COMPANY: D BATTALION: 2/503 IN

TASK	CONDITIONS/ TARGETS/ SITUATION	TOTAL TARGET EXPOSURE TIME	HITS	ENGAGEMENT TIMES	POINTS	CREW/ LDR CUTS	TOTAL POINTS
1. ENGAGE A SINGLE TARGET	FIRED FROM BASELINE. 3,000 - 3,750 METERS 4 - MV TANKS, 2 - ST BMPS	TFTT HMMWV/3:26		MT-33 MT-29 MT-37 MT-31 SB-22 SB-17	95 97 93 96 64 92		90
2. ENGAGE MULTIPLE TARGETS	FIRED FROM PHASE LINE 2. 2,000 - 3,000 METERS 4 - MV BMPS 2 - ST BMPS	TFTT HMMWV/2:26		MB-20 MB-14 MB-23 MB-25 SB-17 SB-0	73 93 96 82 92 0		71
3. ENGAGE MULTIPLE TARGETS	FIRED FROM PHASE LINE 3. 1,000 - 2,000 METERS 2 - ST TANKS, 2 - ST BMPS	TFTT 25 SECS ALL VEHICLES		ST-17 ST-19 SB-21 SB-26	92 79 68 47		72
4. ENGAGE MULTIPLE TARGETS	FIRED FROM PHASE LINE 2. 2,000 - 3,000 METERS 2- MV TANKS, 2 - MV BMPS (NBC)	TFTT BMP/43 SEC TANK/1:03		MT-17 MT-22 MB-23 MB-19	100 99 96 93	-10 INCORRECT FIRE COMMAND	85
5. ENGAGE MULTIPLE TARGETS	FIRED FROM BASELINE. 3,000 - 3,750 METERS 2 - ST TANKS, 2 - ST BMPS (NBC)	TFTT 25 SECS ALL VEHICLES	4	ST-15 ST-21 SB-24 SB-0	100 68 55 0		56
					TOTAL SCORE		374

NOTES:

1. TABLE 11 IS CONDUCTED, TIMED, AND SCORED IN EXACTLY THE SAME MANNER AS TABLES 7 THRU 10.

DA FORM 7340-R, AUG 2003 APD V1.00

Figure 4-14. Example of completed DA Form 7340-R, TOW Gunnery Table 11: Platoon Gunnery Practice.

TOW GUNNERY TABLE 12: PLATOON GUNNERY QUALIFICATION

For use of this form, see FM 3-22.34. The proponent agency is TRADOC.

SECTION 1:	TOW CREW 1:	TC: SSG COOKS	LDR: SPC SMITH	GNR: SPC BROOKS	DVR: PFC MIDDLEBROOK	DATE: 15AUG03
	TOW CREW 2:	TC: CPL YORK	LDR: SPC WALTER	GNR: SPC CARTER	DVR: PFC PORTER	
SECTION 2:	TOW CREW 1:	TC: SSG ASHLEY	LDR: SPC HART	GNR: SPC JORDAN	DVR: PFC JOHNSON	
	TOW CREW 2:	TC: SGT OH	LDR: PFC CLAY	GNR: SPC PRICE	DVR: PFC CARDEN	TIME: 1400
SECTION 3:	TOW CREW 1:	TC: SGT BLAND	LDR: SPC OLIVER	GNR: SPC SELL	DVR: PFC OGLE	
	TOW CREW 2:	TC: SGT WALKER	LDR: PFC HARRIS	GNR: SPC NOEL	DVR: SPC PETMA	

PLATOON: 2D COMPANY: E BATTALION: 5/16 IN

TASK	CONDITIONS/ TARGETS/ SITUATION	TOTAL TARGET EXPOSURE TIME	HITS	ENGAGEMENT TIMES	POINTS	CREW/ LDR CUTS	TOTAL POINTS
1. ENGAGE A SINGLE TARGET	FIRED FROM BASELINE. 3,000 - 3,750 METERS 2 - MV TANKS, 2 - ST BMPS	TFTT BMP/25 SECS TANK/1:13	1, 2, 3, 4	MT-38, MT-0, SB-19, SB-23	92, 0, 79, 59		58
2. ENGAGE MULTIPLE TARGETS	FIRED FROM PHASE LINE 2. 2,000 - 3,000 METERS 3 - ST BMPS, 3 - MV TANKS	TFTT HMMWV/2:28	1, 2, 3, 4, 5, 6	SB-20, SB-17, SB-14, MT-23, MT-33, MT-0	74, 92, 64, 96, 90, 0	INCORRECT FIRE COMMAND -10	59
3. ENGAGE MULTIPLE TARGETS	FIRED FROM PHASE LINE 3. 1,000 - 2,000 METERS 3 - ST TANKS, 3 - MV BMPS	TFTT HMMWV/1:08	1, 2, 3, 4, 5, 6	ST-17, ST-19, ST-26, MB-31, MB-29, MB-0	92, 79, 47, 57, 61, 0		56
4. ENGAGE MULTIPLE TARGETS	FIRED FROM PHASE LINE 2. 1,500 - 2,500 METERS 2 - MV TANKS, 2 - MV BMPS (NBC)	TFTT BMP/34 SECS TANK/1:03	1, 2, 3, 4	MT-23, MT-24, MB-15, MB-0	96, 95, 100, 0		73
5. ENGAGE MULTIPLE TARGETS	FIRED FROM BASELINE. 3,000 - 3,750 METERS 2 - ST TANKS, 2 - ST BMPS (NBC)	TFTT 25 SECS ALL VEHICLES	1, 2, 3, 4	ST-15, ST-17, SB-17, SB-23	100, 92, 92, 59	-10 NO SITREP SENT	76
NOTES:					TOTAL SCORE		322

1. TABLE 12 IS CONDUCTED, TIMED, AND SCORED IN EXACTLY THE SAME MANNER AS TABLES 7 THRU 10.

DA FORM 7341-R, AUG 2003 APD V1.00

Figure 4-15. Example of completed DA Form 7341-R, TOW Gunnery Table 12: TOW Platoon Gunnery Qualification.

MOVING TARGET							
TIME SEC	STATIONARY TARGETS	BMP 65-900 METERS	BMP 901-2000 METERS	BMP 2001-3500 METERS	TANK 65-1500 METERS	TANK 1501-2500 METERS	TANK 2501-3750 METERS
1	100	100	100	100	100	100	100
2	100	100	100	100	100	100	100
3	100	100	100	100	100	100	100
4	100	100	100	100	100	100	100
5	100	100	100	100	100	100	100
6	100	100	100	100	100	100	100
7	100	100	100	100	100	100	100
8	100	100	100	100	100	100	100
9	100	100	100	100	100	100	100
10	100	100	100	100	100	100	100
11	100	92	100	100	100	100	100
12	100	86	100	100	97	100	100
13	100	79	97	100	94	99	100
14	100	74	95	100	91	99	100
15	100	68	93	100	88	99	100
16	100	64	91	98	85	98	100
17	92	59	89	96	81	98	100
18	86	55	87	95	78	98	100
19	79	51	84	93	75	97	100
20	74	47	82	91	72	97	100
21	68	44	80	89	69	97	99
22	64	41	77	88	66	96	99
23	59	38	75	86	63	96	99
24	55	35	73	84	60	95	98
25	51	33	70	82	57	95	98
26	47	30	68	80	54	94	98
27	44	28	66	79	52	94	97
28	41	26	63	77	49	93	97
29	38	24	61	75	46	92	97
30	35	22	59	73	44	92	96
31	33	21	57	71	41	91	96
32	30	19	54	70	39	90	95
33	28	18	52	68	37	90	85
34	26	16	50	66	34	89	94
35	24	15	48	64	32	88	94
36	22	14	46	62	30	87	93
37	21	13	44	61	28	86	93
38	19	12	42	59	27	85	92
39	18	11	40	57	25	84	92
40	16	10	38	56	23	83	91
41	15	9	36	54	22	82	91
42	14	9	35	52	20	81	90
43	13	8	33	51	19	80	90
44	12	7	31	49	18	79	89
45	11	7	30	47	17	78	89
46	10	6	28	46	15	77	88
47	9	6	27	44	14	75	88
48	9	5	26	43	13	74	87
49	8	5	24	41	12	73	86
50	7	4	23	40	12	71	85
51	7	4	22	39	11	70	84
52	6	4	21	37	11	68	83
53	6	3	19	36	9	67	82
54	5	3	18	35	9	65	81

Figure 4-16. Example of TOW live-fire point calculation sheet.

MOVING TARGET							
TIME SEC	STATIONARY TARGETS	BMP 65-900 METERS	BMP 901-2000 METERS	BMP 2001-3500 METERS	TANK 65-1500 METERS	TANK 1501-2500 METERS	TANK 2501-3750 METERS
55	5	3	17	33	8	64	80
56	4	3	16	32	7	62	78
57	4	2	16	31	7	61	77
58	4	2	15	30	6	59	75
59	3	2	14	29	6	58	74
60	3	2	13	28	5	56	72
61	3	2	12	27	5	54	70
62	3	1	12	25	5	53	69
63	2	1	11	24	4	51	67
64	2	1	10	24	4	49	65
65	2	1	10	23	4	48	63
66	2	1	9	22	3	46	61
67	2	1	9	21	3	45	60
68	1	1	8	20	3	43	58
69	1	1	8	19	2	41	56
70	1	1	7	18	2	40	55
71	1		7	18	2	38	53
72	1		6	17	2	37	52
73	1		6	16	2	35	50
74	1		5	15	2	34	49
75	1		5	15	1	32	47
76	1		5	14	1	31	46
77			5	13	1	30	45
78			4	13	1	28	43
79			4	12	1	27	42
80			4	12	1	26	41
81			3	11	1	25	40
82			3	11	1	23	39
83			3	10	1	22	38
84			3	10	1	21	37
85			3	9		20	36
86			2	9		19	35
87			2	9		18	34
88			2	8		17	33
89			2	8		16	32
90			2	7		15	31
91			2	7		15	30
92			2	7		14	29
93			1	6		13	28
94			1	6		12	27
95			1	5		12	26
96			1	5		11	25
97			1	5		10	24
98			1	5		10	23
99			1	5		9	22
100			1	5		9	21
101			1	4		8	20
102			1	4		8	19
103			1	4		7	18
104				4		7	17
105				3		6	16
106				3		6	15
107				3		6	14
108				3		5	13

Figure 4-16. Example of TOW live-fire point calculation sheet (continued).

4-11. FIELD TRACKING

Field tracking (or tracking in a field environment) provides practice and experience in tracking uncooperative targets and should be performed in conjunction with FTXs. It is not an element in the 12 required TOW gunnery tables, but it is useful in preparing for them, especially Tables 7 through 12. Other tactical employment tasks, such as occupying a firing position or completing a range card, should be performed at the same time. Careful planning by unit commanders will result in effective training that makes the best use of time, equipment, and personnel.

a. The tracking range should include an area more than 3,000 meters deep and at least 500 meters wide with hills and valleys, dead space, and covered terrain. Battlefield target conditions should be simulated as follows:

(1) Targets should be a variety of tanks, APCs, and other tactical vehicles. Many training installations now have various types of actual threat vehicles. These vehicles should be used when possible.

(2) Target vehicles should behave as much like real enemy vehicles on the battlefield as possible. They should change directions constantly and quickly (to include backing up for short distances) and should vary their speed. They should move within the gunner's sight picture from left to right, top to bottom, and bottom to top. The vehicles should move at angles to and from the weapon position as well as directly toward and away from the weapon.

(3) Gunners should see tanks and APCs in full and partial frontal, flank, and rear exposures.

(4) Gunners must engage the targets at all ranges with both the daysight tracker and the nightsight. To do so, they must be able to determine if a target is in range. Targets should move in and out of range so gunners can gain experience in target range determination.

(5) Armored vehicles should limit their exposure by moving quickly from one covered area to another to escape ATGM fires. The squad leader must have the gunner acquire and fire at the target while it is exposed. During field tracking, target vehicle exposure time is reduced to a minimum by having targets use covered areas during halts. This gives the gunner experience in tracking targets that disappear and reappear.

b. Field tracking should be performed under conditions that normally occur on the battlefield. These conditions include limited visibility caused by darkness, smoke, or fog; the discomfort of MOPP gear; and the noise and distraction caused by indirect and small-arms fire around the firing positions. Battlefield tracking conditions should be simulated as follows:

(1) The TOW crew members should train to engage targets during reduced visibility conditions using both the daysight tracker and the nightsight. Training should be conducted during daylight using the AN/TAS-4A. The AN/TAS-4A is a thermal sight, ***not*** just a nightsight. It offers many advantages in searching for and tracking targets in all kinds of weather and at any time of day. The AN/TAS-4A should be used extensively in conjunction with the daysight to maximize the abilities of the system.

(2) Enemy capabilities include the use of nuclear, biological, and chemical munitions. These munitions may not affect the weapon, but a gunner's ability to track will be degraded if he has never tracked while wearing a protective mask and protective clothing. Training is more realistic when the entire crew wears protective masks and clothing during part of the field tracking exercises.

(3) The enemy will try to suppress TOW fires with artillery and mortar fires. A gunner's ability to keep the sight crosshairs on a target may be affected under these conditions, especially if he has not been properly trained. Distractors should be used in

training to accustom gunners to such conditions and to minimize flinching. Grenade and artillery simulators are effective distractors. Small-arms fire can also be expected, but gunners should not be seriously distracted by its noise. Therefore, the simulation of small-arms fire is not required.

c. If facilities and equipment are not available to conduct a field tracking exercise, gunners should be allowed to track any uncooperative targets such as military or civilian vehicles traveling on a nearby highway. Any means available should be used to allow gunners to practice tracking targets that vary target speeds and directions of movement. This should be done often and can be accomplished in any location that has vehicle traffic.

d. TOW crews must be proficient in combat vehicle identification to gain the most benefit from field tracking. Combat vehicle identification skills have been standardized by GTA 17-02-011. This GTA includes six modules of 35-mm slides and instructor guides. Each module highlights five different NATO or Threat combat vehicles from five different views: frontal, right side, left side, right oblique, and left oblique.

CHAPTER 5

COMBAT TECHNIQUES OF FIRE

This chapter discusses the techniques and procedures of fire control, helicopter engagement, and firing the TOW under NBC and limited visibility conditions in a combat situation. These techniques and procedures greatly enhance the performance of the TOW weapon system in combat and increase its chances of survival.

Section I. FIRE CONTROL MEASURES

This section discusses fire control measures in combat to include target engagement determination procedures, the elements of fire commands, target tracking procedures, and target engagement procedures with the M220A1 and M220A2 TOW launchers.

5-1. TARGET ENGAGEMENT DETERMINATION

Mechanized units are trained to use terrain driving techniques to conceal movement, and drivers are taught to move vehicles quickly from one concealed position to another. TOW gunners and squad leaders must determine the range to a target and determine if the exposure time—the time a vehicle is in an opening between positions—is long enough to allow a missile to reach its target.

a. **Determine if a Target is Within Range**. The TOW gunner or squad leader can use the nightsight method or the binocular method to determine if a target is within range.

(1) ***Nightsight Method.*** The reticle within the nightsight is marked with stadia lines on each side of the vertical crossline and just above the intersection of the vertical and horizontal crosslines (Figure 5-1, page 5-2). Use these stadia lines to determine if a target is within range.

(a) Check to ensure the nightsight is in the narrow field of view.

(b) Look at the flank (broadside) of a target. If the ends of the target touch or extend beyond the stadia lines, the target is within range (Figure 5-2, page 5-2).

(c) If the target is moving straight or at an angle toward the position, use one-half of the stadia lines. If one side of the vehicle touches one of the stadia lines and the other side of the vehicle touches or extends beyond the vertical line of the crossline, the target is within range (Figure 5-3, page 5-2).

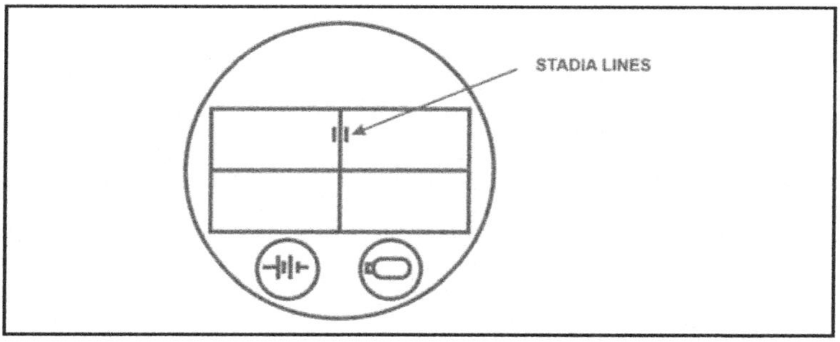

Figure 5-1. Stadia lines.

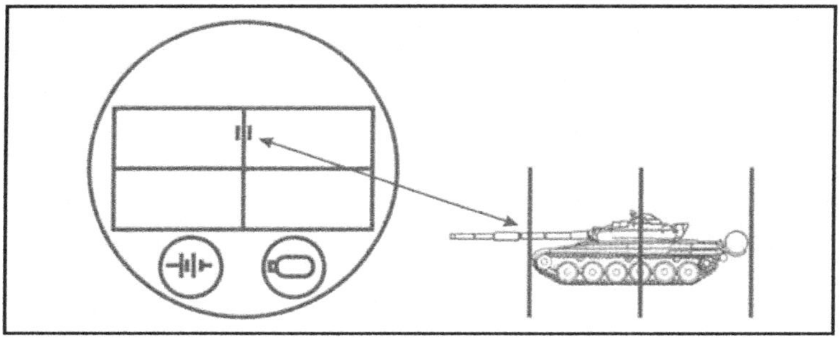

Figure 5-2. Target within range.

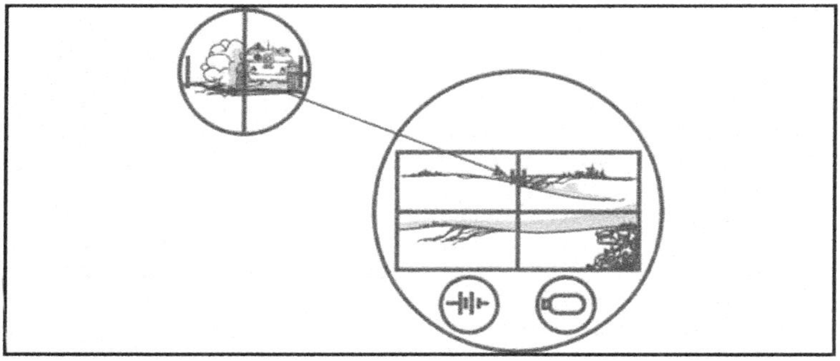

Figure 5-3. Target moving straight toward or away from position within range.

NOTE: When using the stadia lines to determine if a target is within range, keep two things in mind. First, the range is a rough estimate, based on the size of the target. A large vehicle may be out of range when the stadia lines show it to be in range. Second, the stadia lines are designed for a maximum engagement range of 3,000 meters.

(2) ***Binocular Method.*** Use the reticle in binoculars to determine if a target is within range by looking at the length, width, or height of the vehicle. Follow the same procedures when using the M17 and M19 binoculars, even though the reticles differ slightly. The M17 tick marks are only 1.7 mils long while the tick marks on the M19 reticle are 5 mils long (2.5 mils on each side of the horizontal and vertical scales).

(a) To determine if a target is within range at 3,000 meters based on the *length* of the target, place the length of the target on the vertical scale. If one-third or more of the vehicle extends beyond the tick mark, the vehicle is in range (Figure 5-4). A vehicle 6.5 meters long will measure about 2.2 mils at 3,000 meters and about 1.7 mils at 3,750 meters.

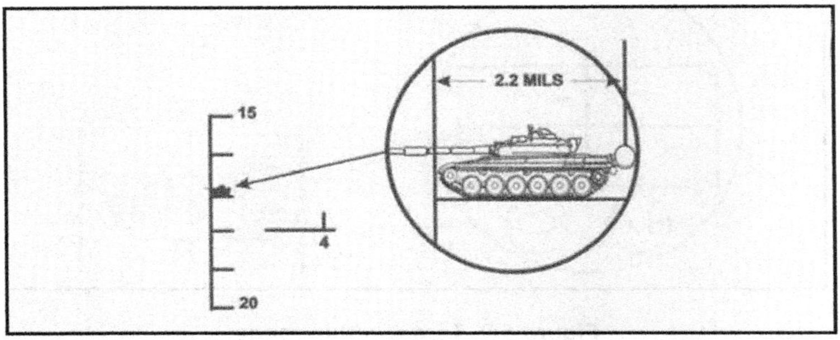

Figure 5-4. Target within range (flank or oblique) using M17 binoculars.

(b) To determine if a target is within range at 3,000 meters based on the *width* of the target, place the target on the small tick mark on the vertical scale. If the target covers two-thirds or more of the tick mark, the vehicle is within range (Figure 5-5, page 5-4). A vehicle 3.4 meters wide will measure 1.1 mils at 3,000 meters and .85 mils at 3,750 meters. (Most Warsaw Pact APCs are less than 3.4 meters wide and can be engaged at smaller mil values.)

(c) To determine if a target is within range at 3,000 meters based on the *height* of the target, place the target on one of the tick marks on the horizontal scale. If the height of the vehicle is one-half or more of the height of one of the tick marks, the vehicle is within range. A vehicle 2.4 meters high (the size of most Warsaw Pact vehicles), will measure .8 mils at 3,000 meters and .6 mils at 3,750 meters (Figure 5-6, page 5-4).

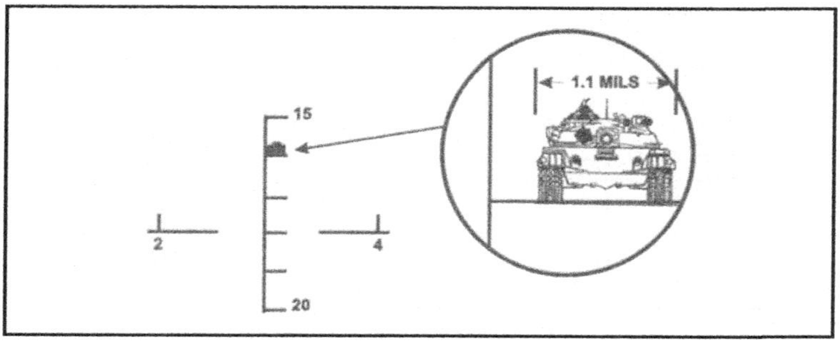

Figure 5-5. Target within range (frontal or rear) using M17 binoculars.

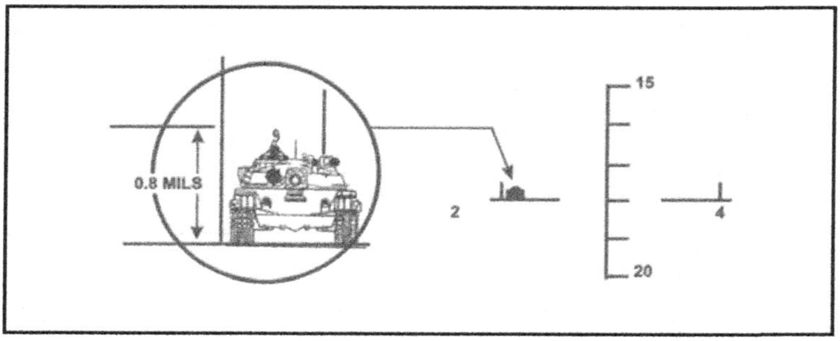

Figure 5-6. Target within range based on height.

NOTE: This method cannot be used if the weapon system is in an elevated firing position or if the lower portion of the target vehicle is hidden by foliage or terrain.

b. **Determine Exposure Time.** The half-sight method of determining exposure time is based on a vehicle speed of 35 kilometers per hour (the expected top vehicle speed of armored vehicles on level or gently sloping dry terrain).

(1) ***Daysight Tracker Method.*** Use the crosslines on the daysight tracker to determine the exposure time of a target.

(a) Place the crosslines of the daysight tracker on the center of the visible mass of the vehicle.

(b) If the area between the vertical crossline and the edge of the field of view in the direction of travel is clear of obstruction, the target is engageable (Figure 5-7).

(c) If obstructions appear between the vertical crossline and the edge of the field of view, the time of exposure would not be long enough for the missile flight before the target moved out of sight (Figure 5-7).

Figure 5-7. Obstruction.

(2) ***Nightsight Method.*** The procedure to determine exposure time is the same for the nightsight as for the daysight tracker, except the nightsight must be set on narrow field of view.

(3) ***Binocular Reticle Method.*** Use the reticle in the binoculars to determine the exposure time of a target.

(a) Place the zero tick mark of the horizontal scale at the center of the vehicle.

(b) If the area between the vehicle and the 50-mil tick mark is clear of obstructions in the direction of travel, the target is engageable (Figure 5-8).

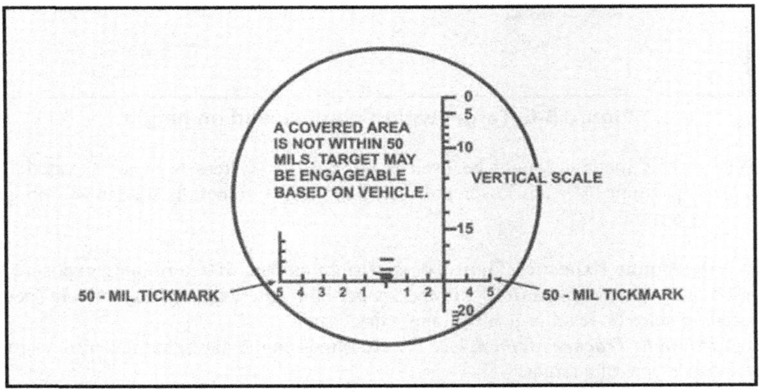

Figure 5-8. Binocular reticle method.

5-2. FIRE COMMANDS

The six elements of a fire command are alert, type of missile, target description, target direction, range, execution, and closing. Whether mounted or dismounted, the elements of the fire command are the same. (Figure 5-9, page 5-6, shows an example of a squad fire command.)

SQUAD LEADER:	SQUAD, TOW2A MISSILE; ONE TANK NORTH OF TRP 1; 2,000 METERS; AT MY COMMAND.
GUNNER:	IDENTIFIED.
SQUAD LEADER:	FIRE.
SQUAD LEADER:	CEASE TRACKING, OUT OF ACTION.

Figure 5-9. Example of a squad fire command.

a. **Alert.** The first element of the fire command alerts the crew for an immediate engagement. The squad leader commands, "Squad," and the gunner begins observing the target area.

b. **Type of Missile.** Because a variety of TOW missiles are used and their capabilities vary, a particular type of missile must be specified. A TOW crew in battle will probably have a mix of different missiles and a wide variety of target arrays to engage (see Chapter 1, TOW missile types and configurations). For example, if the target is a BTR-60 or BMP-1, the squad leader may command, TOW 2; if the target is a T-80 with reactive armor, he may command, TOW 2B.

c. **Target Description.** The second element identifies the target for the gunner. If several similar targets are present, this element tells the gunner which target to engage first. Most targets can be described by using the terms listed in Table 5-1. Targets that are combinations of the ones listed in Table 5-1, such as a truck mounting a missile system, are identified by combining terms—for example, TRUCK MOUNTING ANTITANK. When the gunner sees the target, he announces, "Identified." If multiple targets appear, the commander may specify which target will be engaged by the gunner--for example, FIRST TANK or RIGHT TRACK. (See Appendix H for information on OPFOR/Threat counterpart systems.)

TARGET	ANNOUNCED AS
Any tank or tank-like vehicle	**TANK**
Several tanks	**TANK FORMATION**
Any unarmored vehicle	**TRUCK**
Any halftrack or armored personnel carrier	**TRACK**
Helicopters	**CHOPPER**
All fixed-wing aircraft	**PLANE**
Personnel	**TROOPS**
Any machine gun	**MACHINE GUN**
Any antitank gun or towed artillery piece	**ANTITANK**
Any other target	**A short word or phrase that clearly describes the target.**

Table 5-1. Terms used to describe targets.

d. **Target Direction.** If the target is moving, the direction of movement is given after the description to aid the gunner in locating the target. After the gunner is given the location of a target, he can search for the target in the direction of movement. The following methods are also used to help the gunner locate the target.

(1) ***Target Reference Point.*** A TRP is an easily recognizable feature or point on the ground (either natural or man-made) used for identifying targets and controlling fires. They can be used to designate targets for companies, platoons, sections, and individual weapons. They can also be used to designate the center of an area where the commander plans to distribute or converge the fires of all his weapons. TRPs are usually designated by the company commander or platoon leaders.

(a) Weapons will engage targets from different directions, so compass points (for example, north, east), rather than "right" or "left," are used when giving directions centered on a TRP.

(b) Deflection from the TRP can be estimated, or it can be measured using the binocular mil scale or the circular reticle on the wide field of view–for example, 5.5 degrees at 1,000 meters is about 100 meters, or 200 meters at 2,000 meters.

(2) ***Prominent Features.*** The commander may give the distance and direction from a prominent feature–for example, FROM HILL SEVEN SIX TWO, LEFT TWO HUNDRED, or FROM BRIDGE, RIGHT FOUR HUNDRED.

e. **Range.** The range is given to help the gunner identify his target and to determine its engageability. The squad or section leader can determine the range to the target using the naked eye, binoculars (mil-relation formula), or reference materials (maps, range cards).

(1) ***Naked Eye.*** One method for using the naked eye to determine range is the football field method. The squad or section leader counts in 100-meter increments, estimating the number of football fields that could fit between the firing position and the target.

(2) ***Binoculars.*** Binoculars and the mil-relation formula can be used to determine range. To use this method, the squad or section leader must know the width, height, or length of the target. He determines the width, height, or length with the mil scale on the binoculars; substitutes the mil-relation; and computes the range.

(3) ***Reference Materials.*** Maps can be used to determine range by counting the grid lines between the firing position and the target or by adjusting from a known point. Range cards can also be used to determine the range to the target.

f. **Execution.** Two commands are necessary for execution: a preparatory command and a command of execution.

(1) AT MY COMMAND is a preparatory command that warns the gunner not to fire until given the command of execution.

(2) FIRE is the only command of execution used to fire a missile.

g. **Additional Commands.** In addition to the six elements of the fire command, some other commands are needed.

(1) The command CEASE TRACKING or CEASE TRACKING, OUT OF ACTION is issued after seeing the round detonate or when the squad or section leader wants to halt firing.

(a) CEASE TRACKING tells the crew the squad or section leader intends to stay in position and engage another target immediately or when one appears.

(b) CEASE TRACKING, OUT OF ACTION tells the crew the squad or section leader intends to move to another position.

(2) To determine the method of engagement, the section leader (or above) selects a fire pattern depending on the opposing force's formation. The section leader directs, FRONTAL, DEPTH, or CROSSFIRE, when the gunner is faced with multiple targets.

(3) When the target is identified, the gunner announces, "Identified."

(4) If the gunner cannot see the target, he announces, "Lost."

(5) If the gunner cannot identify the target, he announces, "Cannot identify."

(6) The loader announces, "Backblast clear," *before* the command of execution is given.

h. **Repeating Commands.** When a crew member fails to hear or understand any element of a fire command, he announces the element in question. For example, if the gunner asks, "Location?" the squad leader repeats the location element such as, "From hill seven six two, west two hundred."

i. **Correcting Errors.** To correct an error in a fire command, the squad leader announces, "Correction," and corrects only the element in error. He completes the command by announcing all elements after the corrected element. He does not try to correct an element that has been needlessly included, such as the direction element. He corrects the omission of an element by announcing "Correction" and then the omitted element. After announcing the omitted element, he completes the command.

j. **Commands for the Driver.** Although directions to the driver are not part of the fire command, they are given by the squad leader or gunner in short terms.

5-3. TARGET TRACKING

To track a target, the gunner visually acquires the target through the daysight tracker system of the TOW. He can track the target by either optical or electrooptical means, depending on the system configuration being used and on the visibility conditions.

a. To track the target, the gunner operates the hand controls on the traversing unit to keep the reticles in the launcher sight aligned with the target. The daysight tracker system is attached to and aligned with the launch tube. The launch tube stabilizes the exit of the missile from the launcher for initial alignment during missile flight. On achieving target alignment, the gunner fires the missile by manually depressing the trigger switch. Thereafter, all operations are automatic and the gunner's only task is to maintain alignment of the sight reticle on the target until missile impact.

b. Deviations of the missile from the line-of-sight trajectory are sensed in the launcher sight by infrared means that receive information from infrared radiators attached to the missile. This information is processed in the form of electrical signals to produce error signals proportional to the azimuth and elevation displacements of the missile from the intended trajectory. Correction commands are derived from these error signals and are sent to the missile over the command-link wires, which are dispensed from the missile. The missile performs corrective maneuvers using aerodynamic control surfaces that deflect in response to the command signals from the launcher. On target impact, a high-explosive, shaped-charge warhead is detonated.

5-4. TARGET ENGAGEMENT WITH THE M220A1 (BASIC TOW)

Specific procedures are followed to engage a target with the M220A1.

a. Position the eye well into the rubber eyepiece and place both hands firmly on the control knobs (Figure 5-10). Place the body so there is no contact between the shoulder and the encased missile. The only contact with the launcher is with the hands and eye. Assume a firing position that is comfortable. An uncomfortable position causes muscle tension, which affects the ability to track smoothly.

Figure 5-10. Control knobs.

NOTE: When firing from the tripod, kneel on one or both knees.

b. Raise the trigger protective cover and establish a smooth tracking rate while keeping the crosshairs on the center of visible mass of the target (Figure 5-11).

Figure 5-11. Crosshairs on the center of visible mass of the target.

c. Move the launch tube left or right by applying a smooth, steady force to both control knobs (pushing one and pulling on the other) and rotating the body from the waist up as the launcher moves. Elevate or depress the launch tube by applying a smooth, steady turning force to both control knobs. Applying pressure to only one control knob, or applying uneven pressure, makes it more difficult to track smoothly. Maintain the same arm, shoulder, and head position throughout an engagement. Any change in body position other than leaning with the controls will cause a jerking motion that could result in grounding of the missile.

d. Proper breath control is especially important during the first and last 400 meters of missile flight. Improper breathing will cause poor tracking. Take a deep breath and let part of it out; then, press the trigger. After a 1.5-second delay, the missile will launch. The delayed firing of the launch motor may cause you to flinch or jerk the control knobs if you are not prepared for it. Be prepared for two noises after the trigger is pressed. The first noise is the gyro being activated. While it is not loud, it may cause you to think a misfire has occurred, and you may not be prepared for the next noise. The second noise is the launch motor firing and it *is* loud. The dust, smoke, heat, and debris from the backblast may cause flinching. When the missile appears in the sight picture, ignore it. *Never try to guide the missile.* If distracted, tracking becomes poor and chances of hitting the target are reduced. Continue to track the target at a smooth tracking rate, keeping the crosshairs on the center of visible mass until missile impact (Figure 5-12).

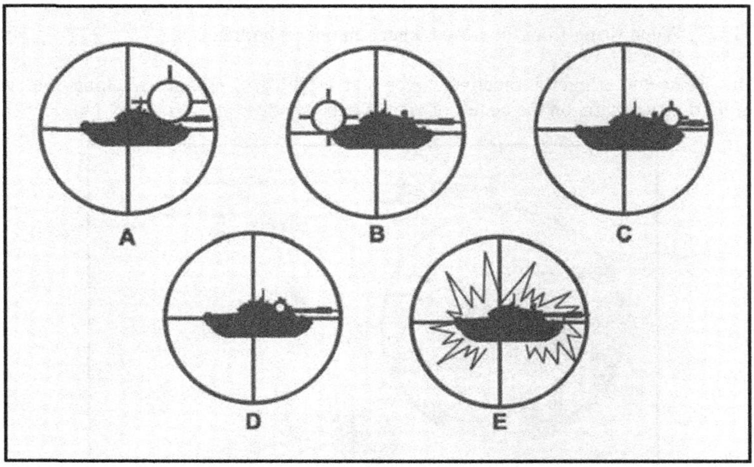

Figure 5-12. Keep crosshairs on the center of visible mass.

5-5. TARGET ENGAGEMENT WITH THE M220A2 (TOW 2)

Specific procedures are followed to engage a target using the M220A2.

a. Looking through the daysight tracker, adjust the focus control until the crosshairs are in focus. (To see the crosshairs clearly, set RETICLE LIGHT switch to ON.) Position the crosshairs on the target, and remove the front lens cover from the nightsight by releasing two latches. Set the ON-OFF-STBY switch to ON. Look through the eyepiece and adjust the diopter adjustment ring to focus the reticle. Ensure the battery monitor light is off. Set the field of view selector to wide field of view, locate the target, and adjust the range focus, contrast, and brightness controls to obtain the best possible sight picture. Set the field of view selector to narrow field of view and adjust the range focus, contrast, and brightness knobs.

(1) Because handoff may occur when limited visibility conditions exist, locate the target with the daysight tracker and fire using the nightsight. During daylight, the battlefield can be obscured by smoke, dust, and so forth. If limited visibility conditions exist, the MGS automatically transfers control of the missile from the daysight tracker to the nightsight. The nightsight allows continuous observation of the target regardless of battlefield visibility conditions.

(2) To view the battlefield, ensure the TOW 2 nightsight ON-OFF-STBY switch is in the ON position. Battlefield conditions requiring the nightsight to be in operation are area target monitoring and target engagement. The ON-OFF-STBY switch on the TOW 2 nightsight allows greater battery life from the BPC when placed in the STBY position. After the TOW 2 nightsight has been turned on and allowed to cool down (about three minutes), place the ON-OFF-STBY switch in the STBY position. The STBY position causes the closed-cycle cooler to cycle OFF for 100 seconds and ON for 20 seconds. Cool down is not required if the nightsight has been operating in the STBY position.

(3) When firing the TOW 2 system, ensure the nightsight is turned ON. Always allow three to five minutes cool down time before operation regardless of which sight is used. This ensures that the enhanced capabilities of TOW 2 are used during firing, flight, and impact. Certain conditions may permit degraded firing. During the system check-out procedure, if the display indicates a nightsight (postamplifier) failure, the TOW 2 can be fired using the daysight tracker only. If limited visibility causes handoff during flight, control of the missile may be lost.

b. To complete the firing sequence, follow the procedures for the basic TOW launcher (paragraph 5-4, page 5-9).

Section II. HELICOPTER ENGAGEMENT

Enemy armor is the primary threat to friendly ground forces employed in forward areas. The primary mission of the TOW is the destruction of these tanks at the greatest possible range. However, TOW gunners can also successfully engage attacking enemy helicopters, which are a significant threat to ground forces.

5-6. OPERATIONAL CONCEPT

Engaging helicopters with the TOW should be considered primarily as a means of self-defense. TOW crews should not consider helicopters as a routine target of opportunity, but they should leave them to conventional ADA assets when possible.

a. TOW positions are selected to cover armor avenues of approach, but these long-range fields of fire also facilitate the engagement of aircraft. The section leader's, squad leader's, and crew's observation from these positions can provide the early warning required to successfully engage aircraft.

b. The engagement of attacking helicopters should be done by TOW sections, not individual weapon systems. TOW sections should automatically engage helicopters that are attacking their positions. If one squad in a section is being attacked by a helicopter, the other squad should engage the helicopter while the first squad seeks cover. TOW crews and sections should be trained to automatically respond to helicopter attacks in this manner.

5-7. GUNNERY TRAINING

Specific gunnery training is required to track a helicopter with a TOW. Crews should perform the training according to the following conditions and standards. In addition to these outdoor training exercises, a number of scenarios for the TGT include helicopters as targets.

a. **Conditions.** Training takes place during daylight on a range. The TOW crew is given either a ground- or vehicle-mounted TOW with a TFTT and a helicopter mounted with TFTT reflectors. The helicopter moves toward the TOW gunner at speeds between 40 and 80 knots and at ranges between 1,000 and 3,500 meters. Three target angles are used for tracking: head-on; approaching at 30 degrees; and an evasive track with a 90-degree turn.

b. **Standards**. Within 5 seconds the gunner must acquire and begin tracking the target. He must get a hit 6 out of 10 times with the TFTT.

Section III. NBC AND LIMITED VISIBILITY CONDITIONS

Specific procedures are followed to operate the TOW during NBC and limited visibility conditions.

5-8. DECONTAMINATION

TOW crewmen must know decontamination procedures and materials. They must know which decontamination materials to use on each type of surface on the weapon system. Using the wrong material can cause damage to the system (for example, using DS2 on rubber surfaces). Decontamination materials and methods are periodically revised. (Refer to Appendix C of FM 3-5 for the most up-to-date information.)

5-9. THERMAL TARGET RECOGNITION, IDENTIFICATION, AND ENGAGEMENT

The nightsight allows the TOW gunner to view targets during limited visibility conditions such as darkness, smoke, fog, rain, and snow. It produces images called *thermal target signatures* or *infrared target signatures*, which are different from the images seen in the daysight tracker. Targets stand out in these infrared images and can be recognized at long ranges on a clear night and at reduced ranges during poor visibility. Recognizing these targets requires trained and experienced gunners.

a. **Temperature and Thermal Images.** Most objects have a radiated temperature either higher or lower than their background. Even if the radiated temperature differences are less than a degree, they appear on the nightsight display. If there is no difference between the temperature of an object and its background, the object will not be seen in the display.

(1) If an object has a high temperature, it will appear bright red in the nightsight. If the object has a low temperature, it will appear black. Usually, targets are easier to identify at night, because their radiated temperature is hotter than their background.

(2) Some targets, such as tanks and APCs, have internal temperature variations that form visible patterns. These patterns are the basis of target signature cues. In a nightsight, the shapes of the hottest vehicle parts, such as engines and exhausts, appear bright red. Objects with a medium temperature, such as the warm tracks, appear a dim red. Objects with a cool temperature, such as the cool hull and other cool parts, appear black.

b. **Sources of Infrared Energy.** Infrared energy comes from different sources such as solar heat, fuel combustion heat, frictional heat, and reflected radiance.

(1) ***Solar Heat.*** Solar heat comes from the sun and affects the exterior surface of objects. This heating highlights the outline of the object, which provides recognition cues to the gunner. These cues are usually similar to the overall appearance of the target. (For example, a solar-heated M113 appears box-like with a sloping front; a solar-heated M60 tank appears as a small oval atop a larger oval.) These shape cues are recognizable out to medium (1,000 to 2,500 meters) and long (beyond 2,500 meters) ranges. Since the sides have more defined contours, the side view shapes are usually easier to recognize than the front view. In addition to atmospheric variables and surface reflections, the solar heating rate is also affected by the object's ability to absorb sunlight. Generally, dark-colored objects are better absorbers of sunlight than light-colored objects.

(2) ***Fuel Combustion Heat.*** Fuel combustion heat comes from operating engines. The heat from operating engines is conducted to the surfaces of the surrounding engine compartment.

(a) Because engine compartment temperatures reach up to 200°F, the surfaces of these compartments radiate features that can be easily detected on the nightsight at long ranges. Heated personnel space is also visible.

(b) Engine muffler and exhaust pipe temperatures are high, providing the gunner with good cues.

(c) Although the engine, heated compartments, and exhaust features themselves do not appear in the nightsight, their cue value is not any less. A trained and experienced gunner can determine much about the vehicle from these cues.

(3) ***Frictional Heat.*** Frictional heat is produced by the moving parts of vehicles. However, these features usually appear a dim red. This heat is less intense than the high temperatures from the engine combustion. Frictional heat is generated only when the vehicle is in motion.

(a) Frictional heat provides long-range cues to classify the vehicle as wheeled or tracked. At medium-range to short-range, these cues can be used to identify the vehicle.

(b) The vehicle's transport systems are the source of most frictional heat cues. Tracked vehicles have frictional heat in the tracks, road wheels, drive sprockets, support rollers, and shock absorbers. The smallest of these features can be identified at longer ranges when they are hot. Wheeled vehicles have frictional heat in the tires, shock absorbers, drive shafts, transmissions, axles, and differentials. The tires, shock absorbers, and differentials can be detected at medium-range to long-range.

(4) ***Reflected Radiance.*** Certain smooth, glossy surfaces, such as windshields and glossy painted fenders, reflect radiation images from other sources. These reflections can produce odd images. For example, the fenders of a T-62 appear black because of thermal reflection. An overcast sky can cause warmer thermal reflections. Generally, surface reflections are diffuse in nature and do not usually cause problems.

5-10. EFFECTS OF WEATHER AND OBSCURANTS

Variations in solar heat, fuel combustion heat, frictional heat, and thermal reflection affect infrared signatures and infrared target recognition cues. In addition, some atmospheric conditions degrade the nightsight, while others can enhance it. Some of these factors are discussed below.

a. **Falling Precipitation.** Infrared energy does not transmit well through falling precipitation (rain, snow, and fog). The temperature of targets and background objects are decreased. The basic signature cues themselves do not change because of atmospheric transmission losses. Falling precipitation restricts nightsight visibility more than precipitation that has fallen.

(1) During rain or snow, background objects and frictionally heated and solar-heated target features lose heat. Frictional heat loss is caused by water and mud accumulating on the tracks, wheels, and other transport system parts. Engine compartment and exhaust temperatures remain high. Landmarks, such as tree lines, trails, and contour features, are often lost. The loss of heat in background objects reduces scene clutter, such as trees and rocks, and can increase target detection. Target recognition cues are usually reduced because of the loss of heat in certain target features.

(2) Because rain and snow have a cooling effect on the target's contrast, the nightsight contrast controls must be increased to compensate for the condition. However, a higher contrast setting produces a "snowy" image.

b. **Fallen Snow.** Fallen snow tends to make all ground temperatures the same. Depth perception by size comparison becomes difficult because of lack of terrain features with which to reference size.

c. **Dust, Diesel Fog, and Oil Smoke.** Dust particles from artillery impact greatly reduce nightsight visibility. Only the hotter objects and target features show through the obscurants.

5-11. COMPENSATION FOR TARGET APPEARANCE VARIABLES

Although vehicles have distinguishing characteristics or cues by which they can be classified and identified, vehicle appearance can be altered by changes in atmospheric and ground conditions. Therefore, the gunner must know to use the control settings of the nightsight to help compensate for these variables.

a. **Contrast and Brightness Control.** The contrast and brightness controls can be set for maximum internal detail of the target. The controls are balanced to give the clearest image of these target recognition cues. The following is general guidance for setting the image brightness and contrast controls. Gunners should be encouraged to experiment with the controls to understand the effects of the image controls on the thermal image and thermal signatures.

(1) ***Brightness, Low; Contrast, Medium to High.*** These settings are for scanning an area in search of targets. Background clutter is suppressed. Cool objects are not visible. Only the hot objects in the field of view are seen. When a possible target has been found, brightness can be increased and contrast can be lowered gradually to reveal more thermal detail in the suspected target. Low brightness and medium-to-high contrast settings are also for nightsight use in light fog, rain, or dust conditions.

(2) ***Brightness, Low to Medium; Contrast, Medium.*** This is the best overall setting for target detail. With medium contrast, brightness can be varied up and down to bring out features and determine the hottest vehicle parts. When brightness is lowered, the cooler parts, such as tracks, darken before hotter parts. Often, small changes in brightness can reveal much about the vehicle. For example, road wheels can sometimes be seen at long range if this technique is used. Experimentation is helpful in learning this target feature extraction technique.

(3) ***Brightness, Medium; Contrast, Medium to High.*** These settings work well in heavy fog or heavy dust when little can be seen with the TOW. The settings increase the snowy effect in the image, but they also increase the apparent sensitivity of the sight. The image appears distorted and is difficult to interpret. These settings are also good for searching a tree line in wet conditions. They are sometimes useful with long-range targets. Beyond 2,500 meters, small target images, such as the front view of a BRDM-2, will have a few recognizable features. At long ranges, the higher contrast setting highlights the vehicle's hull and overall silhouette. This will not provide internal detail, but it will help the target stand out from the background and will aid in target detection.

b. **Focus Controls.** Most nightsight focus controls are sensitive; that is, a small movement of the control knob results in a large change in focal point. If a gunner has difficulty in focusing, he should check the adjustment of the image controls.

(1) The eyepiece focus should be adjusted first. The eyepiece focus is called the *diopter adjustment*. Once a gunner knows his diopter correction number, he can dial it in on any nightsight diopter ring. Correct diopter adjustment can usually be obtained by focusing the eyepiece so that the reticle is focused.

(2) The second focus adjustment is the objective range focus. The image controls should be adjusted at the low-to-medium level before focusing the objective lens. Focusing the objective lens is learned through trial and error. It is made difficult by the fact that infrared heat diffuses on objects and does not usually give clear-cut, straight lines on which to focus. Thus, the gunner must learn to focus by adjusting the control back and forth to get the best image. This is easier to do when the nightsight is aimed at a prominent object. Once the best image is determined, the gunner can experiment by focusing on different objects at different ranges.

5-12. BATTLEFIELD IDENTIFICATION

Battlefield identification using a nightsight is difficult. Although the identification problem is being studied, little is known about the ranges at which high-confidence identification can be expected. In a target-rich environment on a dry, clear night, high-confidence identification requires a thermal image of such features as road wheels, turret shapes, gun tube, and exhaust location. Limited experience with tanks produced in the U.S. and other nations indicates that thermal identification can be made between 1,000 to 2,000 meters in clear weather. When identifying targets, a gunner should ask himself these questions:

- Is the target moving?
- What direction is it moving?
- Where is the engine?
- Where is the exhaust?
- Is the target in the unit's sector?
- Should it be there?
- Is it in a formation?
- Is it firing at the unit?

NOTE: Refer to FM 100-65 for primary recognition cues.

Section IV. ELECTROOPTICAL COUNTERMEASURES

TOW crews may encounter electrooptical countermeasures on the battlefield. Therefore, the following procedures should be used as engagement criteria and not as maintenance checks.

5-13. TOW LAUNCHER

TOW gunners should complete boresight procedures at least as often as prescribed in the operator's manual to ensure the system remains boresighted. When properly boresighted, the boresight meter needles are centered. The sight sensor must point at the ground or away from the sun or any other possible source of IR energy when the boresight procedure is being completed.

a. Once the boresight procedure is completed, the TOW crew leaves the self-test selector switch in position 7 and engages the test/operate switch (move to test position and hold) as the gunner acquires the target. The missile arming lever should not be raised at this time. If the self-test meter needles peg to one side, or become erratic, or both as the gunner aims at the target, electro-optical countermeasures are present and a shot against that target should not be attempted. The gunner waits for a flank shot or shifts targets.

b. The loader releases the test/operate switch as the gunner shifts targets. The loader reengages the test/operate switch (returns it to test position) when the gunner is ready to acquire a second target. Releasing the switch saves wear and tear on the daysight tracker sensor and conserves battery power, if batteries are being used.

c. If the meter needles return to center after the gunner places his crosshairs on a second target, he engages the target. The loader releases the test/operate switch and raises the arming lever, and the gunner fires the missile.

d. If the meter's needles are erratic after the gunner lays on the second target, he does not fire. He selects another target or waits for an oblique or flank shot ensuring the meter needles are centered before firing.

5-14. TOW 2 LAUNCHER

TOW 2 launchers firing TOW 2-series missiles are not vulnerable to electrooptical countermeasures. However, TOW 2 launchers firing non-TOW 2 missiles are vulnerable, and gunners should use the boresight indicators to determine whether a target should be engaged.

a. The loader completes the MGS self-test procedure as target engagement becomes imminent. The loader holds the self-test toggle switch in the test position before the missile is armed, as the gunner places his crosshairs on the target. If the green light stays on indicating electro-optical countermeasures are not present, the loader releases the toggle switch and raises the arming lever, and the gunner engages the target.

b. If the green boresight center light goes off and the red lights begin to flicker on and off indicating electro-optical countermeasures are present, the gunner does not engage the target. The loader holds the self-test toggle switch in the test position as the gunner acquires an alternate target. If the red lights go off and the green light comes on as the gunner lays his crosshairs on a second target, the loader releases the self-test toggle switch and arms the missile, and the gunner engages the target.

c. Because the TOW 2 self-test switch is reactivated each time the toggle switch is placed in the test position and the self-test cycle takes 15 to 20 seconds, the loader does not release the test switch while the gunner is acquiring alternate targets. The leader releases the switch only after the green boresight light appears, signifying that it is safe to engage that target.

APPENDIX A

TOW TRAINING TIPS

Experiences at the National Training Center have produced many useful tips for training and employing TOWs. Following these tips will greatly enhance the performance of TOW units in the field.

1. To reduce the effects of heat scintillation, fire from a higher to a lower elevation whenever possible.

2. Train crewmen to remove the protective covers from the electrical connectors of missiles before loading. This is the most common cause of misfires during live fire.

3. Emphasize the obstacle effects of TOW guidance wires. Wire will impede or stop vehicular and dismounted movement. Plan for their recovery and or destruction. Wire cutters are not fine enough to cut TOW wire. The wire must be cut with low-intensity flame when possible.

4. Perform a system self-test and collimate the nightsight to the daysight tracker every 4 hours, after every 10-degree change in temperature; and after moving cross country.

5. Select flat, level firing positions to improve hit probability.

6. Practice fire control and fire patterns at every opportunity.

7. Identify maximum engagement lines and TRPs as accurately as possible. Use adjacent tanks or FO/FIST laser range finders when available.

8. Perform the 180-day verification checks as often as possible (every 30 days is not too often). Verify in the assembly area immediately before combat or live-fire operations. Allow time to fix deficiencies.

9. Increase the amount of time spent tracking in MOPP4. Tracking in MOPP4 improves tracking skills even under MOPP1 conditions.

10. Ensure TOW live fire is conducted from the narrow field of view only.

11. Ensure M966s and BFVs with TOW 2 weapon systems receive priority issue of TOW 2 missiles.

12. Collimate nightsight to daysight each time the system is boresighted.

13. Know the diopter setting. This information should be posted on the system for ready reference.

14. M966 TOW HMMWV:

a. Shield the nightsight and daysight from direct sunlight. Cool the nightsight as much as possible to reduce system cool-down time.

b. Familiarize gunners with launch obscuration effects:

(1) *White out*—1.5 seconds after trigger pull, block daysight for 2 to 5 seconds.

(2) *Red out*—1.5 seconds after trigger pull, pass a lit cigarette lighter or other IR heat source 6 inches in front of the AN/TAS-4.

c. Remove AN/TAS-4 eyepiece when tracking in a protective mask. Replace with a doughnut-shaped foam cushion made from a sleeping mat or from ammunition packing and taped to AN/TAS-4.

d. Reinforce counter-rotation launch technique to reduce launch tube climb during live fire.

e. Train gunners to bring crosshairs back on target slowly, avoiding jerky movements.

f. Minimize movement of crew members on vehicle when missile is in flight.

g. Minimize blowing of sand and dust from missile launch, by placing a tarp on the ground in front of the launch tube or by wetting the ground down. Blowing dust and sand can cause the weapon system to lose the IR beacon on the missile.

15. Fire Commands/Missile Selection. With the increase in the number of different types of missiles in the TOW inventory, it will sometimes be necessary to tell the gunner what type of missile to engage the target with (see Appendix H, Table H-2). If a missile is not loaded in the weapon system, the squad leader can include the missile type in the target description portion of the fire command (SQUAD, T-80, TOW 2B). If a missile is loaded in the weapon system, the squad leader will have to make the decision whether time and the tactical situation make it desirable to unload and then reload the system. If a missile has just been fired, the missile reload type can be designated as part of the cease tracking command (CEASE TRACKING, TARGET DESTROYED, RELOAD WITH TOW 2A).

16. TOW MILES Gunnery. Because TOW MILES is an engagement simulator adapted as an interim gunnery trainer, it is important to return gunners to a precision tracking system before live-fire or combat operations. Because the tracking simulation of TOW MILES is less realistic and more forgiving of errors, hit success with the MILES simulator comes more easily than hit success with the actual TOW missile. This is especially true with panel targets equipped with the LTID. Exclusive training on TOW MILES sometimes lulls gunners into a false sense of tracking proficiency. As an engagement simulator in the role of tactical trainer, TOW MILES offers adequate tactical realism. Good TOW gunners are usually good MILES gunners. The reverse is not necessarily true.

17. TOW Missile Clearance Requirements. If line-of-sight clearance is less than 30 inches, the probability of the missile hitting the ground or an obstruction is increased. Figure A-1 shows the probability of survival for the TOW. The reason for the deviation is that the missile does not precisely follow a gunner's line of sight.

RANGE TO TARGET	HEIGHT OF LINE OF SIGHT ABOVE THE GROUND		
	18 INCHES	20 INCHES	30 INCHES
200M	.98	.98	1.0
300M	.95	.95	1.0
400M	.91	.91	.98
500M	.86	.91	.98
1,000 TO 3,000/3,750M	.56	.61	.93

Figure A-1. Probability of survival for the TOW (meters).

18. TOW Missile Time-Of-Flight Versus Range. Figure A-2 (page A-4) and Figure A-3 (page A-5) show the time of flight of the TOW missile to various ranges. Using these figures allows the squad leader to accurately calculate the time it takes for the missile to reach its target.

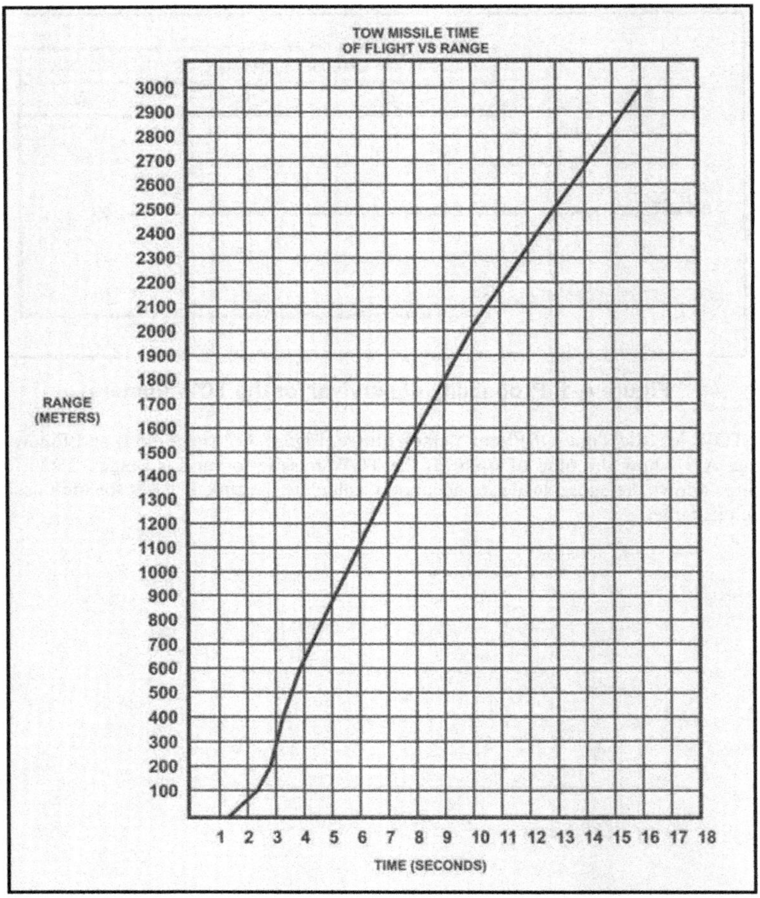

Figure A-2. TOW missile time-of-flight versus range.

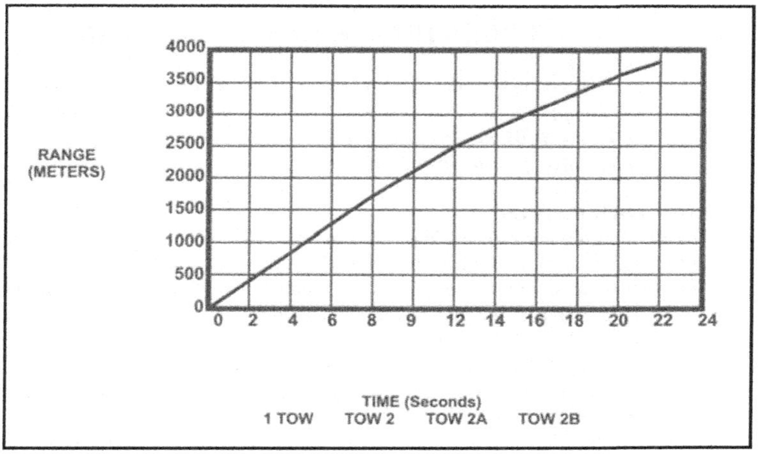

Figure A-3. TOW missile time-of-flight versus range (kilometers).

APPENDIX B

TRAIN-THE-TRAINER PROGRAM

The success of the TOW training program depends on trainers who are competent and well prepared. This appendix provides information to assist the TOW unit's chain of command in preparing and implementing an effective train-the-trainer program.

B-1. OBJECTIVES

The train-the-trainer program has specific objectives. Its purpose is to develop in every TOW trainer the skills, confidence, and willingness to train 11B soldiers to be proficient with their weapon system and effective in combat. An effective trainer must be able to—

- Set up and operate training equipment.
- Perform all individual and collective tasks.
- Effectively coach a gunner during tracking training.
- Accurately assess the training strengths and weaknesses of individuals and crews.
- Plan and conduct training that will correct training weaknesses.
- Instill confidence in the individuals and crews he trains.

B-2. MISSION-ESSENTIAL TASK LIST

Each commander of a TOW unit should carefully examine his unit's wartime mission and develop a METL. He then organizes a unit sustainment training program that supports that METL. The TOW Training Program consists of quarterly and semiannual mandatory training events. That does not mean that no TOW training should be done in the interval between these events. The commander is responsible for assessing his wartime mission and his unit's current training status and then devising a unit sustainment training program. That program should both support his METL and prepare his unit for the next round of TOW gunnery tables.

B-3. TRAINER ASSESSMENT

Trainers in a TOW unit are normally the squad and section leaders and the platoon sergeants. The trainers are not chosen by their proficiency but by their duty position. Therefore, assessing the abilities of all element leaders is critical. Any shortcomings must be corrected. This assessment is primarily the responsibility of the commander and the platoon leaders and sergeants. This can be accomplished in several ways.

- Conduct periodic leader's tests, both written and hands on.
- Have all leaders perform all the tasks that their subordinates are required to take in the quarterly and semiannual mandatory training events (Gunner's Skill Test and TOW Gunnery Tables).
- Hold a competitive skill test each month with the winner being named trainer of the month.

Any of these methods, or any other the commander deems appropriate, are acceptable as long as they provide an accurate assessment. Trainers should be thoroughly refreshed in the skills they are deficient in before they are allowed to train and test subordinates. A soldier must know his superiors are more skilled than he is in order to have confidence in those superiors.

B-4. COMMAND BENEFITS

If a train-the-trainer program is conducted aggressively and regularly, the benefits to the unit are considerable.

a. The base of trainer expertise will remain solid and gradually spread through the unit.

b. Maintaining this base is easier than periodically recreating it.

c. Overall unit proficiency and esprit de corps will steadily rise.

d. A well-trained and proficient unit requires less time for sustainment training and leaves more time for other unit missions.

B-5. TRAINING TASKS

To be successful, the trainer must be thoroughly familiar with the operation of his training devices. All trainers must be able to correctly perform the following tasks:

a. **TOW Gunnery Trainer.**

TASK: Assemble the TGT.

CONDITIONS: Indoors, given a TOW tripod, traversing unit, and launch tube and all components of the TGT.

STANDARDS: Correctly assemble all components of the TGT IAW TM 9-6920-452-10.

TASK: Prepare the TGT for operation.

CONDITIONS: Indoors, given a fully assembled and operational TGT.

STANDARDS: Correctly power on and warm up the TGT, and perform the operational checkout of the TGT IAW TM 9-6920-452-10.

TASK: Conduct pre-operational procedures for the TGT.

CONDITIONS: Indoors, given an operational and fully assembled TGT.

STANDARDS: Correctly load the videodisc, program disk, and the data disk IAW TM 9-6920-452-10.

TASK: Edit a planned group.

CONDITIONS: Indoors, given a fully assembled, operational TGT and matching data and videodiscs.

STANDARDS: Correctly edit a planned group to match the desired conditions in the mission IAW TM 9-6920-452-10.

TASK: Perform operational procedures for the TGT.
CONDITIONS: Indoors, given an assembled, fully operational TGT with matching video and data disks and an edited planned group.
STANDARDS: Correctly brief the student and conduct a mission from the planned group IAW TM 9-6920-452-10.

---

b. **MILES.**
TASK: Install MILES on the TOW weapon system.
CONDITIONS: Outdoors, given a TOW traversing unit, tripod, nightsight, and launch tube and a TOW MILES tracker head, ATWESS tube, and MGS simulator.
STANDARDS: Correctly assemble, install, and test the TOW weapon system with TOW MILES IAW TM 9-1265-368-10-2 (tasks 9 through 12).

TASK: Test the assembled TOW MILES.
CONDITIONS: Outdoors, given an assembled TOW weapon system with TOW MILES installed.
STANDARDS: Correctly test the TOW MILES IAW TM 9-1265-368-10-2 (task 13).

TASK: Fire TOW and ATWESS.
CONDITIONS: Outdoors, given an assembled, tested, and operational TOW MILES system.
STANDARDS: Correctly and safely fire TOW MILES with ATWESS IAW TM 9-1265-368-10-2 (task 15).

TASK: Reset MWLD and TOW MILES.
CONDITIONS: Outdoors, given an MWLD, an assembled, tested, and operational TOW MILES system, and a controller key.
STANDARDS: Correctly reset the MWLD and TOW MILES after they have been "killed" IAW TM 9-1265-368-10-2 17).

---

c. **TOW Field Tactical Trainer.**
TASK: Assemble the TFTT and install the retroreflector.
CONDITIONS: Outdoors, given a complete TOW system and a complete TFTT with retroreflector.
STANDARDS: Correctly install the TFTT and retroreflector IAW TM 9-6920-453-10 (M966).

TASK: Power up the TFTT and align the laser to the retroreflector.
CONDITIONS: Outdoors, given an assembled TFTT and an installed retroreflector.
STANDARDS: Correctly power up the TFTT and align the laser to the retroreflector IAW TM 9-6920-453-10 (M966).

TASK: Conduct operational procedures for the TFTT.
CONDITIONS: Outdoors, given an assembled, operational, aligned TFTT, and an installed retroreflector.
STANDARDS: Correctly conduct operational procedures for the TFTT to included setting target size and obscuration time, loading the M80 blast simulator, and preparing the TFTT for firing IAW TM 9-6920-453-10 (M966).

TASK: Conduct system checks and test procedures for the TFTT.
CONDITIONS: Outdoors, given an assembled, operational, and aligned TFTT.
STANDARDS: Correctly perform all system checks and test procedures for the TFTT to include visual checks, cable checks, power checks, computer checks, alignment checks, and system and operational checks IAW TM 9-6920-453-10 (M966).

B-6. TRAINER CERTIFICATION PROGRAM

The commander may choose whatever method he thinks best to assess trainers, but he must include the following elements in a trainer certification program:

a. **Program Orientation.** When a new trainer first enters the unit, he must be briefed on the unit's instructor certification program so that he understands what is expected of him.

b. **Testing of 11B Individual Soldier Tasks.** Each trainer must be tested on each individual task at least annually. Testing every trainer on every task more frequently is impractical. A more workable method is to conduct monthly or quarterly testing events where some portion of these tasks are tested, and over the course of a year all tasks are included.

c. **Testing of Collective Tasks.** Trainers should be tested in all collective tasks annually. Testing events should be spread out in partial monthly or quarterly events.

d. **Testing of Training Equipment Tasks.** This should be conducted in the same manner as the individual and collective tasks. The testable training equipment tasks are listed in paragraph B-5. These tasks are not found in the soldier's manual. The appropriate TMs must be referenced for the performance measures.

e. **Formal Records Keeping.** Formal records should be carefully maintained. If a trainer fails a particular task, he should be retrained and retested until he receives a go on it and all other tested tasks. No one should be allowed to train troops who has not received a go on all tasks tested to date. The soldiers' full confidence in the abilities of their trainers and leaders is critical to morale.

APPENDIX C

FIGHTING POSITIONS

TOW fighting positions may be mounted or dismounted, depending on the situation. A camouflage net can be used with either configuration.

C-1. DISMOUNTED TOW FIGHTING POSITION

A tripod-mounted TOW fighting position must meet the following specifications:

- Allows the squad to engage the enemy in its assigned sector of fire.
- Provides protection for both the weapon and its crew.
- Blends with its surroundings so that it cannot easily be detected 35 meters to the front and cannot be seen from the air.

a. The steps in constructing the position should be conducted in the following sequence:

(1) Assemble the launcher.

(2) Clear the fields of fire while ensuring the sector is under observation.

(3) Dig the weapon position first, then add overhead protection for the crew.

(4) Build a parapet to the front and flanks of the position.

(5) Disconnect the MGS and place it in the position made for it. Place the launcher into the position. Reconnect the MGS and check the boresight.

(6) Improve the position by adding overhead cover for the crew and system.

(7) Camouflage the position.

(8) Inspect the position and its camouflage. If possible, move at least 35 meters to the front of the position and study it.

b. The initial position should look like Figure C-1.

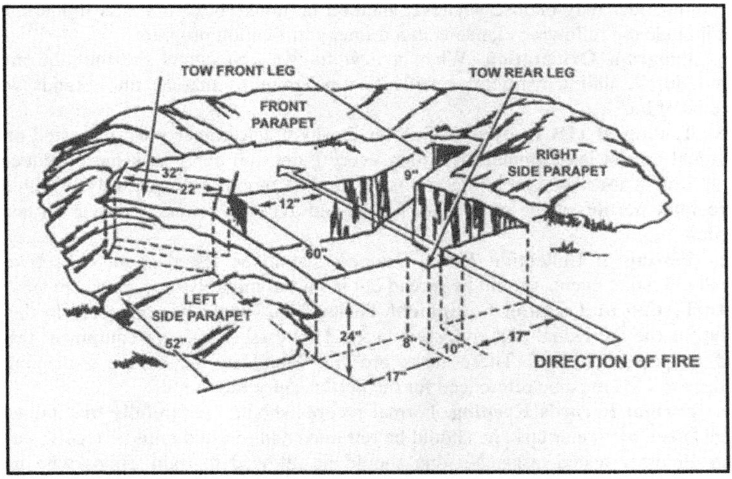

Figure C-1. Tripod-mounted TOW fighting position.

(1) Make the position 24 inches deep.

(2) Make the parapet at least 18 inches thick to keep out small-arms fire and artillery fragments. It should provide 9 inches of muzzle clearance under the launch tube.

(3) Do not place dirt or equipment in the backblast area.

(4) Scoop out a place for the MGS either under or to the front of the tripod.

(5) Dig a storage/protective area for the crew and missiles to one flank at a 90-degree angle to the primary direction of fire. Use the strongest material available for the roof. Put canvas or plastic down before throwing dirt on the roof to keep the ceiling from leaking. Place at least 20 inches of dirt on top of the storage/protective area (Figure C-2).

(6) Ensure the ground behind the TOW is free of leaves and dirt out to 25 meters so the backblast does not leave a signature.

(7) Ensure all the standard principles of camouflage are followed. (For example, cover all fresh dirt with leaves and brush, replace withered foliage, always approach position from rear, and so forth.)

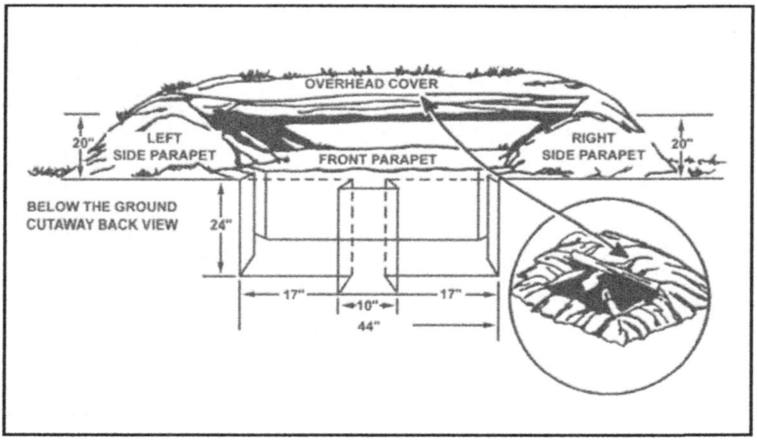

Figure C-2. Storage/protective area.

C-2. MOUNTED TOW FIGHTING POSITION

A mounted TOW fighting position must meet the following specifications:

- Does not restrict the TOW's target engagement.
- Provides concealment from ground and air observation for the crew and vehicle.
- Backblast area is as clear as possible of debris that would increase launch signature.

a. Camouflage and conceal the firing position.

(1) All vehicles travel to a position over the same route to prevent the enemy from detecting the number of vehicles present.

(2) The route taken should follow existing paths, roads, fences, or natural lines in terrain.

(3) Exposed routes should not end at a position.

(4) If only a short portion of the route into a position is exposed, sweep out the tracks with tree branches.

(5) Traffic in and out of a position should be held to essential movements only.

(6) Upon moving into a position, be sure to erase the trail leading into the position.

(7) Ensure all standard principles of camouflage are adhered to (same as for tripod-mounted position).

b. If a hull-down position is not available, use a hide position (Figure C-3).

c. If possible, move at least 35 meters in front of the position to ensure that it looks natural and blends with its surroundings.

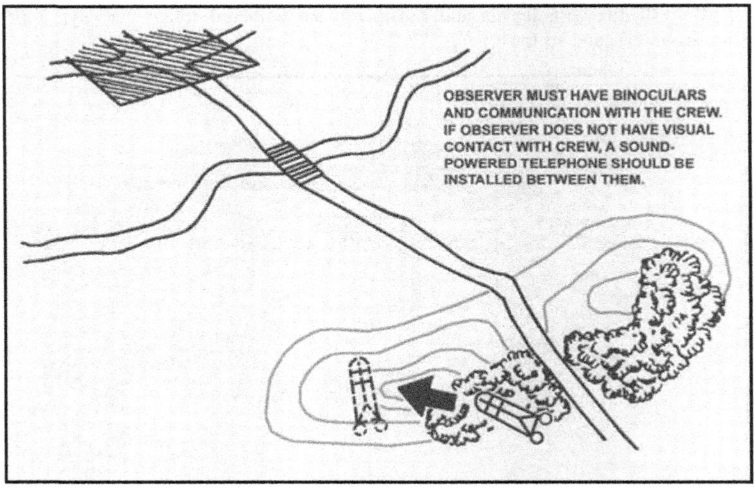

Figure C-3. TOW hide position.

C-3. CAMOUFLAGE NETS

When camouflaging any TOW fighting position, whether mounted or dismounted, the net must be positioned so it does not hang in the backblast area. If any portion of the net is in the backblast area, the net will be damaged and could collapse on top of the launcher when a missile is launched. Also, the net must not interfere with the field of fire.

APPENDIX D

STANDARD RANGE CARD

This appendix describes the various parts of a standard range card. It also provides a detailed discussion of how to complete a range card.

D-1. DESCRIPTION

A range card is a sketch of the terrain a weapon system has been assigned to cover by fire. It contains information that assists in the planning and controlling of fires, the rapid detection and engagement of targets, and the orientation of replacement personnel or units. By using a range card, a gunner can quickly and accurately determine the information he needs to engage targets.

a. A sector of fire is a part of the battlefield within which a gunner is responsible for engaging targets. Sectors of fire are assigned to ensure weapon systems will cover target approaches. Leaders should strive to overlap sectors to cover areas that cannot be engaged by one system. The leader gives a gunner boundaries running between prominent terrain features, or by left and right limits indicated by terrain features or azimuths. If necessary, the leader also assigns a gunner more than one sector of fire, designating one sector as primary and others as secondary.

b. The section or squad leader may also designate anticipated target engagement locations within the sector of fire. Those are recognizable terrain features on or near likely enemy avenues of approach. This information is placed on the range card.

c. Leaders may pick out natural or man-made terrain features that can be used as reference points for locating targets and adjusting direct and or indirect fires. Those features are called target reference points (TRPs). TRPs are requested through the mortar or artillery fire support team (FIST) or fire support officer (FSO). If TRPs are in or near the sector of fire, the leader should point them out and tell the gunner their numbers. If he does, the gunner depicts the TRPs on his range card. Normally, a gunner has at least one TRP, but not more than three, in his sector of fire.

d. Natural or man-made terrain features, such as hills, draws, or buildings, may be within the sector(s) of fire that prevent the gunner from firing in that area. The area blocked by these features is called dead space. All dead space in the sector(s) of fire must be determined so leaders can plan other weapon systems or other types of fire to cover the area (for example, mortars, artillery, or mines). Dead space is indicated on the range card.

e. The length of the sector of fire is normally limited by the maximum engagement range of the antiarmor weapon, but it can be less if any natural or man-made terrain features (trees, fences) prevent the gunner from engaging targets at maximum range. Regardless of what affects it, the maximum engagement range is shown on the card as a maximum engagement line. The squad leader uses a map to determine the distance to the maximum engagement line.

f. All TRPs, anticipated target engagement areas, azimuth and distance to a known point, and left and right limits are numbered on the sector sketch and in the data section with corresponding numbers for quick response.

D-2. PREPARATION PROCEDURES

The section/squad leader's briefing provides the necessary information for the gunner to begin preparing DA Form 5517-R (Standard Range Card). If he is assigned alternate and supplementary firing positions, he prepares a range card for them also. The gunner prepares the range card by performing the following steps.

EXAMPLE OF SECTION OR SQUAD LEADER'S BRIEFING:
"Our mission is to cover a sector of fire that begins at our present position and goes in the direction of the windmill to the maximum engagement range of 3,750 meters; it extends to the right across the high ground behind the houses, and hill, to the right edge of the orchard and returns here. The enemy should approach from the north and will probably use both Marshall Road and Lewis Road to enter our sector. We must plan on engaging the enemy in this area as soon as he is within range. There are two target reference points within your sector; the road intersection of Marshall Road and Duffell Road is TRP-Charlie One and the road junction of Lewis Road and Duffell Road is TRP-Charlie Two. Use the road junction of Campbell Road and Lewis Road to your left as a reference point to locate your position. The distance from the road junction is 633 meters on an azimuth of 85 degrees." (See Figure D-1.)

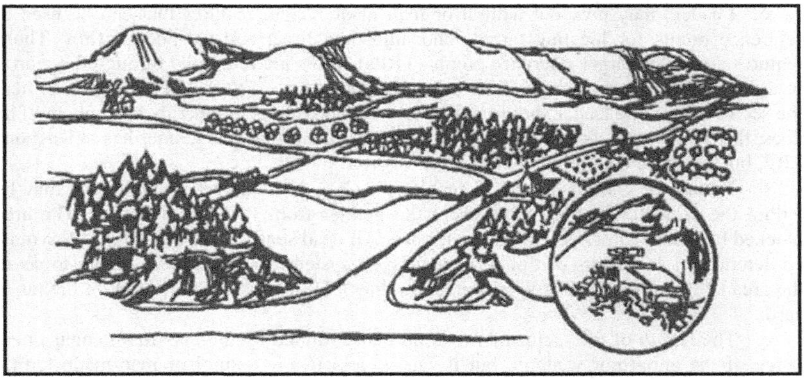

Figure D-1. Section or squad leader's briefing.

a. Draw a sector sketch of the entire sector. Make the sketch as large as possible, not to exceed the largest circle. For a large area covered by trees or woods, draw only the outline and label the area; for example, "woods" or "orchard" (Figure D-2).

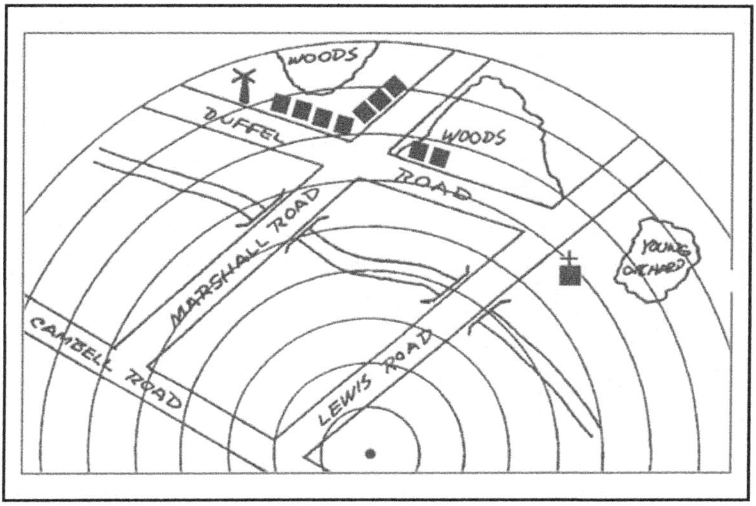

Figure D-2. Sketch of area on range card.

b. Draw lines from the weapon position (indicated by the black dot at lower center of range card) to show the right and left limits. Place a number 1 at the end of the left limit line and draw a circle around the number. Place a number 2 at the end of the right limit line and draw a circle around the number (Figure D-3, page D-4).

NOTE: After drawing left and right limit lines, the weapon symbol can be drawn over the black dot.

c. If there are no limitations, the maximum engagement line is curved and joins the left and right sector of fire boundaries at the maximum engagement range (Figure D-4, page D-4). If there are limitations, the maximum engagement line is drawn in front of the limiting terrain feature.

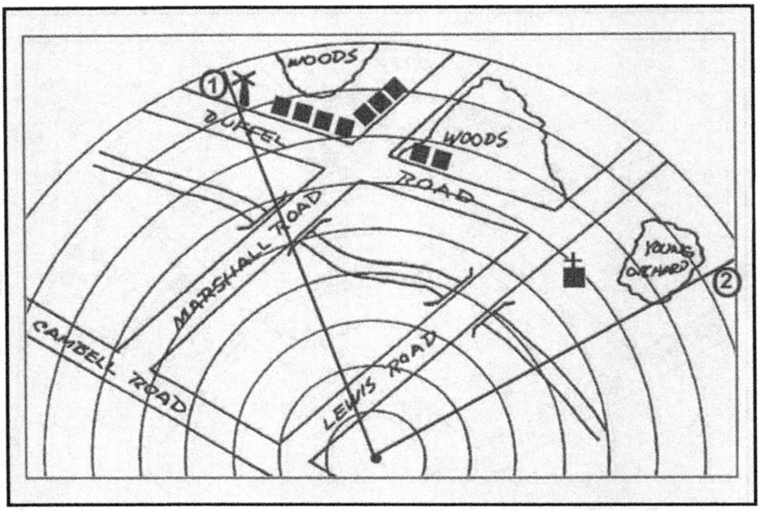

Figure D-3. Labeling sector of fire.

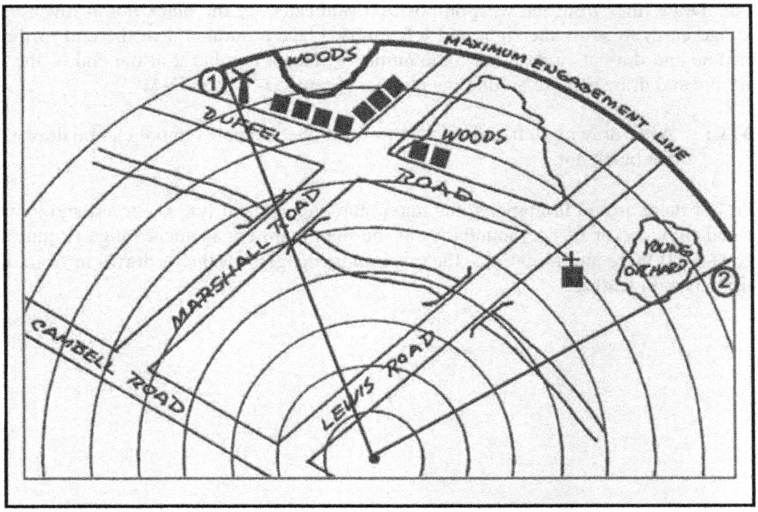

Figure D-4. Placement of maximum engagement line.

d. Number the anticipated target engagement areas (ATEAs) from left to right, starting with number 3. Place a number at the maximum engagement range of the target on the range card and circle the number (Figure D-5, page D-6).

e. Number the TRPs from left to right. Place the number below or next to the TRP on the range card and circle the number (Figure D-6, page D-6).

f. Place diagonal lines, or the words "dead space," where dead space occurs (Figure D-7, page D-7).

g. Use a compass to determine the azimuth from the firing position to the known point. Convert the direction to a back azimuth. Draw a line with multiple arrows from the known point to the firing position. Place a number at the known point and circle the number (Figure D-8, page D-7).

h. Fill in the marginal information at the top of the card.

(1) ***Unit Description–SQD, PLT, CO.*** Never indicate a unit higher than company level.

(2) ***Magnetic North.*** Orient the range card with the terrain. Place the compass on the range card. Determine the direction of magnetic north arrow and mark it on the card.

i. Fill in the data section at the bottom of the card.

(1) ***Position Identification.*** List either primary, alternate, or supplementary.

(2) ***Weapon***. See Figure D-9 (page D-8) for weapon symbols.

(3) ***Date***. List the day and month.

(4) ***Each circle equals ______ meters.*** Write the distance between the circles in meters. To determine the distance, count the intervals from the weapon to the maximum engagement line (as determined by the squad leader). Divide the amount of intervals into the range of the maximum engagement line. This will give the distance between circles (Figure D-10, page D-8).

Example: 9 intervals into 3,750 meters = 416 meters between circles.

(5) ***No. (Number)***. Starting with number 1, list the left limit, the right limit, and locations of ATEAs and TRPs shown on the sector.

(6) ***Direction/Deflection.*** Only degrees or the azimuth from the azimuth bevel ring (improved TOW vehicle) is listed. Line through the word that does not describe the information listed.

(7) ***Elevation***. This is only used with a ground-mounted machine gun using the traverse and elevation mechanism.

(8) ***Range***. Distance in meters from the weapon to the TRP or target engagement area.

(9) ***Ammunition***. List the type of ammunition used, if applicable.

(10) ***Description***. List the name of the object; for example, road, windmill, church. If the item is a TRP, also list the TRP number.

(11) ***Remarks***. Enter the weapon's reference point and any additional information not listed in the range card section. If more space in the data section is needed, use the reverse side of the range card.

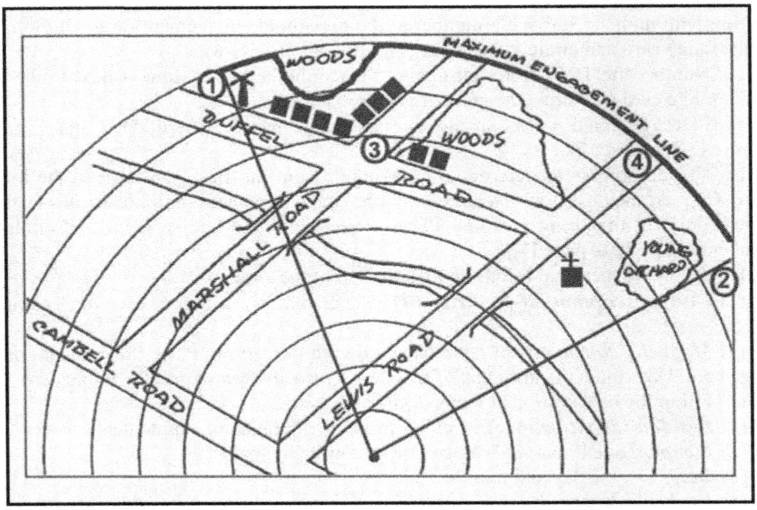

Figure D-5. Numbered target engagement locations.

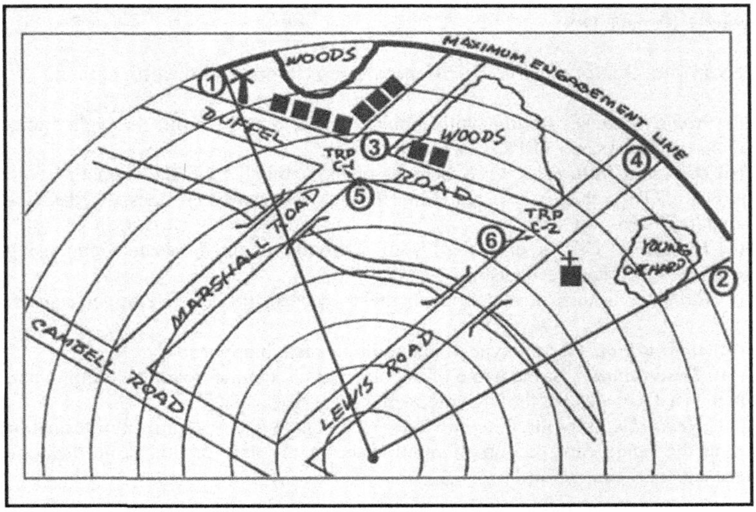

Figure D-6. Target reference points and numbers.

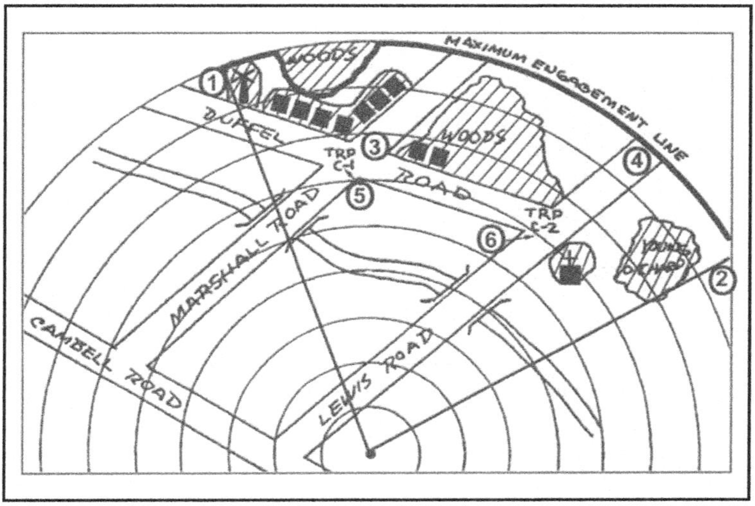

Figure D-7. Dead space indicated.

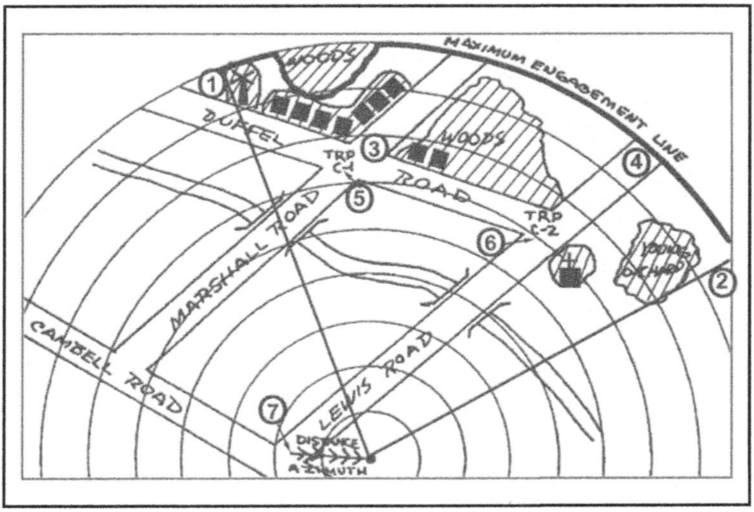

Figure D-8. Placement of distance and azimuth from a known point.

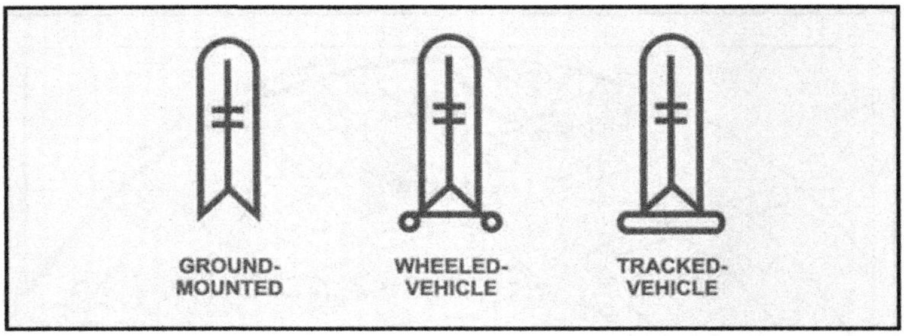

Figure D-9. Weapon symbols.

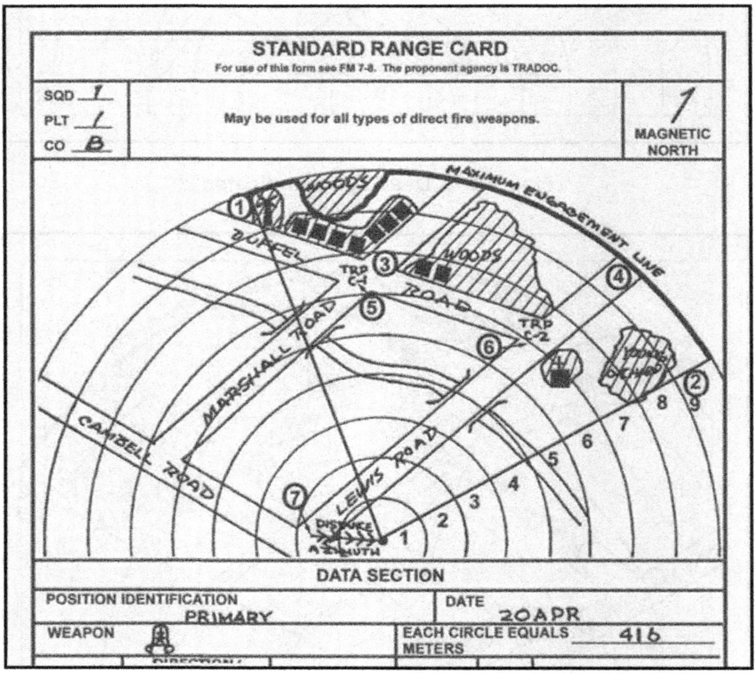

Figure D-10. Determining range interval on range card.

j. Make two range cards. Keep one at the firing position and give one to the squad or section leader for preparation of fire plans and final coordination of fires. (See Figure D-11 for a completed TOW range card.)

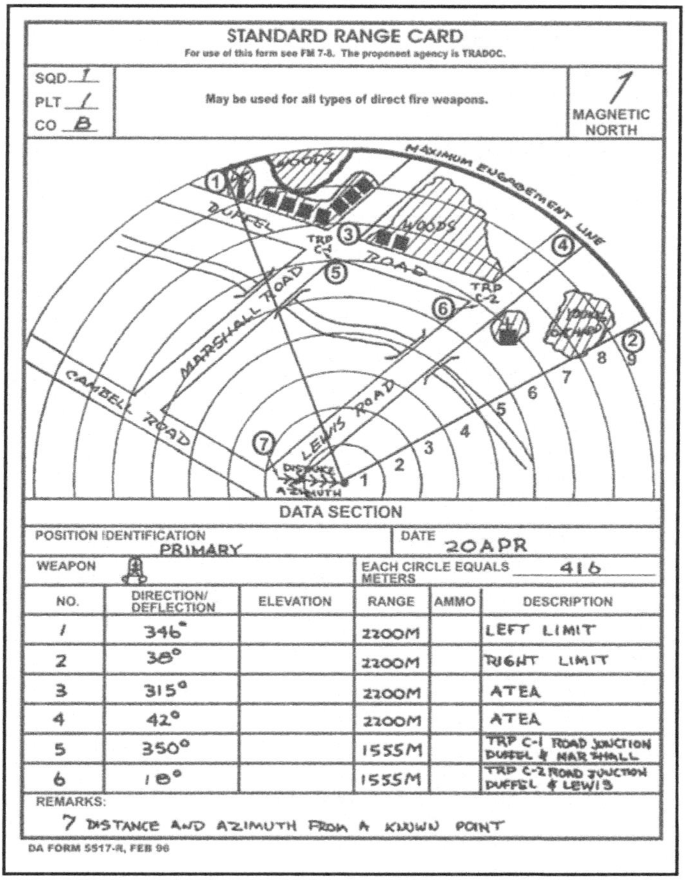

STANDARD RANGE CARD

For use of this form see FM 7-8. The proponent agency is TRADOC.

SQD 1 PLT 1 CO B

May be used for all types of direct fire weapons.

MAGNETIC NORTH

DATA SECTION

POSITION IDENTIFICATION: PRIMARY | DATE: 20APR

WEAPON: | EACH CIRCLE EQUALS 416 METERS

NO.	DIRECTION/ DEFLECTION	ELEVATION	RANGE	AMMO	DESCRIPTION
1	346°		2200M		LEFT LIMIT
2	38°		2200M		RIGHT LIMIT
3	315°		2200M		ATEA
4	42°		2200M		ATEA
5	350°		1555M		TRP C-1 ROAD JUNCTION DUFFEL & MARSHALL
6	18°		1555M		TRP C-2 ROAD JUNCTION DUFFEL & LEWIS

REMARKS:
7 DISTANCE AND AZIMUTH FROM A KNOWN POINT

DA FORM 5517-R, FEB 96

Figure D-11. Example of a completed DA Form 5517-R, Standard Range Card for TOW.

D-3. EXPEDIENT RANGE CARD

In combat, a DA Form 5517-R may not be available. The gunner must then draw a range card on anything available (Figure D-12, page D-10). Preparation of the expedient range card follows the same procedures provided for the standard range card, but the weapon symbol must be used to indicate the location of the weapon position. The range card must include the following eight items:

- Weapon symbol.
- Sector of fire.
- Maximum engagement line.
- Range and azimuth to TRP/ATEA.
- Dead space.
- Distance and azimuth from a known point.
- Magnetic north arrow.
- Data section.

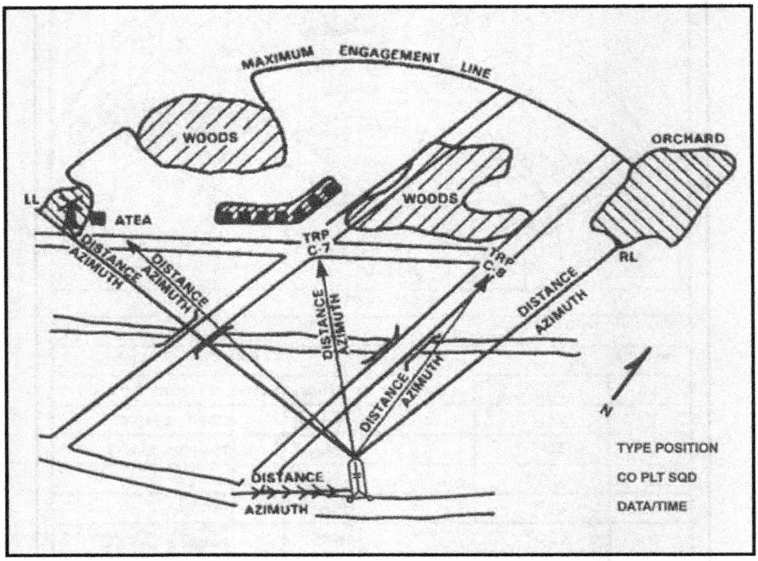

Figure D-12. Expedient range card.

APPENDIX E
SAFETY

Units should develop local directives and SOPs that include individual responsibilities, safety requirements, proximity limits for personnel and explosives, location and sequence of operations, equipment required for handling munitions, and protection for troops. Individual responsibility for operations involving explosives must be designated. Troops must not tamper with the encased missiles other than to remove the forward handling ring and dust cover.

Because of the danger to troops from the backblast, extreme care must be used in all phases of training. Emphasize this danger from the first stage of training. Conduct all crew tasks, position and tracking exercises, and firings with the simulation round as though missiles were being fired.

E-1. SAFETY PRECAUTIONS

The surface danger zone for any firing range consists of a firing area, a target area, impact area, and danger areas surrounding these locations (Figure E-1). An additional area for occupation by personnel during firings may also be required. The shape and size of the surface danger zone varies with the type of missile or rocket being fired. (Refer to DA Pam 385-63 for dimensions.)

a. The primary danger area is a 90-degree cone with a 50-meter radius. The apex of the cone is centered at the rear of the missile launcher. Serious casualties or fatalities are likely to occur to anyone in the area during firing. Hazards include launch motor blast, high noise levels, overpressure, and debris.

b. *Caution area 1* extends radially from each side of the primary danger area to the firing line with a radius of 50 meters. Permanent hearing damage could occur to personnel without adequate hearing protection in this area during firing. The hazards are high noise levels and overpressure.

c. *Caution area 2* is an extension of the primary danger area with the same associated hazards and personnel protection required. The radius of this area is 75 meters.

d. The *200-meter zone* is the danger area for aerial firings 15.25 meters or more above ground level.

e. TOWs will not be fired from buildings or bunkers, or within 100 meters of a vertical or nearly vertical backstop (IAW DA Pam 385-63).

WARNING

All crew members must wear V-51R (or equivalent) earplugs that have been fitted by qualified medical personnel. Failure to use the proper earplugs during missile firing could cause serious injury.

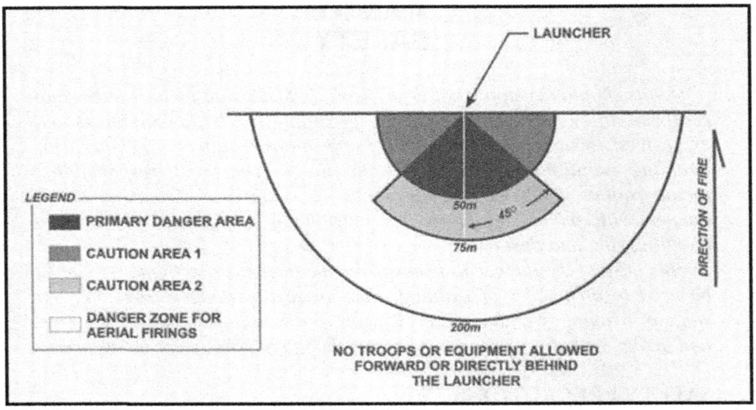

Figure E-1. Surface danger zone for firing basic TOW, TOW 2A, and TOW 2B missiles.

E-2. MISSILE HANDLING PRECAUTIONS

Improper handling of the encased missile may damage the components and cause malfunctions when the missile is launched. If the encased missile is dropped, the end handling rings and the launch container may be damaged. If the missile has been damaged or there is damage to the launch container (other than minor deformation of handling rings), the encased missile should be returned to the ammunition unit for inspection and disposition.

E-3. SIGHTING AND AIMING PRECAUTIONS

Gunners must not look at the sun or bright lights while sighting through the daysight tracker; serious eye burn could result. They must not look through the daysight tracker at an air field test set on the control tower unless the distance between the daysight tracker and the test set is more than 300 feet.

E-4. FIRING ANGLE LIMITATIONS

Azimuth and elevation firing angles are limited by the traversing unit, the vehicle, and other external restrictions. All elevation angles are referenced to the horizontal plane of the traversing unit. All azimuth angles are referenced to the long axis of the vehicle and depend on whether the launch tube points over the front or rear of the vehicle. The other reference line is the line-of-sight from the TOW to the target.

> **WARNING**
> **At angles greater than 20 degrees above ground level, hazards to the gunner may exist in the overpressure waves and debris caused by the backblast during training. Do not use angles greater than 20 degrees.**

a. When the TOW is tripod-mounted, a 360-degree lateral track is possible, because the traversing unit is not restricted in azimuth. Mechanical stops limit the elevation angle coverage to 20 degrees below and 30 degrees above the horizontal plane. Before the missile is fired, the line-of-sight angle should be estimated at the expected time of launch and throughout the expected missile flight time. The firing position should be changed or a different target selected if an expected line-of-sight angle exceeds the firing limitation angle.

b. The firing angle limitations of the TOW carriers are as shown in Figure E-2.

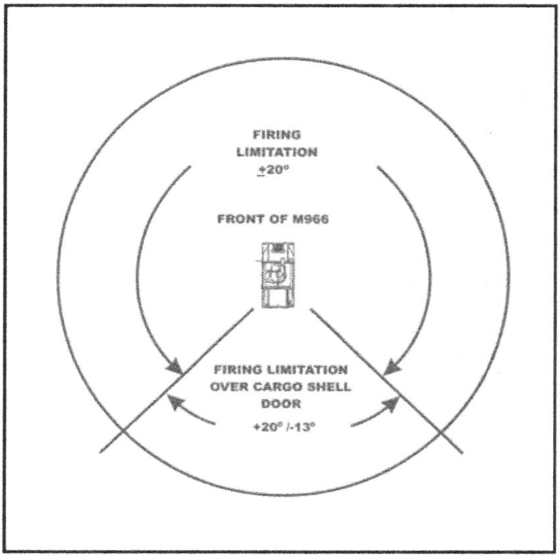

Figure E-2. M966-mounted TOW firing angle limitations.

E-5. SAFETY PRECAUTIONS FOR MILES

The gunner must follow these safety precautions:

a. Do not load an ATWESS cartridge until ready to fire. If the target is lost, remove the ATWESS cartridge from the firing chamber before moving. (Gloves are recommended when loading or unloading the ATWESS cartridge.)

b. Before pulling the PULL-TO-ARM switch, always check to ensure no personnel or equipment (antennas) are in the ATWESS danger zone. This zone extends for 75 meters behind the ATWESS firing chamber and covers an arc 90 degrees wide (Figure E-3).

c. Never view the laser being fired through stabilized optics, such as binoculars or telescopic weapon sights, when within 75 meters of the transmitter.

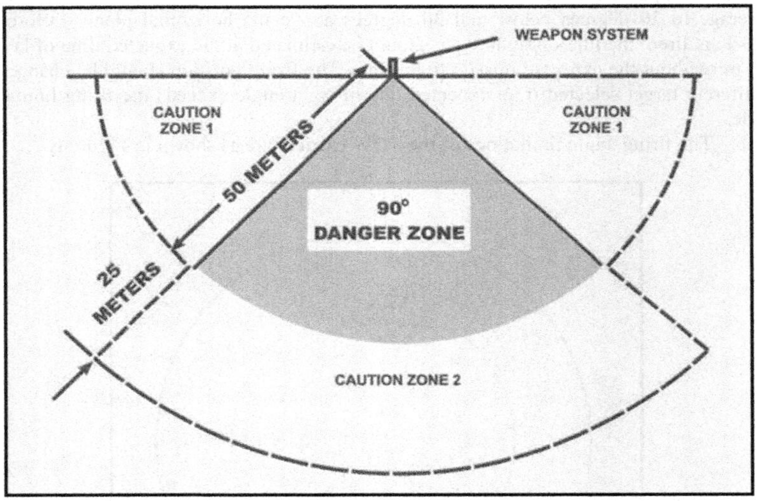

Figure E-3. ATWESS danger zone.

E-6. SAFETY PRECAUTIONS FOR THE TOW GUNNERY TRAINER

The following safety precautions must be observed when using the TGT.

> **DANGER**
>
> **THIS EQUIPMENT USES HIGH VOLTAGE TO OPERATE. NEVER USE UNGROUNDED EXTENSION CORDS, UNGROUNDED ADAPTERS, OR ANY UNGROUNDED OUTLET TO CONNECT THE TGT. DEATH ON CONTACT MAY RESULT IF PERSONNEL FAIL TO OBSERVE SAFETY PRECAUTIONS.**

a. Use two people to lift the instructor console. The console is heavy and lifting with only one man could result in serious injury.

b. Do not attempt to open shipping cases before pressing air pressure release valves on side of cases. Serious injury to personnel could result from opening cases with high pressure inside.

c. Turn off the power to the TGT trainer and disconnect the wall outlet plug before beginning cleaning procedures.

E-7. SAFETY PRECAUTIONS FOR THE TOW FIELD TACTICAL TRAINER

The following safety precautions must be observed when using the TFTT.

a. The laser light emitted by the TFTT is considered eye safe, but suitable precautions must be taken to avoid possible eye damage from overexposure to this radiated energy. The preface to TM 9-6920-453-10 and the laser range safety procedures in DA Pam 385-63 and TB MED-279 discuss these precautions.

b. The M80 blast simulator used with the TFTT can cause death or injury. Observe the precautions listed in the preface of TM 9-6920-453-10.

c. To avoid personnel injury and equipment damage, four people are needed to lift and carry each shipping container.

> **DANGER**
>
> **THIS EQUIPMENT USES HIGH VOLTAGE TO OPERATE. DEATH ON CONTACT MAY RESULT IF PERSONNEL FAIL TO OBSERVE THE SAFETY PRECAUTIONS LISTED IN TM 9-6920-453-10.**

APPENDIX F

TOW TRAINING DEVICES

For many years TOW training devices consisted of the M70-series training sets and MILES TOW equipment. Recently, the Precision Gunnery Training System (PGTS) was fielded. PGTS consists of two systems: an indoor trainer and an outdoor trainer. The TOW Gunner Trainer (TGT) is the indoor system and the TOW Field Tactical Trainer (TFTT) is the outdoor system. The TGT will be used in place of the M70 to qualify individual gunners and for additional sustainment training as needed. The TFTT will be used in place of the M70 for outdoor tracking sustainment and in place of MILES when conducting Tables 5 through 12 of the TOW Training Tables.

MILES TOW equipment is the most realistic device available for simulating tactical engagements. It is valuable in maneuver training exercises and Army training and evaluation programs. However, MILES TOW is not a precision gunnery trainer and should not be used to train gunner tracking skills.

Section I. MULTIPLE-INTEGRATED LASER ENGAGEMENT SYSTEM

MILES is a training system used in force-on-force training and in FTXs. Commanders should use an MPRC to verify MILES equipment before the unit's rotation to the NTC. (See gunnery Tables 7 through 12 in Chapter 4 for target arrays.)

F-1. COMPONENTS AND FEATURES

This paragraph discusses the different components, features, and capabilities of MILES.

a. **Components.**

(1) The following are the nine components in the MILES TOW transit case:

- Operator's manual.
- Two man-worn laser detector sets (halo and harness).
- Two yellow keys.
- TOW laser transmitter/tracker head simulator.
- Missile guidance set adapter.
- Miscellaneous hardware.
- Hook-pile tape pads.
- Control console mounting bracket.
- Collimator adapter bracket.

(2) The following are the ten MILES APC system components packed in the APC transit case.

- Operator's manual.
- Control console.
- Laser transmitter for M2.
- Battery box.
- Hardware.

- Combat vehicle kill indicators.
- Weapon's keys.
- Cable assembly.
- Combat vehicle laser detector belt segments.
- Man-worn laser detectors.

b. **Features.**
Range: 3,750 meters; M60, 1,100 meters.
Tracking time: 10 seconds.

Kill codes:	Visual	Audio
Near miss	2 CVKI flashes	2 tones
Hit	4 to 6 CVKI flashes	4 to 6 tones
Kill	Continuous flash	Continuous tone

Basic load: IAW unit SOP, ATTWESS comes packed 12 to a box.
Backblast area: 75 meters at 90 degrees.
TOW laser tracking: head assembly optics 10 power.
Refire ability: every 10 seconds.

c. **Capabilities.** MILES-equipped weapons have the same range and operational capabilities as the normal weapons. The effective range of the transmitter may be reduced by a dirty transmitter lens and or weak batteries. The M60 machine gun is effective against lightly armored MILES-equipped vehicles and personnel. The TOW 2 is effective against armored MILES-equipped vehicles.

NOTE: Training soldiers to assemble, utilize, and maintain the MILES will greatly decrease down time due to MILES malfunctions.

F-2. SAFETY PRECAUTIONS

The following safety precautions should be observed when using MILES.

a. Do not load an ATWESS cartridge until ready to fire. If the target is lost, remove the ATWESS cartridge from the firing chamber before moving. (Gloves are recommended when loading or unloading the ATWESS.)

b. Before pulling the PULL-TO-ARM switch, always check to ensure no personnel or equipment (antennas) are in the ATWESS danger zone. This zone extends for 75 meters behind the ATWESS firing chamber and covers an arc 90 degrees wide (Figure F-1).

c. Never view the laser being fired through stabilized optics, such as binoculars or telescopic weapon sights, when within 75 meters of the transmitter.

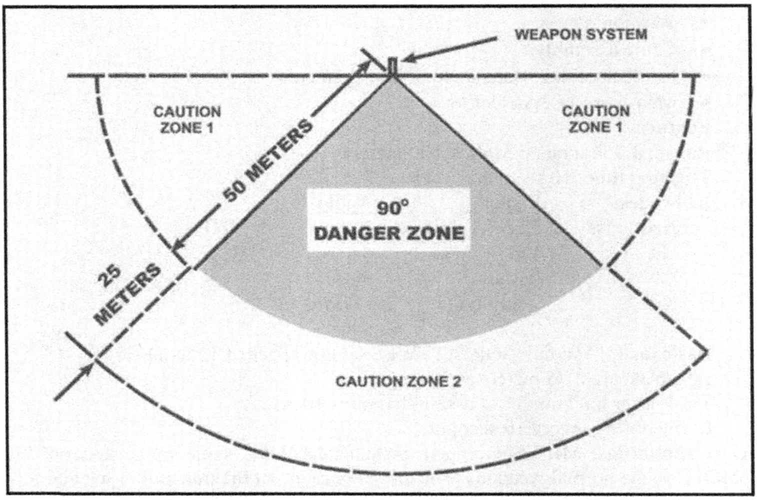

Figure F-1. MILES TOW backblast area.

Section II. TOW GUNNERY TRAINER

The TGT is a new trainer used indoors for individual advanced gunnery tables. It provides realistic gunnery training, not crew drills.

F-3. COMPONENTS AND FEATURES

The TGT consists of TGT-unique components and TOW 2 components. The major TGT-unique components are the instructor station (Figure F-2, page F-4) and the student station (Figure F-3, page F-4). The TOW 2 components are the traversing unit, tripod, and the launch tube.

Figure F-2. Instructor station.

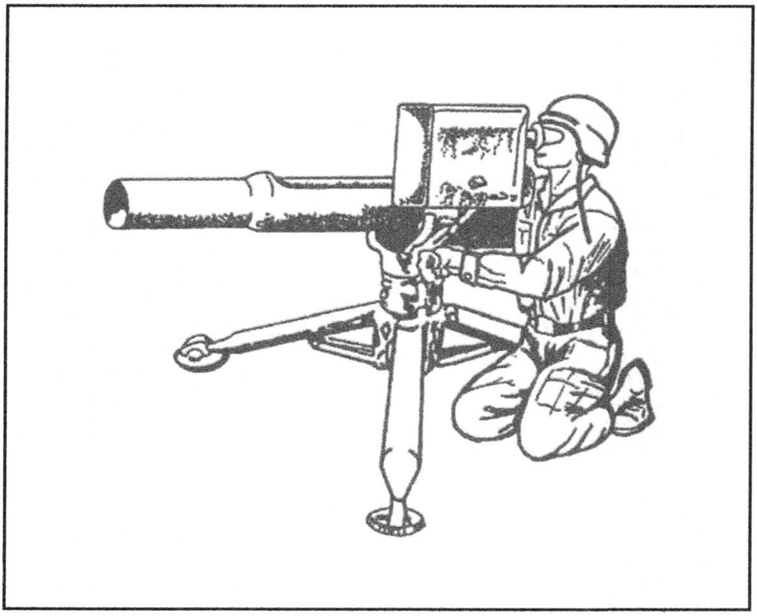

Figure F-3. Student station.

a. The TGT simulates the sight(s), controls, switches and indicators of the M220E4 TOW 2 weapon system. Through the gunner's sight(s) you see battlefield scenes that include both enemy and friendly vehicles. Using the controls and switches, the gunner selects a target, fires, and tracks it. The headsets simulate the blast of the TOW and the singing of the wire. The gunner sees and hears hit-and-miss explosions and hears commands from the instructor and battlefield sounds of small arms and guns. The score is displayed at the end of each mission.

b. TGT trains novice gunnery skill training, gunnery skill progression, and sustainment training. It is used to train the following skills:

- Correct firing position.
- Target identification.
- Target engageability determination.
- Target engagement, including tracking and firing.
- Fire commands.

F-4. ASSEMBLY AND OPERATION

The following manuals are used for proper assembly and operation of the TGT.

a. TM 9-6920-452-10, Chapter 2, Section 1, discusses procedures for unpacking, assembly, disassembly, and repacking for storage and or shipment.

b. TM 9-6920-452-10, Chapter 2, Section 2, discusses preparation for operation, preliminary inspections, warm-up, and operational checkout procedures.

c. TM 9-6920-452-10, Chapter 2, Section 3, discusses operating procedures performed by the instructor in order to conduct training.

F-5. SAFETY PRECAUTIONS

The following safety precautions should be observed when using the TGT.

DANGER

THIS EQUIPMENT USES HIGH VOLTAGE TO OPERATE. NEVER USE UNGROUNDED EXTENSION CORDS, UNGROUNDED ADAPTERS, OR ANY UNGROUNDED OUTLET TO CONNECT THE TGT. DEATH ON CONTACT MAY RESULT IF PERSONNEL FAIL TO OBSERVE SAFETY PRECAUTIONS.

a. Use two people to lift the instructor console. The console is heavy and lifting with only one man could result in serious injury.

b. Do not attempt to open shipping cases before pressing air pressure relief valves on side of cases. Serious injury to personnel could result from opening cases with high pressure inside.

c. Turn off the power to the TGT trainer and disconnect the wall outlet plug before beginning cleaning procedures.

Section III. TOW FIELD TACTICAL TRAINER

The TFTT is used to teach precision gunnery skills to TOW gunners in the field. This training can occur on designated ranges, general outdoor areas, or representative tactical environments. The TFTT trains gunners to adopt a correct firing position, assess target engageability, and to engage and track the target. Missile launch, flight and impact effects are realistically simulated by the TFTT.

F-6. COMPONENTS AND FEATURES

The TFTT consists of TFTT-unique components and a TOW 2 weapon system. The TFTT attaches to the TOW 2 and replaces some of its components. The TFTT can be either tripod mounted or HMMWV mounted and can be configured for the M966 (Figure F-4).

a. The TFTT uses the TOW 2 weapon system equipment to enhance training realism. In addition, most of the TFTT components are designed to resemble actual weapon system equipment. The TFTT uses a retroreflector to designate its target. The retroreflector returns a portion of the laser beam generated by the gunner. The laser beam enables precise measurement of target range and location relative to the gunner. The retroreflector can be mounted on a variety of target vehicles, which can be maneuvered as required during a training mission. Targets equipped with MILES sensors can also be engaged by the TFTT.

b. The TFTT operator loads the M80 blast simulator, sets the duration of the obscuration that simulates the smoke produced at missile launch, and selects the relative size of the target. Following missile launch, the operator monitors gunner performance during missile flight. At the end of each mission, the operator is provided with a readout of mission results.

F-7. ASSEMBLY AND OPERATION

The following manuals are used for proper assembly and operation of the TFTT.

a. TM 9-6920-453-10 (M966), Chapter 2, Section 1, discusses procedures for unpacking, assembly, disassembly, and repacking of the TFTT.

b. TM 9-6920-453-10 (M966), Chapter 2, Section 2, discusses procedures for preliminary inspection, power-up, and operational checkout of the TFTT.

c. TM 9-6920-453-10 (M966), Chapter 2, Section 3, discuss procedures for operating the TFTT.

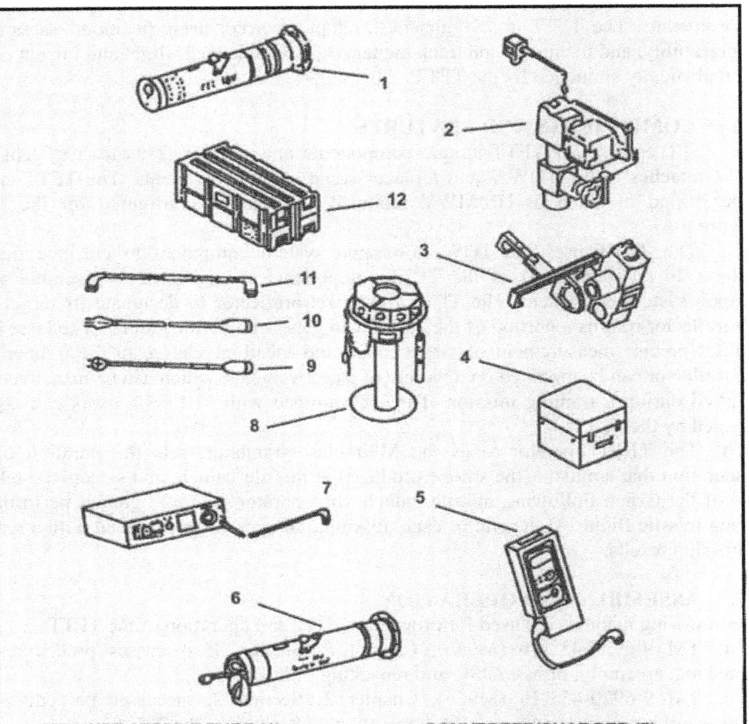

TRAINER MISSILE TUBE (1)
DAYSIGHT TRACKER INTERFACE (2)
THERMAL SIGHT INTERFACE (3)
COMPUTER (4)
REMOTE CONTROL UNIT (5)
BATTERY TUBE (6)
CONVERTER/CHARGER (7)
RETROFLECTOR ASSEMBLY AND PEDESTAL (8)
CABLES (9, 10, 11)
SHIPPING CONTAINERS (12)

Figure F-4. TOW field tactical trainer components (M966).

F-8. SAFETY PRECAUTIONS

The laser light emitted by the TFTT is considered eye safe, but suitable precautions must be taken to avoid possible eye damage from overexposure to this radiated energy. See the preface to TM 9-6920-453-10 and the laser range safety procedures in DA Pam 385-63 and TB MED-279 for these precautions. To avoid personnel injury and equipment damage, four people are required to lift and carry each shipping container.

DANGER

THIS EQUIPMENT USES HIGH VOLTAGE TO OPERATE. DEATH ON CONTACT MAY RESULT IF PERSONNEL FAIL TO OBSERVE SAFETY PRECAUTIONS LISTED IN TMs 9-6920-453-10 (M966).

DANGER

THE M80 BLAST SIMULATOR USED WITH THE TFTT CAN CAUSE DEATH OR INJURY. OBSERVE THE PRECAUTIONS LISTED IN THE PREFACE OF TM 9-6920-453-10.

APPENDIX G

COMBAT VEHICLE IDENTIFICATION

Skill Level 1 11B soldiers must be able to identify combat vehicles according to the conditions and standards listed in this appendix.

CONDITIONS: Given a classroom with a 35-mm slide projector, screen, slide tray with 40 combat vehicle slides, stopwatch, chair, desk or clipboard, and pencil. (See the local Training Support Center for the combat vehicle slides.)

STANDARDS: Given a 12-second exposure time for each slide, the soldier will identify 36 of 40 vehicles by nomenclature.

EVALUATION PROCEDURES: This task requires the crewman to identify combat vehicles from the front, side, oblique, or rear using day and thermal sights. Views of these combat vehicles should replicate ranges between 800 and 1,200 meters.

a. **Administrative Process.** At the station, the evaluator logs the soldier's information on a roster. The evaluator provides the soldier with all materials and equipment and displays them in a manner outlined in this guide. The evaluator must use the criterion performance checklist for this task. The evaluator informs the soldier of his performance on this task and directs him to the next station or to further training. The number of soldiers to be tested at one time depends on the classroom seating without compromising test procedures.

b. **Administrative Procedures for Personnel Receiving a NO-GO.** If the soldier does not achieve the standard indicated on the checklist, he receives a NO-GO. When a soldier receives a NO-GO, he must be critiqued, to include an explanation of vehicles that were incorrectly identified. The soldier will be retested with a second set of slides showing the same vehicles in a different view or a different order.

PERSONNEL, EQUIPMENT, AND MATERIAL REQUIRED:

- Qualified 11B sergeant or above.
- Forty 35-mm slides of combat vehicles. The 30 required vehicles for this test are included in the list in Figure H-1 and are indicated by an asterisk. The balance will be taken from the other vehicles listed in Figure G-1 (page G-4). (Recognition of Combat Vehicles [ROC-V] compact discs may also be used.) (For more information, see GTA 17-02-013.)
- Classroom.
- Slide projector with screen.
- Stopwatch.
- Performance checklist (one per soldier).
- Chair, desk or clipboard (one per soldier).
- Pencil (one per soldier).

PRETEST PREPARATION: Thirty vehicles will be taken from the prescribed list and ten will be selected by the commander based on the unit's contingency area. The station evaluator must ensure the slide projector is operational and the slides can be recognized from all locations in the classroom. Some projectors have an automatic setting for exposure of each slide; if used, this feature must be checked for accuracy.

a. Select 30 required vehicle slides to be used along with 10 slides selected by the commander.

b. Ensure the slide projector is operational.

c. Ensure the slides can be recognized from all locations in the classroom.

PRETEST CONDITIONS FOR EACH EXAMINEE: The slide projector will be on, but slides will not be exposed.

TEST PLANNING TIME:

Administrative: 5 minutes
Test: 15 minutes
Total: 20 minutes

INSTRUCTIONS TO EXAMINEES:

"LET ME HAVE YOUR ATTENTION. AT THIS STATION, YOU WILL BE TESTED ON YOUR ABILITY TO IDENTIFY COMBAT VEHICLES. YOU MUST CORRECTLY IDENTIFY THIRTY-SIX OF FORTY VEHICLES BY NOMENCLATURE. YOU WILL HAVE TWELVE SECONDS TO VIEW EACH SLIDE AND TEN SECONDS BETWEEN EACH SLIDE TO WRITE YOUR ANSWER ON THE PERFORMANCE CHECKLIST PROVIDED. BE SURE YOU DO NOT GET OUT OF SEQUENCE OR YOUR ANSWERS WILL BE INCORRECT. DO YOU UNDERSTAND THESE INSTRUCTIONS?"

Pause 5 seconds, then say, "YOU WILL HAVE TWENTY-TWO SECONDS TO COMPLETE EACH TASK." Pause, then say, "BEGIN."

Begin timing when the command "BEGIN" is given. If the examinee has not completed each task after the appropriate time, announce loud enough for the examinee to hear, "STOP."

PERFORMANCE CHECKLIST

TASK: Identify combat vehicles by nomenclature.

PERFORMANCE MEASURES:

1. Identified vehicle slide No. 1.
2. Identified vehicle slide No. 2.
3. Identified vehicle slide No. 3.
4. Identified vehicle slide No. 4.
5. Identified vehicle slide No. 5.
6. Identified vehicle slide No. 6.
7. Identified vehicle slide No. 7.

8. Identified vehicle slide No. 8.
9. Identified vehicle slide No. 9.
10. Identified vehicle slide No. 10.
11. Identified vehicle slide No. 11.
12. Identified vehicle slide No. 12.
13. Identified vehicle slide No. 13.
14. Identified vehicle slide No. 14.
15. Identified vehicle slide No. 15.
16. Identified vehicle slide No. 16.
17. Identified vehicle slide No. 17.
18. Identified vehicle slide No. 18.
19. Identified vehicle slide No. 19.
20. Identified vehicle slide No. 20.
21. Identified vehicle slide No. 21.
22. Identified vehicle slide No. 22.
23. Identified vehicle slide No. 23.
24. Identified vehicle slide No. 24.
25. Identified vehicle slide No. 25.
26. Identified vehicle slide No. 26.
27. Identified vehicle slide No. 27.
28. Identified vehicle slide No. 28.
29. Identified vehicle slide No. 29.
30. Identified vehicle slide No. 30.
31. Identified vehicle slide No. 31.
32. Identified vehicle slide No. 32.
33. Identified vehicle slide No. 33.
34. Identified vehicle slide No. 34.
35. Identified vehicle slide No. 35.
36. Identified vehicle slide No. 36.
37. Identified vehicle slide No. 37.
38. Identified vehicle slide No. 38.
39. Identified vehicle slide No. 39.
40. Identified vehicle slide No. 40.

TANKS

1. AMX13
2. AMX30
3. AMX30 SA
4. JP Kanone
5. Centurion
6. Centurion 15
7. Centurion RE
*8. Challenger
*9. Chieftain
10. Leopard 1
*11. Leopard 2
*12. M1 Abrams
13. M48A5
14. M551 Sheridan
15. M60A1
*16. M60A3
*17. Merkava
18. PT76
19. PT76 Type 60
20. T8OU
21. T34 Medium
22. T54
23. T55
24. T55K
*25. T62
26. T64
27. T64A
28. T64B
29. T64K
*30. T72
31. T72 M84
32. T72M1
*33. T80
34. Type 59
35. Type 62
36. Type 69
37. Type 79
38. Type 80
39. Type 90
*40. Le Clerc

LIGHT-ARMOR—APC/IFV/RECONNAISSANCE

1. AML
2. AMX10
3. AMX10P
4. AMX10RC
5. BMD1
6. BMD 1979/3
*7. BMD2
*8. BMP1
9. BMP1K
10. BMP1KSH
*11. BMP2
*12. Bradley M2/M3
13. AMX VCI
14. Bradley M2A2
15. LVTP7 (AAV7A1)
16. Bradley M3A1
17. Bradley M3A2
*18. BRDM1
*19. BRDM2
20. BRDM2 AT3
21. BRDM2 AT5
22. BRDM2 RKH
23. BRDM2 SA9
24. BRDM2 U
25. BRM
26. BRM1
27. BTR152
28. KM900
29. BTR50PK
30. BTR60PA Cmd
*31. BTR60PB
*32. BTR70
33. BTR80
34. BTRD
35. Charrua
36. Condor
37. Cougar
38. EE-9 Cascavel
39. EE-11
40. Fox
41. FV432
42. IPR
43. IRM
44. Jaguar 1
45. Jaguar 2
*46. Luchs
*47. M113
48. M113A3
49. M3 APC (Fr)
*50. Marder
51. MCT (S)
52. Ferret MK 1
53. Ferret MK 2
54. MTLB
55. NFV-1
56. NVH-1
57. OT64
58. OT65
59. PSZH-IV
60. Ratel
61. Rooikat
62. Saxon
*63. Scimitar
*64. Scorpion
65. Striker
*66. VAB
*67. Warrior
68. WZ523
69. WZ551
70. YW531
71. YW534

*Required for test.

Figure G-1. Combat vehicle slide list.

ANTITANK SYSTEMS

1. ASU57
2. ASU85
3. AT1 Snapper
4. AT2 Swatter
5. AT3 Sagger
6. AT4 Spigot
7. AT5 Spandrel
8. AT6 Spiral
9. B10
10. PANHARD VCR HOT
*11. ITV M901
12. Milan ATGM
13. MT12
14. RPG2
15. RPG7
16. SPG9
17. SS11 Harpoon ATGM
18. Swingfire ATGM
19. Type 52 75-mm Recoilless Rifle
20. Type 56 RPG-2
21. Type 70-1 62-mm Rocket Launcher
22. Type 86

ANTIAIRCRAFT

1. AMX DCA 30-2
2. Chaparral
*3. Gepard
4. M163A1 VULCAN
5. Roland
6. S60
7. ZPU4
8. 2S6
*9. ZSU23-4
10. ZSU57-2
11. ZU23

SELF-PROPELLED MORTAR

1. M106 107-mm
2. M125 81-mm
3. T54 160-mm
4. Type 85 120-mm
5. Type 85 82-mm
6. YW304 82-mm
7. YW381 120-mm

*Required for test.

Figure G-1. Combat vehicle slide list (continued).

SELF-PROPELLED AND TOWED ARTILLERY

1. 2S1
2. 2S3
3. 2S4
4. 2S5
5. 2S7
6. 2S9
7. AL FAO 210-mm
8. Astros (MRL) 127-mm
9. Astros (MRL) 180-mm
10. Astros (MRL) 300-mm
11. BM21
12. CGT F1
13. Dana 152-mm (SP)
14. G-5 155-mm towed
15. G-6 155-mm (SP)
16. GCT 155-mm (SP)
17. GHN-45 155-mm towed
18. M107
19. M109A1
20. M109A2
21. M109A3
22. M109A6
23. M110A2
24. M1972 122-mm
25. M1973 152-mm
26. M1974 152-mm (SP)
27. M1975 130-mm (SP)
28. M1977 122-mm (SP)
29. M1978 170-mm (SP)
30. M1981 122-mm (SP)
31. M1985 MRL 122-mm (SP)
32. M1985 MRL 240-mm (SP)
33. M1989 170-mm
34. Majnoon 155-mm (SP)
35. MKF3
36. MLRS
37. Type 54-1 122-mm
38. Type 59 130-mm
39. Type 85 122-mm

MISCELLANEOUS

1. Artillery Command and Reconnaissance Vehicle
2. AVLB
3. Fox, NBC Reconnaissance M93
4. * LAV-25
5. M728 CEV
6. * M88A1
7. M9 ACE
8. PRP-3

*Required for test.

Figure G-1. Combat vehicle slide list (continued).

APPENDIX H

OPFOR/THREAT COUNTERPART SYSTEMS

Recent years have seen rapid and revolutionary changes in the field of armor and antiarmor warfare. The most radical changes have been in the design of armor itself, the fire control systems (especially in thermal sights), the warheads and kill mechanisms of the missiles, and the various countermeasures that have been developed. The rate of change is increasing, and antiarmor crewmen and trainers must make a continual effort to keep abreast of recent developments. What was true two years ago, or even one year ago, is not necessarily true today. Leaders who take their men into battle and make decisions based on outmoded information and assumptions are dooming those men to failure and death.

H-1. OPFOR/THREAT ARMORED VEHICLES

Armored vehicle identification in the past was a simple matter of distinguishing between vehicles manufactured by Eastern Bloc and Rest of the World (ROW) nations and those manufactured by NATO nations. Even then, many soldiers had difficulty distinguishing between vehicles on a simple "friend or foe" basis. Changes in international relations have made it less likely that we will meet nations of the former Soviet Union in battle, but more likely that we will fight the military of a third world nation equipped with a wide variety of equipment. From the standpoint of vehicle identification, this greatly complicates the problem. Most third world nations do not manufacture their own vehicles, but buy them from other countries. As a result, many of them have vehicles from nearly every manufacturing nation in the world in their inventories. Twenty nations in the world today have more than 500 main battle tanks each. Two possible examples of many illustrate this point (Table H-1).

COUNTRY	TYPE	ORIGIN
Iran	Type 59 Type 69 Chieftain M48A5 M60A1 T62 T72 EE-9 Cascavel M113	China China United Kingdom USA USA Russia Russia Brazil USA
Iraq	Type 59 AMX-30 Chieftain T62 T72 EE-11 Urutu Walid MOWAG Roland M113A1	China France United Kingdom Russia Russia Brazil Egypt Switzerland USA

Table H-1. Select armored vehicle types and countries of origin.

a. The task of armored vehicle identification has become much more complex. The antiarmor crewman must now become familiar with a much wider variety of vehicles than before. He must also be well informed about what vehicles a potential enemy possesses. In the future, the antiarmor crewman will probably face armored vehicles produced in the U.S. as well as in other friendly nations. The trend of the U.S. to fight with a coalition means that, on future battlefields, both friendly and enemy forces may have identical equipment.

b. This manual is not a complete FM on vehicle identification. However, the antiarmor crewman should be readily familiar with a few of the most recently fielded vehicles and the particular problems and capabilities they present.

(1) ***British Main Battle Tank, Chieftain Mk 5*** (Figure H-1).

- Date of introduction: 1967 original Chieftain
- Proliferation: At least six countries
- Automotive performance:
  Engine type: 750-hp diesel
  Cruising range: 400 to 500 km
- Speed:
  Max road: 48 km/h
  Max off-road: INA
  Average cross country: 30 km/h
  Max swim: INA
- Fording depths: 1.1 meters unprepared
- Main armament ammunition:
  Caliber type name: 120-mm high L11A5
  Rate of fire: 8 to 10 rounds per minute first minute; 6 sustained
  Maximum aimed range: 5,000 meters
  Maximum effective range:
    Day: 3,000 meters
    Night: INA
  Armor penetration: INA

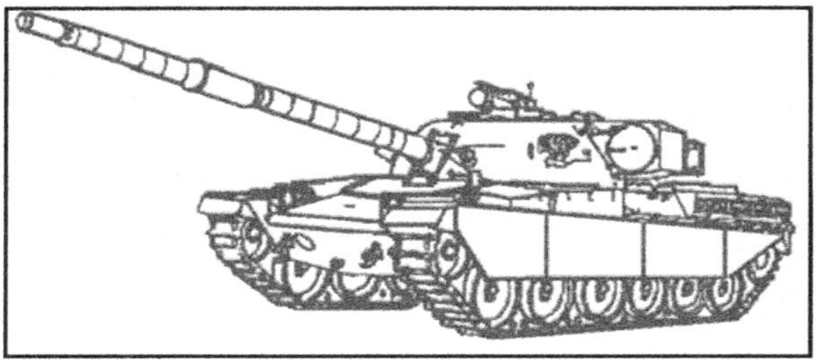

Figure H-1. British main battle tank, Chieftain Mk 5.

(2) ***German Main Battle Tank, Leopard 2*** (Figure H-2).

- Date of introduction: 1979
- Proliferation: At least seven countries
- Automotive performance:
  Engine type: 750-hp diesel
  Cruising range: 400 to 500 km
- Speed:
  Max road: 48 km/h
  Max off-road: INA
  Average cross country: 30 km/h
  Max swim: INA
- Fording depths: 1.1 meters unprepared
- Main armaments ammunition:
  Caliber type, name: 120-mm APFSDS-T, DM43
  Rate of fire: INA
  Maximum aimed range: 3,500 meters
    Day: INA
    Night: INA
  Armor penetration: 450 to 2,000 mm

Figure H-2. German main battle tank, Leopard 2.

(3) ***Polish/Czechoslovakian Main Battle Tank, T-72M1*** (Figure H-3, page H-4).

- Date of introduction: 1975
- Proliferation: At least seven countries
- Automotive performance:
  Engine type: 780-hp diesel
  Cruising range: 460 to 700 km with extra tanks
- Speed:
  Max road: 60 km/h
  Max off-road: 45 km/h
  Average cross country: 35 km/h
  Max swim: INA
- Fording depths: 1.2 meters unprepared; 5 meters with snorkel

- Main armaments ammunition:
  Caliber type, name: 125-mm APFSDS-T, BM-42M
  Rate of fire: INA
  Maximum aimed range: 2,000 to 3,000 meters
    Day: 2,000 to 3,000 meters
    Night: 850 to 1,300 meters
  Armor penetration: 590 to 630 rounds per min at 2,000 meters

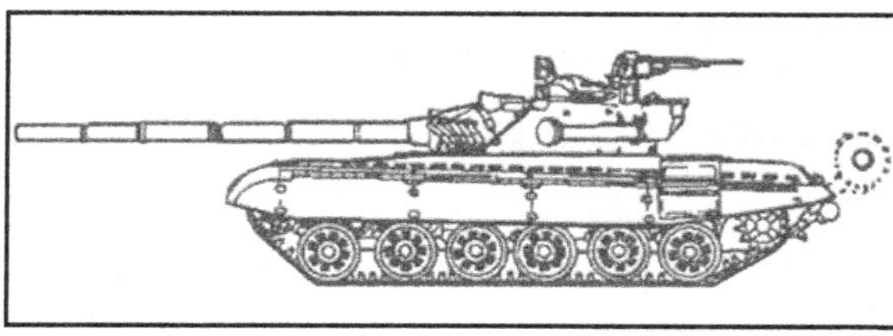

Figure H-3. Polish/Czechoslovakian main battle tank, T-72M1.

(4) ***Russian Main Battle Tank, T-80U*** (Figure H-4).

- Date of introduction: 1987
- Proliferation: At least three countries
- Automotive performance:
  Engine type: 1250-hp gas turbine (multi-fuel); diesel on T-80UD
  Cruising range: 335 km; 600 km with extra tanks
- Speed:
  Max road: 70 km/h
  Max off-road: 48 km/h
  Average cross country: 40 km/h
  Max swim: INA
- Fording depths: 1.8 meters unprepared; 5 meters with snorkel; 12 meters with BROD-M system
- Main armaments ammunition:
  Caliber type, name: 125-mm APFSDS-T, BM-42M
  Rate of fire: INA
  Maximum aimed range: 2,000 to 3,000 meters
    Day: 2,000 to 3,000 meters
    Night: 800 to 1,300 meters
  Armor penetration: 560 to 630 mm at 2,000 meters

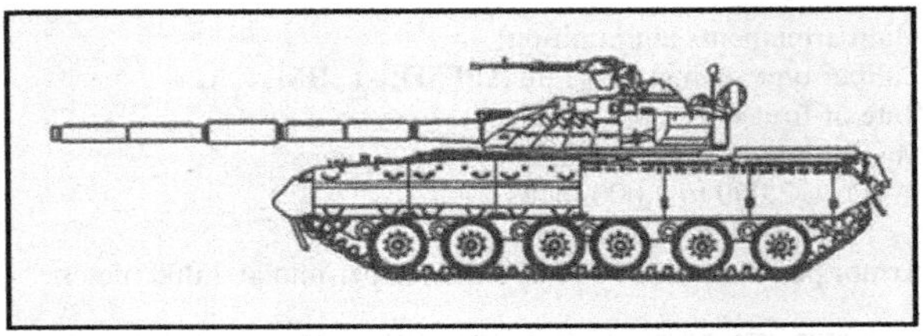

Figure H-4. Russian main battle tank, T-80U.

(5) ***Chinese Main Battle Tank, Type 85-IIM*** (Figure H-5).

- Date of introduction: 1991
- Proliferation: At least two countries
- Automotive performance:
  Engine type: 750-hp diesel
  Cruising range: 700 km; 900 km with external tanks
- Speed:
  Max road: 57 km/h
  Max off-road: 45 km/h
  Average cross country: 35 km/h
  Max swim: INA
- Fording depths: 1.4 meters unprepared; 2.4 meters with snorkel
- Main armaments ammunition:
  Caliber type, name: 125-mm HEAT-MP, BK-29M
  Rate of fire: INA
  Maximum aimed range: 3,000 meters
    Day: INA
    Night: 850 to 1,300 meters
  Armor penetration: 560 to 750 mm

Figure H-5. Chinese main battle tank, Type 85-IIM.

H-2. THREAT RIFLE-PROPELLED GRENADES

The most prolific threat now is the hand-held RPG. The following are some of the RPGs that U.S. forces will encounter.

a. **Russian 40-mm Antitank Grenade Launcher, RPG-7V** (Figure H-6).

- Date of introduction: 1962
- Proliferation: At least 40 countries
- Ammunition: PG-7V
  Caliber: 85 mm
  Type: Heat
  Range:
    Effective: 500 meters
    Minimum: N/A
    Moving: 300 meters
  Armor penetration: 330 mm

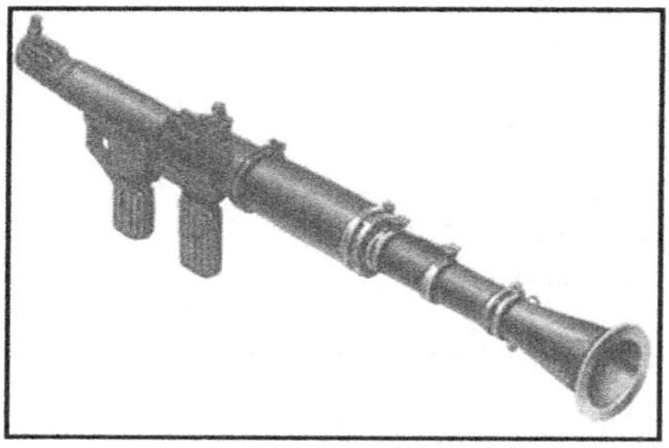

Figure H-6. Russian 40-mm antitank grenade launcher, RPG-7V.

b. **Swedish 84-mm Recoilless Rifle, Carl Gustav M2** (Figure H-7).

- Date of introduction: INA
- Proliferation: At least 20 countries
- Ammunition: FFV 751
  Type: Heat (tandem)
  Range:
    Effective: 500 meters
    Minimum: N/A
  Armor penetration: +500 mm

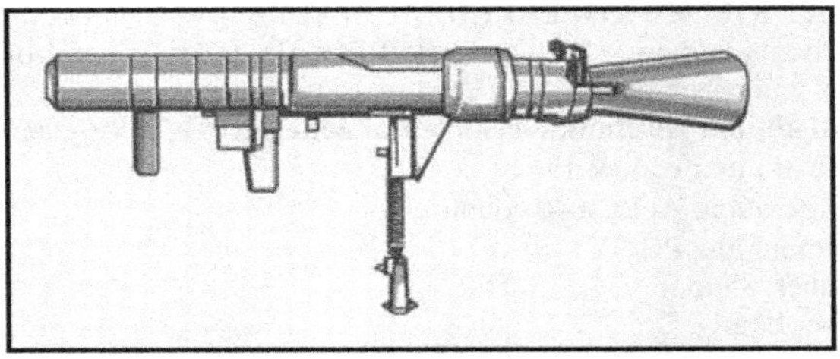

Figure H-7. Swedish 84-mm recoilless rifle, Carl Gustav M2.

c. **German 60-mm Antitank Grenade Launcher, Panzerfaust-3** (Figure H-8).

- Date of introduction: 1990
- Proliferation: At least eight countries
- Ammunition: INA
  Caliber: 110 mm
  Type: Heat
  Range:
    Effective (moving): 300 meters
    Effective (stationary): 500 meters
  Armor penetration: 500 mm

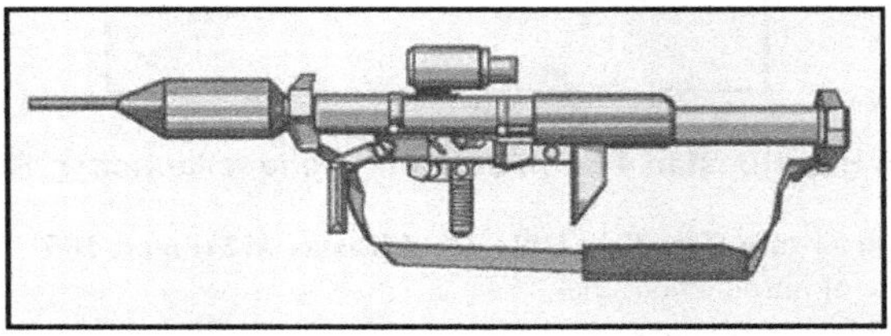

Figure H-8. German 60-mm antitank grenade launcher, Panzerfaust-3.

d. **German 67-mm Disposable Antitank Grenade Launcher, Armbrust** (Figure H-9, page H-8).

- Date of introduction: INA
- Proliferation: At least seven countries
- Ammunition: INA
  Type: Heat
  Range:
    Maximum: 1500 meters
    Effective AT: 300 meters

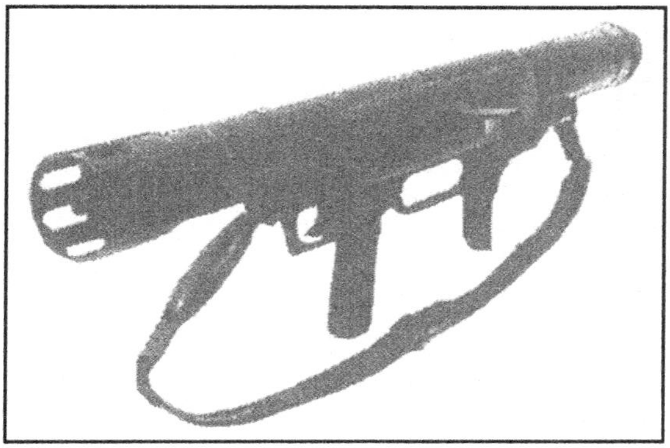

Figure H-9. German 67-mm disposable antitank grenade launcher, Armbrust.

e. **Russian 105-mm Antitank Grenade Launcher, RPG-29** (Figure H-10).

- Date of introduction: Late 1980s
- Proliferation: Former Soviet Union
- Ammunition: PG-29V
  Caliber (warhead): 105 mm
  Type: Heat (tandem)
  Range:
    Effective: 500 meters
    Minimum: N/A
  Armor penetration: +750 mm

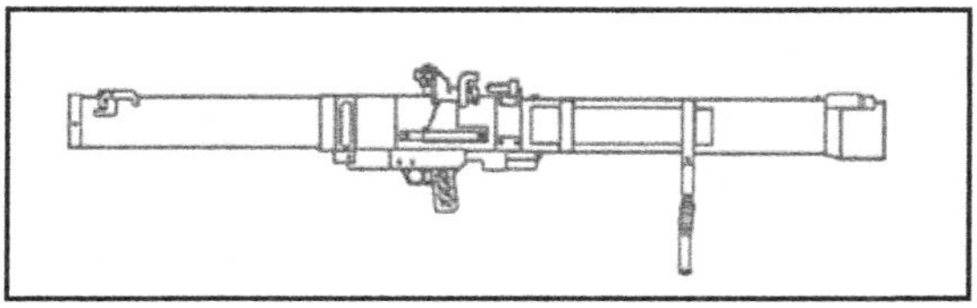

Figure H-10. Russian 105-mm antitank grenade launcher, RPG-29.

H-3. ANTITANK GUIDED MISSILE LAUNCHER VEHICLES

Although there are unique ATGM launcher vehicles with unique ATGMs, most launcher vehicles are military and commercial vehicles adapted with pintel mounts for portable ground launchers, with ATGMs manually loaded and launched. The systems described in this paragraph are the more common threat systems or represent the spectrum of antitank systems that can threaten U.S, Army forces in the world today.

a. **French ATGM Launcher Vehicle, AMX-10 HOT** (Figure H-11).

- Date of introduction: INA
- Proliferation: At least one country
- Guidance system: SALCOS
- Command link: Wire
- Ammunition: ATGM
  Name: HOT
    Warhead type: shaped charge (HEAT)
    Armor penetration: 800 mm
    Minimum/maximum range: 75 to 4,000 meters
  Name: HOT 2
    Warhead type: tandem shaped charged (HEAT)
    Armor penetration: 900 mm
    Minimum/maximum range: 75 to 4,000 meters
  Name HOT 2
    Warhead type: tandem shaped charged (HEAT)
    Armor penetration: 1,250 mm
    Minimum/maximum range: 75 to 4,000 meters
- Probability of hit: INA
- Time of flight to max range: 17.3 seconds

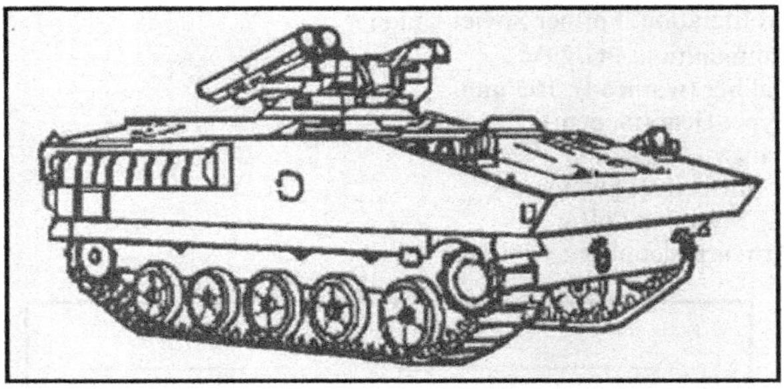

Figure H-11. French ATGM launcher vehicle, AMX-10 HOT.

b. **Russian ATGM Launcher Vehicle, 9P148** (Figure H-12, page H-10).

- Date of introduction: 1977
- Proliferation: At least six countries
- Guidance system: SALCOS
- Command link: Wire
- Ammunition
- ATGM:
  Name: AT-5/SPANDREL
  Warhead type: shaped charge (HEAT)
  Armor penetration: 650 mm
  Minimum/maximum range: 75 to 4,000 meters

- Probability of hit: 90 percent
- Time of flight to max range: 20 seconds

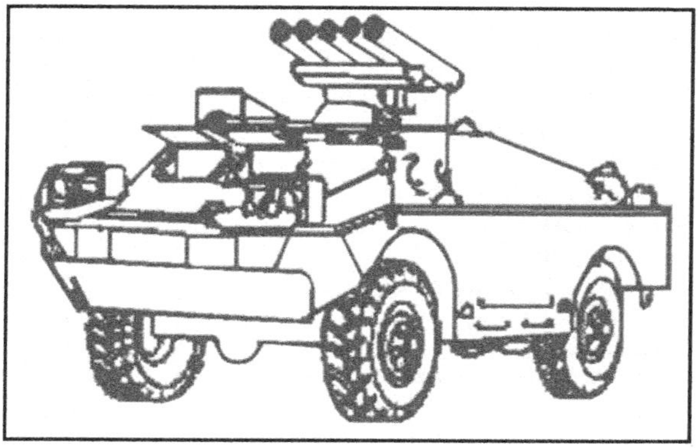

Figure H-12. Russian ATGM launcher vehicle, 9P148.

c. **Russian ATGM Launcher Vehicle, 9P149** (Figure H-13).

- Date of introduction: 1990
- Proliferation: At least nine countries
- Guidance system: SALCOS
- Command link: radio frequency
- Ammunition: ATGM
  Name: AT-56a/SPIRAL
    Warhead type: shaped charge (HEAT)
    Armor penetration: 750 mm
    Minimum/maximum range: 400 to 5,000 meters
- Probability of hit: 90 percent
- Time of flight to max range: 14.5 seconds

Figure H-13. Russian ATGM launcher vehicle, 9P149.

H-4. ANTITANK GUIDED MISSILE LAUNCHER

As the levels of armor protection and the lethality of antitank weapons continue to rise, armor protection for many modern tanks has outpaced most AT weapons. However, ATGMs have been able to increase their range and warhead configurations to threaten even the heaviest tanks. Among notable trends in ATGMs is the worldwide proliferation and variety of man-portable and portable antitank ballistic missiles (ATBMs).

a. **French ATGM Launcher, Eryx** (Figure H-14).

- Date of introduction: 1991
- Proliferation: At least five countries
- Guidance: SACLOS
- Command link: Wire
- Beacon type: infrared laser diode
- Ammunition:
  Name: Eryx
  Warhead type: tandem shaped charge (HEAT)
  Armor penetration: 900 mm
  Minimum/maximum range: 50 to 600 meters
- Probability of hit: 90 percent
- Time of flight to max range: 3.7 seconds

Figure H-14. French ATGM launcher, Eryx.

b. **Russian ATGM Launcher, AT-4/AT-5** (Figure H-15, page H-12).

- Date of introduction: 1973
- Proliferation: At least 25 countries
- Guidance: SACLOS
- Command link: Wire
- Beacon type: Incandescent infrared bulb
- Ammunition:

Name: AT-5B/SPANDREL-B
- Warhead type: tandem shaped charge (HEAT)
- Armor penetration: 925 mm
- Minimum/maximum range: 75 to 4,000 meters
- Probability of hit: 90 percent
- Time of flight to max range: 10 seconds

Name: AT-5B/SPANDREL
- Warhead type: shaped charge (HEAT)
- Armor penetration: 650 mm
- Minimum/maximum range: 75 to 4,000 meters
- Probability of hit: 90 percent
- Time of flight to max range: 20 seconds

Name: AT-4/SPIGOT
- Warhead type: shaped charge (HEAT)
- Armor penetration: 480 mm
- Minimum/maximum range: 70 to 4,200 meters
- Time of flight to max range: 11 seconds

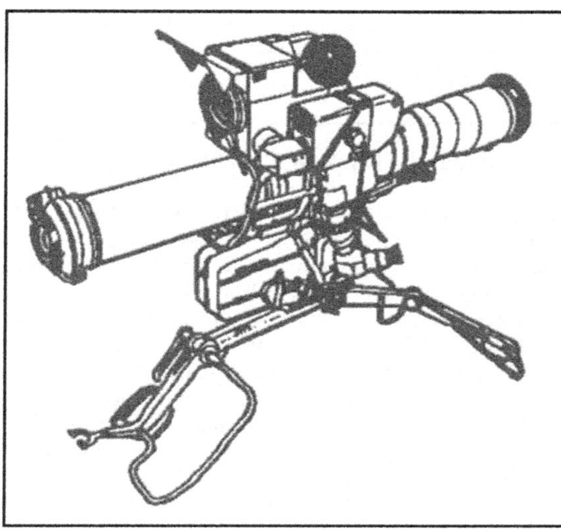

Figure H-15. Russian ATGM launcher, AT-4/AT-5.

c. **Russian ATGM Launcher, AT-3** (Figure H-16).

- Date of introduction: 1963
- Proliferation: At least 45 countries
- Guidance: MCLOS
- Command link: Wire

- Beacon type: Incandescent infrared bulb (SACLOS)
- Ammunition:
  Name: AT-3, 3a, 3b/SAGGER
    Warhead type: tandem shaped charge (HEAT)
    Armor penetration: 400 mm
    Minimum/maximum range: 500 to 3,000 meters
    Probability of hit: 70 percent
    Time of flight to max range: 26 seconds

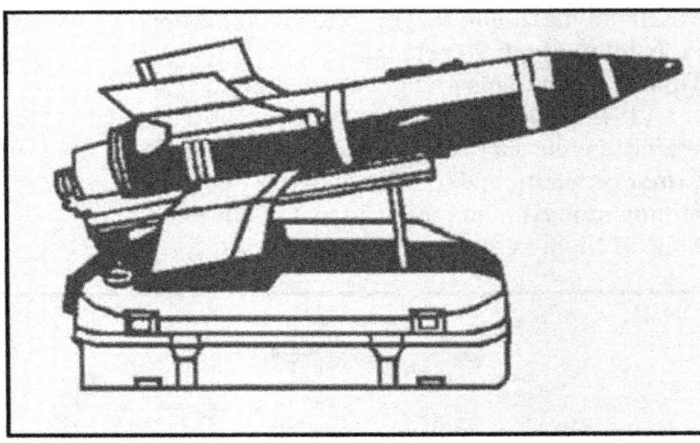

Figure H-16. Russian ATGM launcher, AT-3.

H-5. THREAT INFANTRY ARMOR COUNTERMEASURES

Historically, the deployment of thick belts of ATGMs has been the foundation of U.S. antiarmor defense planning. Potential enemies have developed countermeasures to these antiarmor measures, which any enemy we may face may employ.

a. **Reconnaissance.** A high priority for enemy reconnaissance elements is to locate our ATGM sites before their main body enters our engagement areas. Methods of reconnaissance can include aircraft, drones, signals intelligence, and conventional ground reconnaissance elements. Leaders must learn to emphasize careful camouflage and concealment and to conduct effective counterreconnaissance operations. Lessons learned at NTC show that the commander who loses the counterreconnaissance phase of the battle loses the battle. For enemy reconnaissance to be effective, their reconnaissance elements must spot our weapon systems and survive to report their location.

b. **Artillery.** Once enemy reconnaissance elements have spotted our ATGM positions, the weapon of choice to destroy them is artillery. Since ATGMs are a high-priority target for artillery, these positions will be subjected to intense indirect fire. The TOW is not designed to withstand such barrages. The best way to survive is to remain undetected and to avoid terrain that is likely to attract heavy artillery fire. Positions should be as well protected against artillery as possible. Move, if possible, when positions come under artillery fire.

c. **Helicopters.** Attack helicopters routinely follow artillery preparation. Again, ATGMs are high-priority targets.

d. **Direct Fire.** Enemy forces will attempt to suppress or destroy friendly ATGM systems with direct fire.

e. **Grenade Launcher.** This weapon is very effective against a tripod- or HMMWV-mounted TOW, and is somewhat effective against lightly armored vehicles such as the Bradley.

f. **Lasers.** Most armies use laser range finders and target designators. Many of these are quite powerful and can cause temporary or permanent damage to a gunner's eyes when he is using direct-view magnifying optics. Some of these devices can even damage the sights themselves. Thermal sights are less vulnerable to lasers. Laser weapons, such as the PRC-produced ZM-87, or even frequency agile laser weapons may be encountered on the future battlefield. Many countries, such as China, Russia, and France, have extensive laser development and procurement programs.

g. **Battlefield Obscurants.** The AN/TAS-4 series of thermal sights is able to penetrate many kinds of battlefield obscurants, but not all. Bispectral smoke and battlefield dust, among others, are largely opaque to thermal sights if they are thick enough. In addition, when using a basic or ITOW, the target may be seen, but the system is unable to guide the missile through the obscurants.

h. **Combined Arms.** Enemy doctrine prefers to fight mounted, but if they run into strong ATGM defenses, they will dismount their infantry and advance at foot pace, with the infantry engaging personnel and ATGMs. The tanks will advance first, at about 6 kilometers per hour, and the infantry will closely follow the tanks. The fighting vehicles will follow the infantry and provide supporting fire.

H-6. INFANTRY ANTITARMOR COUNTER-COUNTERMEASURES

United States forces must have counter-countermeasures against reconnaissance, artillery, helicopters, direct fire, grenade launchers, lasers, obscurants, and combined arms.

a. **Reconnaissance.** The best counter to reconnaissance elements is effective counterreconnaissance operations by defending forces. Equally important is effective and rigidly enforced camouflage and concealment.

b. **Artillery.** The best defense against artillery is not to be discovered by the enemy's reconnaissance elements. This may not be possible, so positions must be as well prepared against heavy artillery fire as time permits. When all else fails, it may be necessary to move positions. Secondary and alternate positions should be preselected for this eventuality.

c. **Helicopters.** The best defense against helicopters is not to be spotted in the first place. TOW positions should only engage helicopters in self-defense when they are already spotted and under attack. Helicopters not engaging TOW positions should be left to regular ADA assets. (See Chapter 5 for more detailed information.)

d. **Direct Fire.** TOW squads must maximize their chances of winning engagements with tanks and fighting vehicles by choosing positions with good cover and concealment, striving for flank shots, striving for surprise, exercising good fire control, and carefully prioritizing their targets, to ensure that targets posing the most immediate threat are engaged first. This is, in fact, precisely the type of combat the TOW was intended to engage in.

e. **Grenade Launcher.** TOW squads must try to choose positions where it is difficult for enemy grenade launcher teams to get in range without being detected and engaged. Where the terrain is too close to permit this, TOW squads must emplace behind friendly infantry, and let the infantry engage the grenade launcher team.

f. **Lasers.** Protective lenses should be installed on optical devices. The protective covers or doors on these devices should be closed when protective lenses are not available. Soldiers should wear laser-proof goggles at all times. (Soldiers are vulnerable to eye damage even when not looking through sighting devices.)

g. **Battlefield Obscurants.** TOW crews should train under realistic battlefield obscurant conditions whenever possible, so that they better understand the limitations of their thermal sights. Alternate positions that view the battlefield from a different angle and that may afford a better view should be identified and prepared. The battlefield should be carefully examined and marked to give gunners a better perspective of where things are when the field becomes partially or largely obscured. Enemy smokescreens normally end about 1,000 meters from our positions, so gunners should be prepared to engage targets immediately when they emerge from the smokescreen.

h. **Combined Arms.** TOW positions are ill-equipped to defend themselves against dismounted infantry. To survive a properly executed combined arms attack, TOW squads should be integrated into a combined arms defense that includes overwatching dug-in infantry, fighting vehicles, and tanks. The TOW was never intended as a stand-alone weapon system, and leaders who attempt to employ it as such guarantee their destruction.

H-7. DIRECTED-ENERGY WEAPONS

Directed-energy weapons include lasers, microwave radiation emitters, and particle beam generators. By far the most likely type that ground troops will see in combat is the laser. This can take the form of a powerful laser range finder used as a weapon or as a specially designed laser weapon. The common perception of the laser is as a science fiction weapon that can vaporize a tank with a single blast. This is far from the truth. In fact, the laser beam will normally be invisible and is generally able only to temporarily blind a soldier who is hit in the eyes with it. If the soldier is using an optical sight of some kind, the effect is magnified. More powerful lasers may cause permanent blindness, but these are rare. Some lasers are also powerful enough to damage the lenses of optical sights or the electronics of electrooptical devices (generally thermal sights and image intensifiers). (For means of defense, see paragraph H-4 of this appendix. See FM 7-91, Appendix F for a more detailed discussion.)

H-8. IMPROVED ARMOR TECHNOLOGY

In the past decade, there has been a revolution in armor technology. Research and new developments have come from Europe, the United States, and Israel. These improvements are continuing worldwide, and the antiarmor crewman must keep up with them. These improvements are also becoming much more common in third world armies, so the antiarmor crewman must expect to meet them wherever he goes. In addition, many older tanks and other AFVs are being retrofitted with improved armor protection, so it is not just the newer vehicles we need be concerned with. These advanced armor configurations improve the vehicles' survivability against all weapons, but for the most part they are specifically designed to protect against HEAT warheads. This is a complex technical subject, and growing more so every year, but essentially these improved types of armor fall into four categories.

a. **Reactive Armor**. Originally fielded by the Israeli Army, reactive armor appeared shortly after on enemy tanks. Reactive armor comes in several varieties, but the principle is essentially the same on all. The armor consists of blocks of explosives sandwiched between two metal plates and bolted on the outside of the vehicle. Small-arms and artillery shrapnel will not set off the blocks. However, when a HEAT round strikes the block, the explosive ignites and blows outwards. The blast and the moving steel plates disperse and deflect the jet of the HEAT warhead, dramatically reducing its ability to penetrate armor. Many countries are now fielding different versions of reactive armor. One of its advantages is that it can easily be retrofitted onto older vehicles.

b. **Laminated Armor.** Laminated armor was developed in the West, most specifically by England, the United States, and West Germany. It consists of flat layers of steel armor plate with layers of ceramics, fiberglass, or other nonmetallic materials in between. This armor is highly effective against all types of weapons, but is difficult and expensive to manufacture. Vehicles with laminated armor are characterized by flat, slab sides, such as on the M1 and Leopard II.

c. **Composite Armor.** Composite armor consists of a nonmetallic core (usually some kind of ceramic) around which the rest of the steel of the hull or, more commonly, the turret, is molded. This is much more effective than conventional steel armor against all types of weapons, but less so than laminated armor. However, it is less difficult and expensive to manufacture. Hulls made of composite armor do not have to be slab sided, like those made of laminated armor.

d. **Applique Armor.** Applique armor is essentially extra plates mounted or welded on top of the hull or turret of a vehicle. They can be made of any material, but are frequently made of ceramic or laminated materials. Like reactive armor, applique armor is an easy and cost-effective way of improving the protection of older vehicles.

H-9. COUNTERMEASURES TO IMPROVED ARMOR TECHNOLOGY

The long-term answer to improved armor technology is primarily a technical issue: improving the kill mechanisms of our missiles to defeat the improved armor.

a. Two responses to improved armor have already been fielded: the TOW 2A and the TOW 2B. Other improved kill mechanisms will be fielded as time passes.

(1) The TOW 2A has an explosive charge in the tip of the extensible probe to prematurely detonate reactive armor and minimize its effect on the penetration abilities of the missile.

(2) The TOW 2B employs a new kill mechanism. It flies about 6 or 7 feet above the line of sight and detonates above the hull or turret of the tank. When it detonates, it creates two explosively formed penetrators (EFPs), which are metallic slugs travelling at very high speeds. These EFPs strike the top of the vehicle where the armor is thinnest and there are fewer reactive armor plates.

b. TOW crews can expect to be issued a mix of TOW missile types on the battlefield, with widely varying capabilities. Gunners and leaders must be familiar with the different missile types and their respective capabilities. The proper type of missile must be chosen for each type of target (Table H-2).

c. TOW crews must strive harder than ever to find positions where they can engage enemy vehicles from the flank. Modern tanks with reactive armor have become increasingly difficult to kill from the front.

THREAT VEHICLE TYPE TARGETS	SELECTION PRIORITY			
	FIRST	SECOND	THIRD	FOURTH
Tanks with applique armor.	TOW 2B	TOW 2A	TOW 2	ITOW
Tanks with explosive reactive armor.	TOW 2B	TOW 2A	TOW 2	ITOW
Tanks without applique/ reactive armor.	TOW 2B	TOW 2A	TOW 2	ITOW
Light armored personnel carriers.	TOW 2	TOW 2A	TOW 2B	ITOW
Light armored wheeled vehicles.	TOW 2	TOW 2A	TOW 2B	ITOW
Antiaircraft vehicles.	TOW 2	TOW 2A	TOW 2B	ITOW
Armored vehicles in hull defilade positions.	TOW 2B	TOW 2A	TOW 2	ITOW
Bunkers/fortifications.	TOW 2	TOW 2A	ITOW	TOW 2B

Table H-2. Missile selection priority chart.

GLOSSARY

ACCP	Army Correspondence Course Program
ADA	air defense artillery
AFV	armored fighting vehicle
ANCOC	Advanced noncommissioned officer course
APC	armored personnel carrier
AR	Army regulation
ARTEP	Army training and evaluation program
ATEA	anticipated target engagement area
ATGM	antitank guided missile
ATTN	attention
ATWESS	antitank weapon effect signature simulator
AZ	azimuth
BFV	Bradley fighting vehicle
BNCOC	Basic noncommissioned officer course
BPC	battery power condition
CO	company
cm	centimeter
CVC	combat vehicle crewman (helmet)
CVKI	combat vehicle kill indicator
DA	Department of the Army
DS	direct support
EFP	explosively formed projectiles
ELEV	elevation
EMER PWR	emergency power
EOD	explosive ordnance disposal
FIST	fire support team
FM	field manual
FO	forward observer
FSO	fire support officer
FTX	field training exercise

GCP	gunner's control panel
GHC	gunner's hand control
GST	Gunner's Skill Test
HMMWV	high-mobility, multipurpose wheeled vehicle
IAW	in accordance with
IFV	infantry fighting vehicle
IMT	infantry moving target
IR	infrared
ITA	image transfer assembly
ITOW	improved TOW
IVD	interactive video disk
km	kilometers
km/h	kilometers per hour
LOS	line of sight
LTD	laser target detector
LTID	laser target interface device
MCLOS	manual command to line-of-sight guidance
METL	mission-essential task list
METT-T	mission, enemy, terrain, troops, and time available
MGS	missile guidance set
MILES	Multiple Integrated Laser Engagement System
MOPP	mission-oriented protective posture
MOS	military occupation specialty
MPRC	multipurpose range complex
MSR	missile simulation round
MV	moving
MWLD	man-worn laser detectors
MWO	modified work order
NATO	North Atlantic Treaty Organization
NBC	nuclear, biological, and chemical
NCO	noncommissioned officer
NTC	National Training Center

OCL	operator's checklist
OPFOR	opposing force
OSUT	one station unit training
PGTS	Precision Gunnery Training System
PLT	platoon
PMCS	preventive maintenance checks and services
psi	pounds per square inch
ROW	rest of the world
rpm	revolutions per minute
SACLOS	semiautomatic command-to-line-of-sight guidance
SDZ	surface danger zone
SFC	sergeant first class
SGT(P)	sergeant (promotable)
SL	skill level; squad leader (figures only)
SLP	squad leader's periscope
SM	soldier's manual
SOP	standing operating procedure
SQD	squad
ST	stationary
STBY	standby
STRAC	Standards in Training Commission
STP	soldier training publication
STX	situational training exercise
TFTT	TOW Field Tactical Trainer
TG	trainer's guide
TGT	TOW Gunnery Trainer
TOW	tube-launched, optically tracked, wire-guided
TM	technical manual (training manual)
TRP	target reference point
TVPC	TOW vehicle power conditioner
US	United States
VDC	volts direct current
VPC	vehicle power conditioner

REFERENCES

DOCUMENTS NEEDED

These documents must be available to the intended users of this publication.

AR 75-1	Malfunctions Involving Ammunition and Explosives. 23 April 2001.
ARTEP 7-91-Drill	Drills for the Antiarmor (TOW) Platoon, Section, and Squad. 26 July 2002.
ARTEP 7-91-MTP	Mission Training Plan for Antiarmor Company/Platoon/Section. 12 October 1989.
DA Form 7331-R	Scorecard for Gunner's Skill Test-Part 1. August 2003.
DA Form 7332-R	Scorecard for Gunner's Skill Test-Part 2-Carrier-Specific Tasks. August 2003.
DA Form 7333-R	TOW Gunnery Tables 1 and 2: Individual Gunnery Practice and Qualification Scorecard. August 2003.
DA Form 7334-R	TOW Gunnery Tables 3 and 4: Advanced Gunnery Practice and Qualification Scorecard. August 2003.
DA Form 7335-R	TOW Gunnery Tables 5 and 6: Baseline TFTT Gunnery Practice and Qualification Scorecard. August 2003.
DA Form 7336-R	TOW Gunnery Table 7: Squad Gunnery Practice Scoresheet. August 2003.
DA Form 7337-R	TOW Gunnery Table 8: Squad Gunnery Qualification Scoresheet. August 2003.
DA Form 7338-R	TOW Gunnery Table 9: Section Gunnery Practice Scoresheet. August 2003.
DA Form 7339-R	Tow Gunnery Table 10: Section Gunnery Qualification Scoresheet. August 2003.

DA Form 7340-R	TOW Gunnery Table 11: Platoon Gunnery Practice Scoresheet. August 2003.
DA Form 7341-R	TOW Gunnery Table 12: Platoon Gunnery Qualification Scoresheet. August 2003.
DA Pam 350-38	Standards in Weapons Training. 01 October 2002.
DA Pam 385-63	Range Safety. 10 April 2003.
DA Pam 611-21	Military Occupational Classification and Structure. 31 March 1999.
FM 3-21.91(FM 7-91)	Tactical Employment of Antiarmor Platoons, Companies, and Battalions. 26 November 2002.
FM 3-37.3(FM 3-5)/ with change	NBC Decontamination. 28 July 2000.
FM 5-250	Explosives and Demolitions. 30 July 1998.
FM 10-500-29	Airdrop of Supplies and Equipment: Rigging Heavy Antitank Assault Weapon System (TOW). 14 April 2000.
FM 25-101	Battle Focused Training. 30 September 1990.
GTA 17-02-011	Combat Vehicle Identification (CVI) Training Cards. 1984.
GTA 17-02-013	Armored Vehicle Recognition. 01 January 1997.
STP 7-11BCHM24-SM-TG	Soldier's Manual and Trainer's Guide for MOS 11B, 11C, 11H, and 11M Infantry Skill Levels 2/3/4. 19 June 2000.
TB MED 524	Occupational and Environmental Health: Control of Hazards to Health From Laser Radiation. 20 June 1985.
TC 25-8/with change	Training Ranges. 25 February 1992.

TM 9-1265-368-10-2	Operator's Manual for Multiple Integrated Laser Engagement System (MILES), Simulator System, Firing Laser: M64 (NSN 1265-01-077-6083) for TOW Standard, TOW Long Range, TOW II, TOW II Extended Weapon System. 15 July 1988.
TM 9-1410-470-34	Direct Support and General Support Maintenance Manual for Surveillance of Guided Missile Surface Attack, BGM-71A, BGM-71A-1, BGM-71A-2, BGM-71A-2A, BGM-71A-3, BGM-71A-3A, BGM-71C, BGM-71C-1, BGM-71C-1A, BGM-71D, or BGM-71E and Guided Missile Practice BTM-71A, BTM-71A-1, BTM-71A-2, BTM-71A-2A, BTM-71A-3, or BTM-71A-3A (TOW Heavy Antitank/Assault Weapon System). 01 March 1988.
TM 9-1425-450-12	Operator and Organizational Maintenance Manual for TOW 2 Weapon System, Guided Missile System M220A2 (NSN 1440-014-9034) (TOW 2 Heavy Antitank/Assault Weapon System). 25 May 1983.
TM 9-2320-280-10	Operator's Manual for Truck, Utility: TOW Carrier, Armored, 1-1/4 Ton, 4X4, M966. 31 January 1996.
TM 9-6920-452-10	Trainer, Gunnery, Antitank: AN/TWQ-T1 (TOW Gunnery Trainer). 01 August 1991.
TM 9-6920-453-10	Trainer, Guided Missile: AN/TWQ-T3 (TOW Field Tactical Trainer). 20 March 1991.
TM 750-244-6	Procedures for Destruction of Tank-Automotive Equipment to Prevent Enemy Use (US Army Tank-Automotive Command). 03 October 1972.

INTERNET WEB SITES

Army Publishing Directorate
http://www.usapa.army.mil

Reimer Digital Library
http://www.adtdl.army.mil

INDEX

GUNNER'S SKILL TEST - PART 1

For use of this form, see FM 3-22.34. The proponent agency is TRADOC.

NAME	RANK	UNIT	DATE

TEST#	TASK NUMBER	TASK TITLE	GO/NO GO
1	071-056-0004	MAINTAIN AN M220-SERIES LAUNCHER SYSTEM	
2	071-056-0007	LOAD AN M220-SERIES LAUNCHER SYSTEM	
3	071-035-0008	UNLOAD AN M220-SERIES LAUNCHER SYSTEM	
4	071-056-0009	ENGAGE TARGETS WITH AN M220-SERIES LAUNCHER SYSTEM	
5	071-056-0010	PERFORM IMMEDIATE ACTION FOR AN M220-SERIES LAUNCHER SYSTEM	
6	071-056-0005	CONDUCT A SYSTEM CHECK OUT ON AN M220A1 LAUNCHER SYSTEM	
7	071-056-0013	CONDUCT A SYSTEM CHECK OUT ON AN M220A2 LAUNCHER SYSTEM	
8	071-056-0030	DETERMINE M220-SERIES LAUNCHER SYSTEM FIRING LIMITATIONS	
9	NO TASK NUMBER	IDENTIFY COMBAT VEHICLES	
10	NO TASK NUMBER	DETERMINE IF A TARGET CAN BE ENGAGED BY AN M220-SERIES LAUNCHER SYSTEM (SEE FM 3-22.34, CH5)	
11	071-317-0000	PREPARE AN ANTIARMOR RANGE CARD	
12	NO TASK NUMBER	ASSEMBLE AN M220-SERIES LAUNCHER SYSTEM	

DA FORM 7331-R, AUG 2003

APD V1.00

GUNNER'S SKILL TEST - PART 2

For use of this form, see FM 3-22.34. The proponent agency is TRADOC.

NAME	RANK	UNIT	DATE

TEST#	TASK NUMBER	TASK TITLE	GO/NO GO
1	NO TASK NUMBER	PLACE THE M966 VEHICLE IN THE READY-TO-FIRE CONFIGURATION (SEE FM 3-22.34 CH 3)	
2	071-056-9932	CONDUCT DISMOUNTING AND REMOUNTING OF AN M220A1 LAUNCHER SYSTEM ON AN M966 VEHICLE (M220A1- EQUIPPED UNITS ONLY)	
3	071-056-0034	CONDUCT DISMOUNTING AND REMOUNTING OF AN M220A2 LAUNCHER SYSTEM ON AN M966 VEHICLE (M220A2 - EQUIPPED UNITS ONLY)	
4	071-034-0001	LOAD AN M243 OR M259 SMOKE GRENADE LAUNCHER	
5	071-034-0002	UNLOAD AN M243 OR M259 SMOKE GRENADE LAUNCHER	

DA FORM 7332-R, AUG 2003

APD V1.00

TOW GUNNERY TABLES 1 AND 2: INDIVIDUAL GUNNERY

☐ PRACTICE ☐ QUALIFICATION

For use of this form, see FM 3-22.34. The proponent agency is TRADOC.

DATA REQUIRED BY PRIVACY ACT OF 1974

AUTHORITY: 10USC 30129g. Executive Order 9397. **PRINCIPAL PURPOSE(S):** Records individual's performance on record fire range. **ROUTINE USE(S):** Evaluation of individual's proficiency and basis for determination of award of proficiency badge; SSN is used for positive identification purposes only. **MANDATORY OR VOLUNTARY DISCLOSURE AND EFFECT ON INDIVIDUAL NOT PROVIDING INFORMATION:** Voluntary; individuals not providing information cannot be rated/scored on a mass basis.

NAME	RANK	SSN	UNIT

DATE	DISK#

MISSION#	INDIVIDUAL GUNNERY TASK TESTED	SIGHT	SCORE	SUCCESSFUL	
#				YES	NO
#				YES	NO
#				YES	NO
#				YES	NO
#				YES	NO
#				YES	NO
#				YES	NO
#				YES	NO
#				YES	NO
#				YES	NO
	TOTAL TRACKING SCORE			SUCCESSFUL MISSIONS: ____	

GUNNER CLASSIFICATION

EXPERT / 875 - 1,000 1ST CLASS / 750 - 874 2ND CLASS / 625 - 749

SCORING INDIVIDUAL GUNNERY TABLES

IN THE INDIVIDUAL GUNNERY TABLES, THE GUNNER'S RATING IS DETERMINED BY HIS TOTAL POINT SCORE FOR 10 MISSIONS WITH AT LEAST 6 SUCCESSFUL MISSIONS. A SUCCESSFUL MISSION IS ONE WHERE THE GUNNER CORRECTLY FOLLOWS THE TRAINERS FIRE COMMANDS AND ENGAGES AND DESTROYS THE CORRECT TARGET.

TOW GUNNERY TABLES 3 AND 4: ADVANCED GUNNERY

☐ **PRACTICE** ☐ **QUALIFICATION**

For use of this form, see FM 3-22.34. The proponent agency is TRADOC.

DATA REQUIRED BY PRIVACY ACT OF 1974

AUTHORITY: 10USC 30129g. Executive Order 9397. **PRINCIPAL PURPOSE(S):** Records individual's performance on record fire range. **ROUTINE USE(S):** Evaluation of individual's proficiency and basis for determination of award of proficiency badge; SSN is used for positive identification purposes only. **MANDATORY OR VOLUNTARY DISCLOSURE AND EFFECT ON INDIVIDUAL NOT PROVIDING INFORMATION:** Voluntary; individuals not providing information cannot be rated/scored on a mass basis.

NAME	RANK	SSN	UNIT

DATE	DISK#

MISSION#	INDIVIDUAL GUNNERY TASK TESTED	SIGHT	SCORE	SUCCESSFUL	
#				YES	NO
#				YES	NO
#				YES	NO
#				YES	NO
#				YES	NO
#				YES	NO
#				YES	NO
#				YES	NO
#				YES	NO
#				YES	NO
	TOTAL TRACKING SCORE			SUCCESSFUL MISSIONS: ______	

GUNNER CLASSIFICATION

EXPERT / 875 - 1,000 1ST CLASS / 750 - 874 2ND CLASS / 625 - 749

SCORING ADVANCE GUNNERY TABLES

IN THE INDIVIDUAL GUNNERY TABLES, THE GUNNER RATING IS DETERMINED BY HIS TOTAL POINT SCORE FOR 10 MISSIONS, PLUS HE MUST HAVE AT LEAST 6 SUCCESSFUL MISSIONS. A SUCCESSFUL MISSION IS ONE WHERE THE GUNNER CORRECTLY FOLLOWS THE TRAINERS FIRE COMMANDS AND ENGAGES AND DESTROYS THE CORRECT TARGET. IF THE MISSION IS A FRATRICIDE MISSION, THE GUNNER SHOULD **NOT** ENGAGE THE TARGET. IF HE **DOES NOT** ENGAGE THE FRIENDLY VEHICLE, HE SCORES 100 POINTS AND A SUCCESSFUL MISSION. IF HE **DOES** ENGAGE THE FRIENDLY VEHICLE, HE SCORES 0 POINTS AND AN UNSUCCESSFUL MISSION. IN TARGET ENGAGEABILITY AND EVASIVE TARGET SCENARIOS IT DOES NOT MATTER WHICH TARGET IS ENGAGED.

TOW GUNNERY TABLES 5 AND 6: BASELINE TFTT GUNNERY			
☐ PRACTICE ☐ QUALIFICATION			
For use of this form, see FM 3-22.34. The proponent agency is TRADOC.			
SQUAD	SECTION	PLATOON	COMPANY
DATE		DAY	NIGHT

ENGAGEMENT	TARGET DESCRIPTION	HIT/MISS/TIME	GO/NG GO
TARGET 1	STATIONARY T72/FLANK/ 1,000 - 1,500 METERS	HIT MISS TIME:	
TARGET 2	STATIONARY T72/FLANK/ 2,100 - 3,750 METERS	HIT MISS TIME:	
TARGET 3	STATIONARY T72/FRONTAL/ 2,100 - 3,750 METERS	HIT MISS TIME:	
TARGET 4	STATIONARY T72/OBLIQUE/ 2,100 - 3,750 METERS	HIT MISS TIME:	
TARGET 5	STATIONARY T72/DEFILADE/ 800 - 1,500 METERS/NBC	HIT MISS TIME:	
TARGET 6	MOVING T72/FLANK/ L-R, 30 KPH 1,600 - 2,000 METERS/NBC	HIT MISS TIME:	
TARGET 7	MOVING T72/FLANK/ L-R, 30 KPH 1,600 - 2,000 METERS/NBC	HIT MISS TIME:	
TARGET 8	MOVING T72/FLANK/ L-R, 5 KPH 1,000 - 1,500 METERS	HIT MISS TIME:	
TARGET 9	MOVING T72/FLANK/ L-R, 10 KPH 2,100 - 3,750 METERS	HIT MISS TIME:	
TARGET 10	MOVING T72/FLANK/ L-R, 40 KPH 1,000 - 1,500 METERS	HIT MISS TIME:	
		OVERALL GO/NO GO:	

TO RECEIVE A "GO" ON ANY TARGET ENGAGEMENT, THE GUNNER MUST HIT THE TARGET WITHIN 30 SECONDS AFTER THE TARGET IS EXPOSED. TO RECEIVE A "GO" ON TABLE 5, THE CREW MUST RECEIVE A "GO" ON "INSTALL TFTT" AND A "GO" ON AT LEAST 6 OUT OF 10 TARGET ENGAGEMENTS, BOTH DAY AND NIGHT.

TOW GUNNERY TABLE 7: SQUAD GUNNERY PRACTICE

For use of this form, see FM 3-22.34. The proponent agency is TRADOC.

TC: __________ LDR: __________ GNR: __________ DVR: __________ DATE: __________ TIME: __________

PLATOON: __________ COMPANY: __________ BATTALION: __________

TASK	CONDITIONS/ TARGETS/ SITUATION	TOTAL TARGET EXPOSURE TIME	HITS	ENGAGEMENT TIMES	POINTS	CREW/ LDR CUTS	TOTAL POINTS
1. ENGAGE A SINGLE TARGET	FIRED FROM BASELINE. 3,000 - 3,750 METERS 1 STATIONARY TANK	TFTT 25 SECS ALL VEHICLES	1				
2. ENGAGE MULTIPLE TARGETS	FIRED FROM PHASE LINE 2. 2,000 - 2,500 METERS 1 - ST TANK, 1 - MV TANK, 1 - MV BMP	TFTT HMMWV/2:30	1 2 3				
3. ENGAGE MULTIPLE TARGETS	FIRED FROM PHASE LINE 3. 2,500 - 3,000 METERS 1 - MV TANK, 1 - ST BMP	TFTT HMMWV/1:40	1 2				
4. ENGAGE MULTIPLE TARGETS	FIRED FROM PHASE LINE 2. 2,500 - 3,000 METERS 1 ST TANK, 1 - MV TANK (NBC)	TFTT HMMWV/1:40	1 2				
5. ENGAGE MULTIPLE TARGETS	FIRED FROM BASELINE. 2,500 - 3,000 METERS 1 - MV TANK, 1 - MV BMP (NBC)	TFTT HMMWV/1:40	1 2				
					TOTAL SCORE		

NOTES:

1. ENGAGEMENT TIMES ARE CALCULATED FROM THE TIME THE TARGET IS EXPOSED TO THE TIME THE TARGET IS HIT. USE THE LIVE-FIRE POINT CALCULATION SHEET FOR SECOND AND THIRD ENGAGEMENTS FROM THE SAME POSITION. THE TIME BEGINS WHEN THE VEHICLE REOCCUPIES ITS FIRING POSITION (STOPS MOVING) AND ENDS WHEN THE TARGET IS HIT.

2. EACH TASK IS WORTH 100 POINTS. EACH TARGET IS ALSO WORTH 100 POINTS FOR TASKS WITH MULTIPLE ENGAGEMENTS. TOTAL THE NUMBER OF POINTS AND DIVIDE BY THE NUMBER OF TARGETS PRESENTED.

3. CREWLEADER CUTS: SEE FM 3-22.34, CH 4.

4. SCORING TABLE 7: TABLE 7 IS PERFORMED TWICE, ONCE DURING THE DAY AND ONCE AT NIGHT. EACH EVENT IS WORTH A MAX OF 500 POINTS. ADD THE TWO SCORES TOGETHER FOR A POSSIBLE TOTAL OF 1,000 POINTS. SQUADS MUST ACHIEVE A MINIMUM TOTAL OF 700 POINTS TO VERIFY ON TABLE 7 AND PROGRESS TO TABLE 8.

DA FORM 7336-R, AUG 2003

APD V1.00

TOW GUNNERY TABLE 8: SQUAD GUNNERY QUALIFICATION

For use of this form, see FM 3-22.34. The proponent agency is TRADOC.

TC: ______ LDR: ______ GNR: ______ DVR: ______ DATE: ______ TIME: ______

PLATOON: ______ COMPANY: ______ BATTALION: ______

TASK	CONDITIONS/ TARGETS/ SITUATION	TOTAL TARGET EXPOSURE TIME	HITS	ENGAGEMENT TIMES	POINTS	CREW/ LDR CUTS	TOTAL POINTS
1. ENGAGE A SINGLE TARGET	FIRED FROM BASELINE. 1,500 - 2,000 METERS 1 STATIONARY TANK	TFTT 25 SECS ALL VEHICLES	1				
2. ENGAGE MULTIPLE TARGETS	FIRED FROM PHASE LINE 1. 2,000 - 2,500 METERS 1 - MV TANK, 1 - ST BMP	TFTT HMMWV/1:40	1 2				
3. ENGAGE MULTIPLE TARGETS	FIRED FROM PHASE LINE 2. 3,000 - 3,750 METERS 2 - MV TANK, 1 - ST BMP	TFTT HMMWV/2:45	1 2 3				
4. ENGAGE MULTIPLE TARGETS	FIRED FROM PHASE LINE 1. 1,500 - 2,000 METERS 1 - ST TANK, 1 - MV TANK (NBC)	TFTT HMMWV/1:40	1 2				
5. ENGAGE MULTIPLE TARGETS	FIRED FROM BASELINE. 2,500 - 3,000 METERS 1 - MV TANK, 1 - MV BMP (NBC)	TFTT HMMWV/1:40	1 2				
					TOTAL SCORE		

NOTES:

1. TABLE 8 IS CONDUCTED AND TIMED IN EXACTLY THE SAME MANNER AS TABLE 7.
2. TABLE 8, UNLIKE TABLE 7, IS FOR QUALIFICATION. THE SQUAD IS RATED ON ITS FINAL SCORE, RATHER THAN SIMPLY HAVING TO MEET A MINIMUM STANDARD.
3. CREW/LEADER CUTS: SEE FM 3-22.34, CH 4.
4. SCORING TABLE 8: TABLE 8 IS PERFORMED TWICE, ONCE DURING THE DAY AND ONCE AT NIGHT. EACH EVENT IS WORTH A MAX OF 500 POINTS. ADD THE TWO SCORES TOGETHER FOR A POSSIBLE TOTAL OF 1,000 POINTS. SQUADS ARE RATED ON THE FOLLOWING BASIS:

DISTINGUISHED: 900 - 1,000 POINTS SUPERIOR: 800 - 899 POINTS QUALIFIED: 700 - 799 POINTS UNQUALIFIED: 0 - 699 POINTS

DA FORM 7337-R, AUG 2003

APD V1.00

TOW GUNNERY TABLE 9: SECTION GUNNERY PRACTICE

For use of this form, see FM 3-22.34. The proponent agency is TRADOC.

SECTION TC: ________ LDR: ________ GNR: ________ DVR: ________ DATE: ________

SQUAD TC: ________ LDR: ________ GNR: ________ DVR: ________ TIME: ________

PLATOON: ________ COMPANY: ________ BATTALION: ________

TASK	CONDITIONS/ TARGETS/ SITUATION	TOTAL TARGET EXPOSURE TIME	HITS	ENGAGEMENT TIMES	POINTS	CREW/ LDR CUTS	TOTAL POINTS
1. ENGAGE A SINGLE TARGET	FIRED FROM BASELINE. 3,000 - 3,750 METERS 1 - ST TANK, 1 - ST BMP	TFTT 40 SECS ALL VEHICLES	1 2				
2. ENGAGE MULTIPLE TARGETS	FIRED FROM PHASE LINE 2. 2,000 - 2,750 METERS 1- ST BMP	TFTT 25 SECS ALL VEHICLES	1				
3. ENGAGE MULTIPLE TARGETS	FIRED FROM PHASE LINE 3. 1,000 - 1,500 METERS 2 - MV BMPS, 1 - ST TANK	TFTT HMMWV/1:40	1 2 3				
4. ENGAGE MULTIPLE TARGETS	FIRED FROM PHASE LINE 2. 2,500 - 3,000 METERS 2 - MV TANKS, 2 - ST TANKS (NBC)	TFTT HMMWV/2:38	1 2 3 4				
5. ENGAGE MULTIPLE TARGETS	FIRED FROM BASELINE. 1,000 - 1,500 METERS 1 - ST TANK, 1 - ST BMP (NBC)	TFTT 25 SECS ALL VEHICLES	1 2				
					TOTAL SCORE		

NOTES:

1. TABLE 9 IS CONDUCTED AND TIMED IN EXACTLY THE SAME MANNER AS TABLES 7 AND 8.
2. SCORING TABLE 9: TABLE 9 IS PERFORMED TWICE, ONCE DURING THE DAY AND ONCE AT NIGHT. EACH EVENT IS WORTH A MAX OF 500 POINTS. ADD THE TWO SCORES TOGETHER FOR A POSSIBLE TOTAL OF 1,000 POINTS. TABLE 9 IS A PRACTICE TABLE. TO VERIFY ON TABLE 9 AND PROGRESS TO TABLE 10, THE SECTION MUST ACHIEVE A MINIMUM COMBINED SCORE OF AT LEAST 700 POINTS, AFTER CREW AND LEADER CUTS.

TOW GUNNERY TABLE 10: SECTION GUNNERY QUALIFICATION

For use of this form, see FM 3-22.34. The proponent agency is TRADOC.

SECTION TC: ______ LDR: ______ GNR: ______ DVR: ______ DATE: ______

SQUAD TC: ______ LDR: ______ GNR: ______ DVR: ______ TIME: ______

PLATOON: ______ COMPANY: ______ BATTALION: ______

TASK	CONDITIONS/ TARGETS/ SITUATION	TOTAL TARGET EXPOSURE TIME	HITS	ENGAGEMENT TIMES	POINTS	CREW/ LDR CUTS	TOTAL POINTS
1. ENGAGE A SINGLE TARGET	FIRED FROM BASELINE. 3,000 - 3,750 METERS 2 - MV TANK, 2 - ST BMPS	TFTT HMMWV/2:38	1 2 3 4				
2. ENGAGE MULTIPLE TARGETS	FIRED FROM PHASE LINE 2. 2,000 - 2,500 METERS 1 - MV TANK	TFTT 40 SECS ALL VEHICLES	1				
3. ENGAGE MULTIPLE TARGETS	FIRED FROM PHASE LINE 3. 1,000 - 1,500 METERS 1 - MV BMP, 1 - ST TANK	TFTT 25 SECS ALL VEHICLES	1 2				
4. ENGAGE MULTIPLE TARGETS	FIRED FROM PHASE LINE 2. 2,500 - 3,000 METERS 1 - ST BMP, 2 - MV TANKS (NBC)	TFTT HMMWV/2:40	1 2 3				
5. ENGAGE MULTIPLE TARGETS	FIRED FROM BASELINE. 3,000 - 3,750 METERS 1 - ST TANK, 1 - ST BMP (NBC)	TFTT 25 SECS ALL VEHICLES	1 2				
					TOTAL SCORE		

NOTES:

1. TABLE 10 IS CONDUCTED AND TIMED IN EXACTLY THE SAME MANNER AS TABLE 8 AND 9.
2. SCORING TABLE 10: TABLE 10 IS PERFORMED TWICE, ONCE DURING THE DAY AND ONCE AT NIGHT. EACH EVENT IS WORTH A MAX OF 500 POINTS. ADD THE TWO SCORES TOGETHER FOR A POSSIBLE TOTAL OF 1,000 POINTS. TABLE 10 IS A QUALIFICATION TABLE. TO PROGRESS TO TABLE 11, THE SECTION MUST ACHIEVE A MINIMUM COMBINED SCORE OF AT LEAST 700 POINTS.

TOW GUNNERY TABLE 11: PLATOON GUNNERY PRACTICE

For use of this form, see FM 3-22.34. The proponent agency is TRADOC.

SECTION 1:	TOW CREW 1:	TC: ______	LDR: ______	GNR: ______	DVR: ______	DATE: ______
	TOW CREW 2:	TC: ______	LDR: ______	GNR: ______	DVR: ______	
SECTION 2:	TOW CREW 1:	TC: ______	LDR: ______	GNR: ______	DVR: ______	
	TOW CREW 2:	TC: ______	LDR: ______	GNR: ______	DVR: ______	TIME: ______
SECTION 3:	TOW CREW 1:	TC: ______	LDR: ______	GNR: ______	DVR: ______	
	TOW CREW 2:	TC: ______	LDR: ______	GNR: ______	DVR: ______	

PLATOON: ______ COMPANY: ______ BATTALION: ______

TASK	CONDITIONS/ TARGETS/ SITUATION	TOTAL TARGET EXPOSURE TIME	HITS	ENGAGEMENT TIMES	POINTS	CREW/ LDR CUTS	TOTAL POINTS
1. ENGAGE A SINGLE TARGET	FIRED FROM BASELINE. 3,000 - 3,750 METERS 4 - MV TANKS, 2 - ST BMPS	TFTT HMMWV/3:26	1 2 3 4 5 6				
2. ENGAGE MULTIPLE TARGETS	FIRED FROM PHASE LINE 2. 2,000 - 3,000 METERS 4 - MV BMPS 2 - ST BMPS	TFTT HMMWV/2:26	1 2 3 4 5 6				
3. ENGAGE MULTIPLE TARGETS	FIRED FROM PHASE LINE 3. 1,000 - 2,000 METERS 2 - ST TANKS, 2 - ST BMPS	TFTT 25 SECS ALL VEHICLES	1 2 3 4				
4. ENGAGE MULTIPLE TARGETS	FIRED FROM PHASE LINE 2. 2,000 - 3,000 METERS 2 - MV TANKS, 2 - MV BMPS (NBC)	TFTT BMP/43 SEC TANK/1:03	1 2 3 4				
5. ENGAGE MULTIPLE TARGETS	FIRED FROM BASELINE. 3,000 - 3,750 METERS 2 - ST TANKS, 2 - ST BMPS (NBC)	TFTT 25 SECS ALL VEHICLES	1 2 3 4				
					TOTAL SCORE		

NOTES:

1. TABLE 11 IS CONDUCTED, TIMED, AND SCORED IN EXACTLY THE SAME MANNER AS TABLES 7 THRU 10.

TOW GUNNERY TABLE 12: PLATOON GUNNERY QUALIFICATION

For use of this form, see FM 3-22.34. The proponent agency is TRADOC.

SECTION 1:	TOW CREW 1:	TC: ______	LDR: ______	GNR: ______	DVR: ______	DATE: ______
	TOW CREW 2:	TC: ______	LDR: ______	GNR: ______	DVR: ______	
SECTION 2:	TOW CREW 1:	TC: ______	LDR: ______	GNR: ______	DVR: ______	
	TOW CREW 2:	TC: ______	LDR: ______	GNR: ______	DVR: ______	TIME: ______
SECTION 3:	TOW CREW 1:	TC: ______	LDR: ______	GNR: ______	DVR: ______	
	TOW CREW 2:	TC: ______	LDR: ______	GNR: ______	DVR: ______	

PLATOON: ______ COMPANY: ______ BATTALION: ______

TASK	CONDITIONS/ TARGETS/ SITUATION	TOTAL TARGET EXPOSURE TIME	HITS	ENGAGEMENT TIMES	POINTS	CREW/ LDR CUTS	TOTAL POINTS
1. ENGAGE A SINGLE TARGET	FIRED FROM BASELINE. 3,000 - 3,750 METERS 2 - MV TANKS, 2 - ST BMPS	TFTT BMP/25 SECS TANK/1:13	1 2 3 4				
2. ENGAGE MULTIPLE TARGETS	FIRED FROM PHASE LINE 2. 2,000 - 3,000 METERS 3 - ST BMPS, 3 - MV TANKS	TFTT HMMWV/2:28	1 2 3 4 5 6				
3. ENGAGE MULTIPLE TARGETS	FIRED FROM PHASE LINE 3. 1,000 - 2,000 METERS 3 - ST TANKS, 3 - MV BMPS	TFTT HMMWV/1:08	1 2 3 4 5 6				
4. ENGAGE MULTIPLE TARGETS	FIRED FROM PHASE LINE 2. 1,500 - 2,500 METERS 2- MV TANKS, 2 - MV BMPS (NBC)	TFTT BMP/34 SECS TANK/1:03	1 2 3 4				
5. ENGAGE MULTIPLE TARGETS	FIRED FROM BASELINE. 3,000 - 3,750 METERS 2 - ST TANKS, 2 - ST BMPS (NBC)	TFTT 25 SECS ALL VEHICLES	1 2 3 4				
					TOTAL SCORE		

NOTES:

1. TABLE 12 IS CONDUCTED, TIMED, AND SCORED IN EXACTLY THE SAME MANNER AS TABLES 7 THRU 10.

FIELD MANUAL
NO. 3-23.24

*FM 3-23.24 (FM 23-24)
HEADQUARTERS
DEPARTMENT OF THE ARMY
WASHINGTON, DC, 30 August 2001

M47 DRAGON
MEDIUM ANTITANK WEAPON SYSTEM

CONTENTS

DISTRIBUTION RESTRICTION: Approved for public release; distribution is unlimited.

* This publication supersedes FM 23-24, 3 April 1990.

PREFACE

This manual provides guidance for using and employing the M47 Dragon, a medium-range, antiarmor, surface-attack, guided missile weapon system. It provides information about the M47 Dragon's characteristics, nomenclature, and functions. The training program provided helps develop, evaluate, and maintain the proficiency of Dragon gunners. Tactical employment procedures described herein do not cover all situations.

Leaders and members of units equipped with the M47 Dragon should use this manual to train and test individual gunners or teams. This manual also discusses five key elements for maintaining combat proficiency with the Dragon:

- Technical and tactical information.
- A training program that stresses command involvement.
- Centralized training.
- Gunner selection.
- Development of qualified trainers.

The proponent of this publication is the United States Army Infantry School. Send comments and recommendations on DA Form 2028 (Recommended Changes to Publications and Blank Forms) directly to the Commandant, US Army Infantry School, ATTN: ATSH-INB-O, Fort Benning, Georgia 31905-5594 or to lusanoh@benning.army.mil.

Unless otherwise stated, whenever the masculine gender is used, both men and women are included.

CHAPTER 1

INTRODUCTION

This chapter describes the M47 Dragon and its components and provides pertinent technical data. The Dragon is a surface-attack, wire-guided, man-portable, shoulder-fired, medium antitank weapon system. It can defeat armored vehicles, fortified bunkers, concrete gun emplacements, and other hardened targets. The individual soldier or a two-man team can operate this weapon. Mechanized infantry gunners can use the Dragon with the M175 guided missile launcher mount. This mount provides a stable platform on either the M113 APC or on the M3 or M122 machine gun tripod. The Dragon can engage targets in daylight (with the daysight) and in limited visibility conditions, such as smoke, fog, or darkness (with the nightsight).

1-1. MAJOR COMPONENTS

The Dragon weapon system consists of a daysight, a nightsight, and a round of ammunition.

a. **Daysight**. The daysight is an electro-optical sight that determines the position of the missile relative to the gunner's line of sight (LOS).

(1) The daysight generates signals that travel by wire to the missile control system. The missile control system fires the appropriate rocket thrusters to keep the missile heading along the gunner's LOS (Figure 1-1).

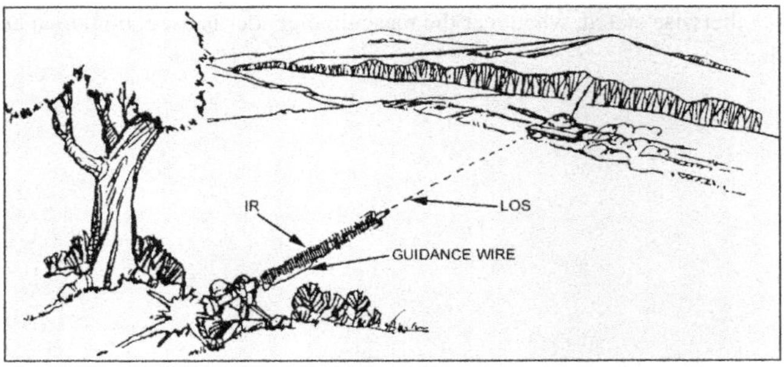

Figure 1-1. Daysight line of sight.

(2) The daysight attaches to and removes quickly from the round (Figure 1-2, page 1-2). The daysight's aluminum housing assembly contains the lens, infrared receiver assembly, and control signal comparator. The firing mechanism, located on the right side of the daysight's housing, consists of a trigger safety plunger and trigger lever bar. A rubber boot seals out moisture from the firing mechanism and safety plunger.

(3) A connector cover provides both mechanical and moisture protection for the electrical connector. A lanyard attaches the cover to the daysight.

(4) Shock absorbers surround the forward and rear ends of the daysight to protect it and its lens. The lens cover is attached to the daysight by a lanyard.

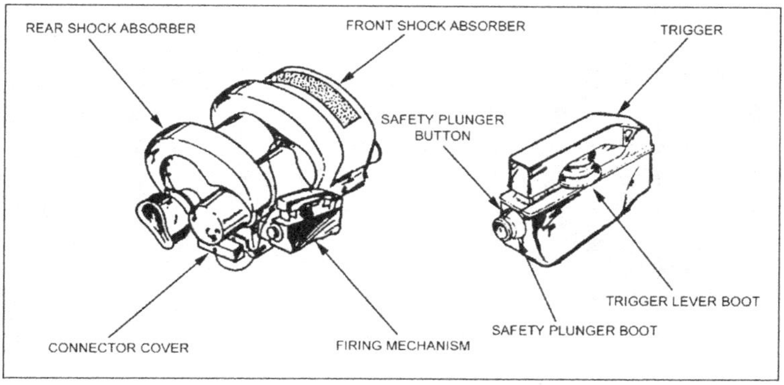

Figure 1-2. Daysight parts.

(5) The daysight's lens is a 6-power telescope. It magnifies the image across a 6-degree field of view (FOV) to help the gunner locate, identify, and track targets. A fitted and molded rubber eye guard protects the gunner's eyes and allows him to adjust rapidly to the sight (Figure 1-3). The infrared receiver consists of an objective lens assembly, graded filter, infrared detector, and nutator mirror. The sensor detects the missile's infrared output while discriminating against background signals. It detects any deviations or excursions of the missile from the gunner's LOS.

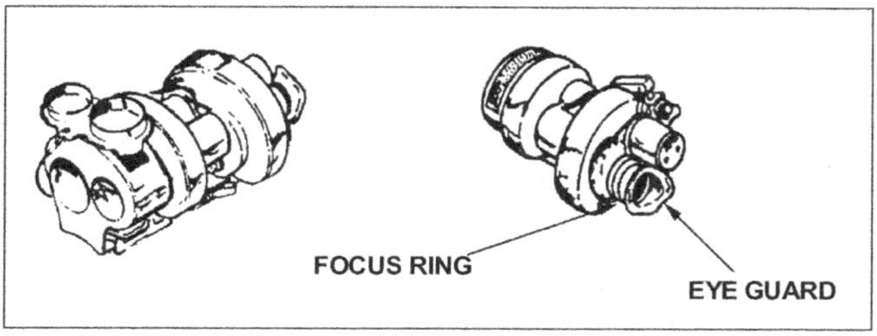

Figure 1-3. Daysight lens and rubber eye guard.

(6) The daysight's telescopic lens and infrared receiver align to define the LOS to the target. The tapered cross hair helps the gunner concentrate on the center of his FOV. Then, he can quickly place the cross hairs on the target's center of mass. He uses the stadia lines to determine the range to the target (Figure 1-4). For example, the stadia lines show how large a standard 3-meter by 6-meter Soviet-built vehicle would appear at the Dragon's maximum engagement range of 1,000 meters.

b. **Nightsight**. At dusk, or anytime the light level drops below what is needed to use the daysight, Dragon gunners switch to the nightsight (Figure 1-5). It attaches to the round the same as the daysight.

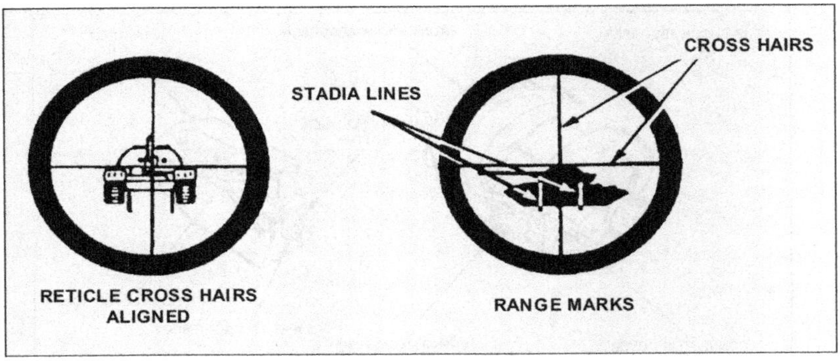

Figure 1-4. Daysight reticle and stadia lines.

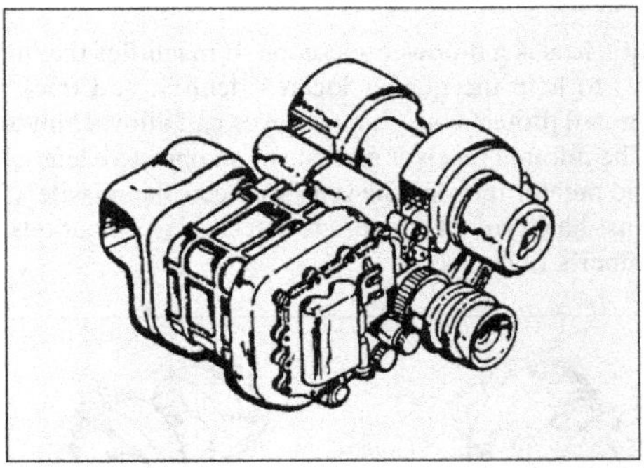

Figure 1-5. Nightsight.

(1) The passive, 4-power AN/TAS-5 nightsight has a 3.4-degree by 6.8-degree FOV. The nightsight is larger and heavier than the daysight. The nightsight converts heat emissions (infrared energy) from a target area into electrical signals, then to visible light (Figure 1-6, page 1-4). This enables the nightsight to display a real-time scene. Objects warmer than the ambient (surrounding) temperature appear in shades of red; cooler objects appear black. Like the daysight, the nightsight monitors the missile and sends commands to keep it on the gunner's line of sight.

(2) The firing mechanism, control system, and guidance system are the same as those on the daysight.

(3) The externally mounted 4.8-volt DC battery and coolant cartridge enable the nightsight to function without an external power source. If the appropriate accessory equipment is available, the nightsight can use military vehicle power.

(4) The nightsight has more controls and operates differently than the daysight.

Figure 1-6. Nightsight's infrared (heat) capability.

(a) The ACTUATOR switch (Figure 1-7) turns the system on and off. This four-position switch allows the gunner to evaluate the operating condition of the nightsight. It is normally set to the OFF-LOCK position. Rotating the switch to the AIR-BATT-CHECK position allows the gunner to check the condition of the coolant cartridge and battery. If the indicator light is off and the coolant cartridge and battery conditioner are good, the gunner rotates the switch to the ON position to operate normally.

CAUTION

Use the fourth position, RELEASE, to release an expended coolant cartridge.

(b) The gunner uses the focusing ring to adjust the focus. The brightness (*BRT*) and contrast (*CTRS*) controls work the same as those on a television set (Figure 1-8, page 1-5).

c. **Ammunition**. The Dragon's ammunition is an expendable component consisting of both the missile and the launcher itself (Figure 1-9, page 1-6). The missile is installed in the launcher at the factory and is shipped ready to fire. The launcher serves as a storage and carrying case for the missile before it is launched. The Dragon has two types of missiles: the M223 practice round, which has an inert warhead, and the M222 HEAT round. The launcher consists of—

- A smooth-bored fiberglass tube.
- A breech and gas pressure generator assembly.
- A wiring harness.
- A sling.
- Forward and rear shock absorbers.
- A sight battery, which provides power to operate the sight and fire the missile.
- A sight support bracket, which houses the Dragon's electrical connectors.
- A bipod, which supports the weapon system during dismounted firing.

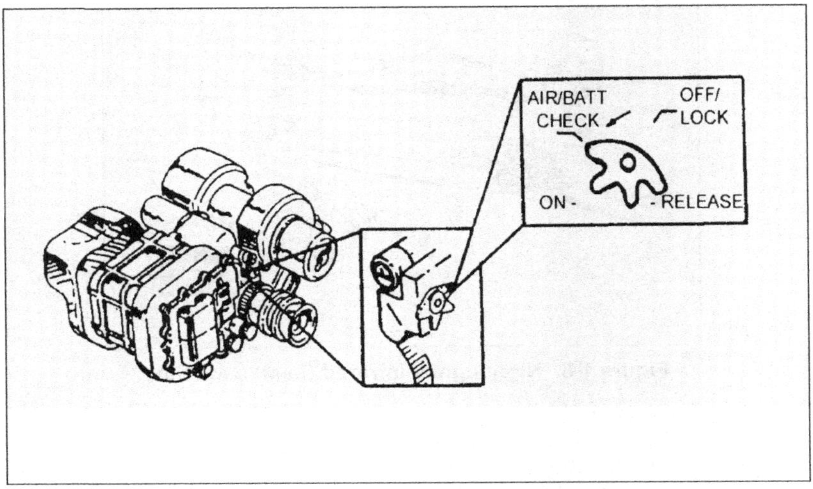

Figure 1-7. Nightsight actuator switch.

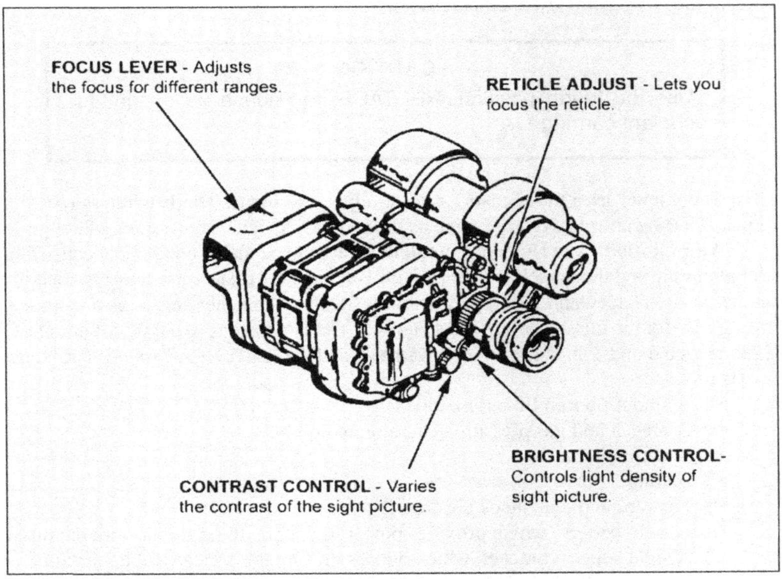

Figure 1-8. Nightsight controls.

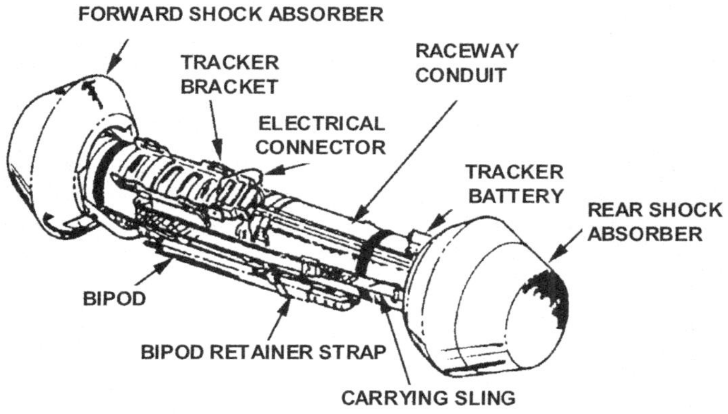

Figure 1-9. External components of the M222 HEAT (tactical) round.

1-2. TECHNICAL DATA

Table 1-1 shows technical data for the Dragon.

AMMUNITION	M222 HEAT round	Weight	14.60 kg	(25.29 lb)
		Missile Length	744 mm	(29.39 in)
		Launcher Length	1,154 mm	(44.10 in)
	M223 inert practice round	Weight	4.70 kg	(10.40 lb)
		Missile Length	846 mm	(33.32 in)
		Launcher Length	1,154 mm	(44.10 in)
OPERATIONAL CHARACTERISTICS		Maximum range	1,000 m	
		Minimum range	65 m	
DAYSIGHT--Infrared Tracker, Guided Missile, SU-36/P		Weight	3.10 kg	(6.75 lb)
		Length	196 mm	(7.72 in)
NIGHTSIGHT--Night Vision Sight, Tracker, Infrared, AN/TAS-5		Weight	9.82 kg	(21.65 lb)
		Length	368 mm	(14.50 in)

Table 1-1. Technical data for the M47 Dragon.

CHAPTER 2
OPERATION AND FUNCTION

This chapter discusses normal operation and function procedures for the Dragon in the man-portable mode. (For a detailed discussion on the operation of Dragon support equipment, refer to TM 9-1425-484-10.)

2-1. INSPECTION

Before placing the Dragon into operation, the gunner visually inspects the Dragon for physical damage and checks its function. During offensive operations, the gunner inspects the Dragon before leaving the assembly area. During defensive operations, the gunner inspects the Dragon before and after preparing the position. (See TM 9-1425-484-10 for information about PMCS.) The gunner inspects the daysight, nightsight, and the round.

2-2. CARRYING TECHNIQUES

Gunners can use either the long-distance or short-distance carry with the Dragon.

a. **Long-Distance Carry**. The Dragon gunner normally carries the round and the sight separately. This carrying technique is most often used during long-distance moves such as foot marches or cross-country travel, or when contact is not likely. For long-distance travel, the gunner uses carrying technique 1, 2, or 3, as shown in Figure 2-1 (page 2-2).

b. **Short-Distance Carry**. The gunner can carry the Dragon short distances with the sight mated to the round. When using this type of carry, the gunner must keep the forward protective shock absorber and lens covers in place until the weapon is likely to be used. The protective lens covers prevent possible damage to the sight lens and keep foreign objects out of the launcher. Carry technique 4 and 5, also shown in Figure 2-1 (page 2-2), can be used with the round and the sight mated. The soldier should only use one of these carry techniques when firing is likely. Otherwise, he should avoid carrying the weapon system in a ready-to-fire configuration.

2-3. PREPARATION FOR FIRING

Improper or careless handling of the round can damage its components and cause the missile to malfunction when launched. If there is any reason to believe the missile has been damaged, the round must be returned to the responsible ammunition personnel for inspection and disposition. Before using the nightsight, the gunner must conduct a preoperational check IAW TM 9-1425-484-10.

a. **Releasing the Bipod**. Unsnap the bipod-retaining strap (A, Figure 2-2, page 2-3). Push the bipod forward until you feel resistance. Snap off the forward shock absorber. Ensure that the desiccant bag, attached to the forward shock absorber, is not lodged in the launch tube (B, Figure 2-2, page 2-3). While still holding the round, depress the bipod friction lock. Push the bipod legs downward to number 4 or 5, then release the friction lock (C, Figure 2-2, page 2-3).

b. **Selecting a Firing Position**. Select the firing position that best meets the situation. For training, use the sitting position (discussed later in this chapter). In a combat situation, you can use any one of the three positions: sitting, kneeling, or standing supported.

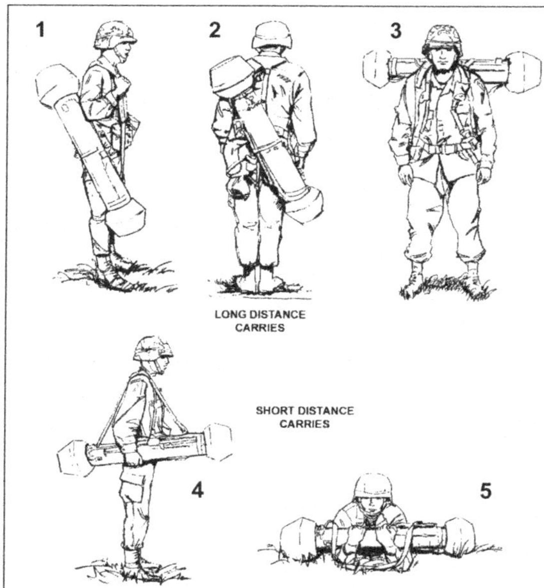

Figure 2-1. Dragon carry techniques.

> **CAUTION**
>
> 1. Do not use the shock absorbers as handles to lift the sight. The shock absorbers might tear off.
> 2. Do not touch the lenses, as this can easily damage them.

c. **Removing the Sight from the Carrying Bag or Rucksack**. Open the carrying bag or rucksack. Grasp the sight by the trigger mechanism or telescoping barrel. Remove the sight and remove the cover from the electrical connector. Secure the cover to the hook-pile tape on the forward shock absorber.

d. **Removing the Lens Cover**. Secure the lens cover to the top of the forward shock absorber (Figure 2-3). Visually inspect the lenses for damage or obstruction. When using the nightsight, make sure the ACTUATOR switch is in the ON position.

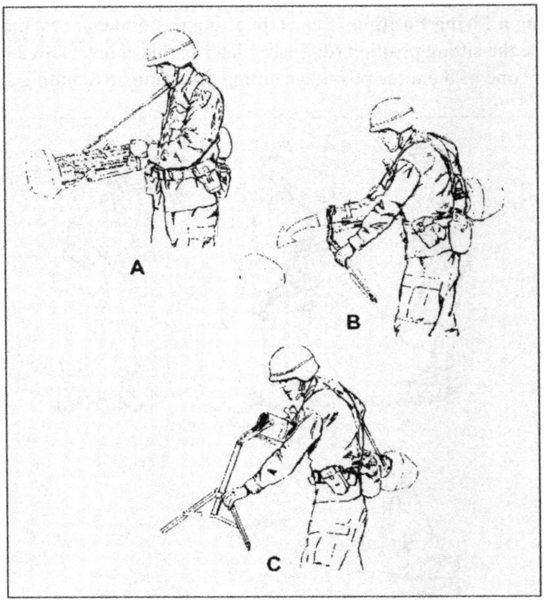

Figure 2-2. Releasing the bipod.

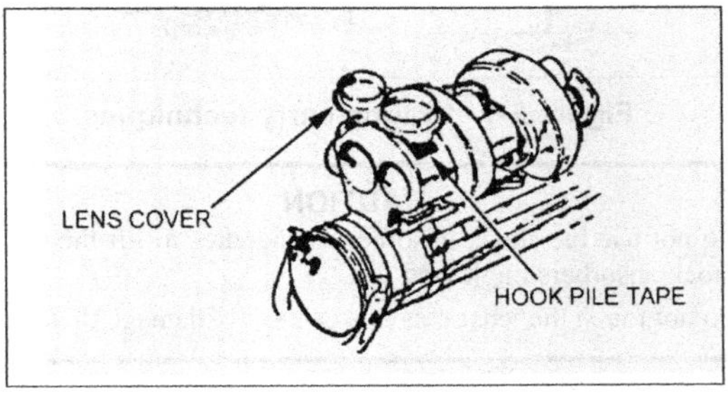

Figure 2-3. Cover secured to forward shock absorber.

e. **Mating the Sight to the Round**. Place the sight guide pins in the slots of the sight bracket guide rails (Figure 2-4, page 2-4). This example uses the daysight, but you will use the same procedure for the nightsight. Use both hands and slide the sight firmly to the rear until the spring clip locks the guide pin in place (Figure 2-5, page 2-4). If the sight does not mate, notify your squad or section leader. Assume the firing position and put the round on your shoulder.

WARNING

Avoid pressing the safety and the trigger while mating the sight to the round.

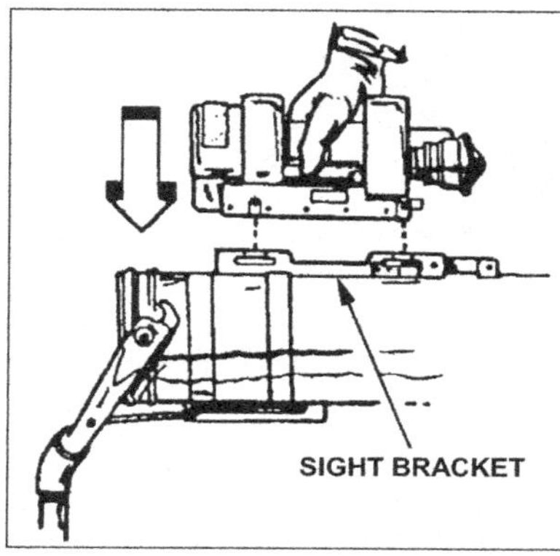

Figure 2-4. Aligning the daysight guide pins.

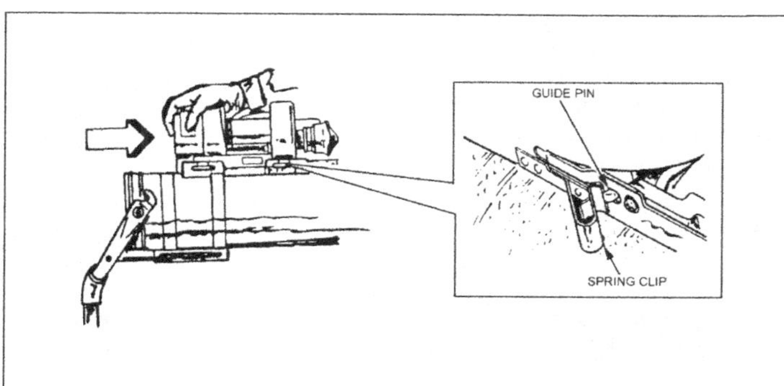

Figure 2-5. Seating the daysight on the round.

f. **Adjusting for Height and a Level Sight Picture**. Look through the sight to see if the sight picture is level. An unlevel sight picture (more than 6 degrees off level) will cause the sight to send bad commands to the missile. The missile will fly erratically and the missile will hit the ground. To obtain a level sight picture, adjust the bipod friction lock and the foot adjustment lever (Figure 2-6, page 2-5).

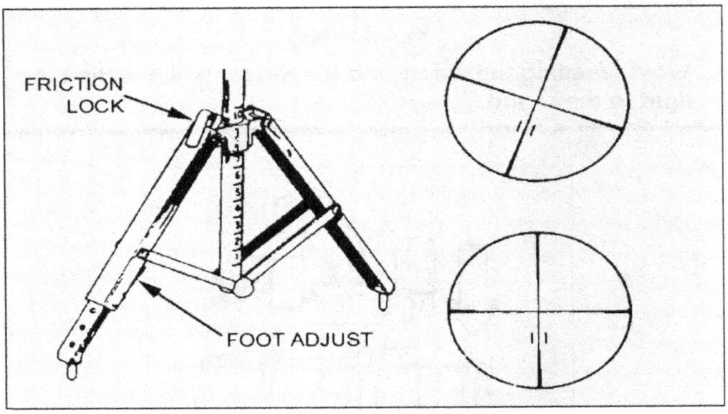

Figure 2-6. Leveling the sight picture.

2-4. FIRING POSITIONS

The Dragon can be fired from any one of the three basic firing positions: sitting, standing supported, or kneeling. When the M175 mount is installed on the M3 or M122 machine gun tripods, the gunner can use a modified sitting or standing supported position. Chapter 5 discusses firing positions in detail.

2-5. TARGET ACQUISITION AND EVALUATION

After assuming a firing position, the gunner looks through the sight and acquires the target (Figure 2-7). The gunner must evaluate the target to determine whether or not it is within range and engageable. By moving the launcher, the gunner adjusts his sight picture, placing the target within the stadia lines (Chapter 6).

a. Acquire and track the target.

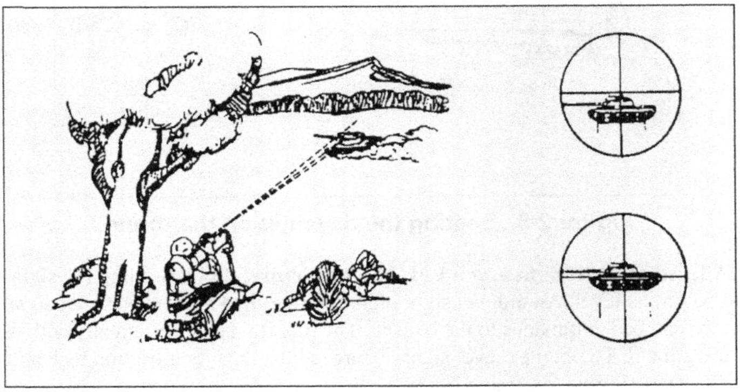

Figure 2-7. Target acquisition.

b. To fire the weapon, push the safety plunger in with the thumb of the right hand, depress and hold the trigger, and maintain the sight picture until impact (Figure 2-8). When engaging a moving target, move the upper part of your body left or right to adjust for azimuth; move up and down to adjust for elevation. Jerking, bumping, or moving the sight while the missile is in flight may cause you to miss the target. To establish a smooth tracking rate, track the target for about two seconds before squeezing the trigger.

c. Ignore the missile when it appears in the sight picture. Do not try to fly the missile. Keep the cross hairs on the target—let the sight guide the missile.

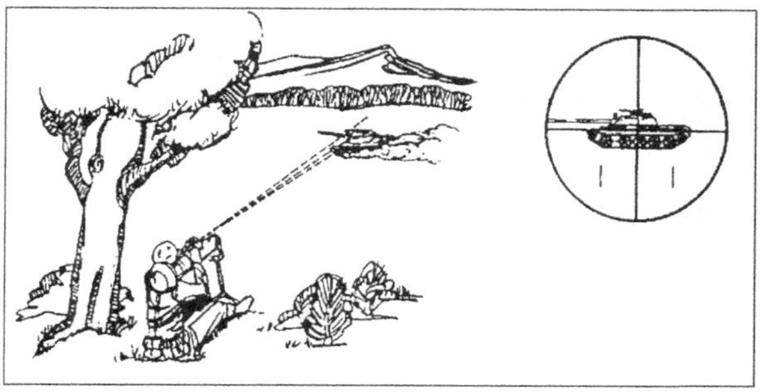

Figure 2-8. Correct sight picture.

2-6. MISSILE FLIGHT AND CONTROL

As the missile leaves the launch tube, the infrared flares begin to operate, and the command-link wire is dispensed from the missile. The missile fins fold out and lock into place. The safety and arming device sequence is completed after the missile has traveled about 65 meters (Figure 2-9).

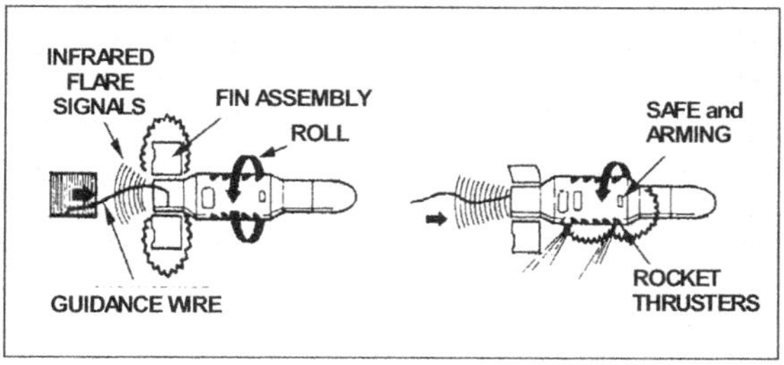

Figure 2-9. Dragon functioning concept.

a. The sight's electronic controls compare the missile's position with the gunner's LOS. It sends electronic position information to the missile through the command-link wire.

b. Based on information received from the sight, the missile generates commands to correct its position. The missile may store the commands until it rolls to the proper position for its rocket motors to fire, or it may fire the rocket motors immediately to correct the missile's position relative to the gunner's LOS.

c. The missile's rocket motors fire to accelerate it or to correct its position. The gunner keeps the cross hairs on the selected aiming point until impact.

d. To fire a second round, the gunner releases the spring clip at the rear left side of the sight support assembly, slides the sight forward, and lifts it clear of the support. The gunner expends or destroys and discards the round IAW unit SOP. The gunner either prepares another round or he secures the sight and any extra rounds and prepares to displace on order.

2-7. FAILURE TO FIRE

A failure to fire occurs whenever the missile does not leave the launcher after the gunner depresses the trigger lever bar. When the missile fails to fire, the gunner immediately resqueezes the trigger lever bar and continues to track the target for 15 seconds. If after 15 seconds the missile has not fired, the gunner announces "Misfire," releases the trigger lever bar, and carefully feels near the sight battery. If the sight battery is hot, the gunner performs hangfire procedures; if it is cold, the gunner performs misfire procedures. Ideally, a two-man team operates the Dragon, and the assistant gunner secures the new round. In the normal tactical situation, a moving armored vehicle closes on the position and moves to within 1,000 meters. To preserve forces, leaders should have an emergency action SOP in place that covers what to do if a Dragon fails to fire.

WARNING

A hangfire is a delayed firing. It is difficult to distinguish immediately from a misfire. The difference lies in when they occur and what causes them.

- **A misfire occurs when a gunner tries to fire a round from a cold weapon and the round jams for mechanical reasons.**
- **A hangfire occurs when a round jams in a hot weapon.**

a. **Hangfire Procedures (Combat Only).** In a typical tactical situation, a moving armored vehicle closes on the position and moves to within 1,000 meters.

(1) Remove the sight from the round and place it on the ground away from the firing site. Keep the round pointed toward the enemy; advise nearby soldiers of the round's failure to fire.

(2) Obtain a new round and mount the sight to it.

(3) Acquire the previous target and continue with the mission.

b. **Hangfire Procedures (Training Only).**

(1) The gunner notifies the range officer or NCOIC about the hangfire.

(2) The gunner removes the round from his shoulder and places the round flat on the ground. The gunner ensures the bipod legs are pointing to the side, with the round pointing downrange.

(3) The gunner moves away from the position.

(4) The range officer or NCOIC clears all personnel within 50 meters of the round and notifies the EOD.

c. **Misfire Procedures**.

(1) Loosen the sight from the round, then remate it to the round.

(2) Track the target and try to fire again.

(3) If the round again fails to fire, feel carefully near the sight battery once more.

(4) If the battery still feels cold, remove the sight from the round. Place the round on the ground away from the firing site. Keep the round pointed toward the enemy; advise soldiers near you of the misfire.

(5) Mount the sight on another round and continue with the mission.

(6) If the new round also fails to fire, carefully feel near the sight battery. If that battery is also cold, the sight is probably defective.

(7) If you can find another sight, replace the existing sight and continue with the mission, using previously unfired rounds.

2-8. RESTORATION TO CARRYING CONFIGURATION

The gunner uses a long-distance or short-distance carry to move a Dragon round that he has prepared for firing but that he no longer needs. To use the long-distance carry, the gunner performs Steps 1 and 2. To use the short-distance carry, the gunner performs only Step 2.

a. **Step 1**. Remove sight from round.

(1) Replace sight lens cover.

(2) Remove sight from round; and replace connector cover on sight and round.

(3) Return sight to carrying bag or rucksack.

b. **Step 2**. Restore round to carrying configuration.

(1) Replace connector cover.

(2) Replace forward shock absorber.

(a) Set round in an upright position with rear shock absorber resting on the ground.

(b) While lightly pushing down on bipod, press bipod brace toward round to lift it from locking slot.

(c) Once bipod brace is unlocked, push down on bipod until upper part of bipod that attaches to round is parallel with muzzle of launcher.

(d) Align cutout portion of forward shock absorber with bipod at launcher muzzle.

(e) Push down on shock absorber while lowering bipod against round. Ensure upper part of bipod engages shock absorber.

(f) Retract bipod legs and secure them to round with retainer strap.

(3) Lift up on forward shock absorber to ensure it is secure. Round is now in safe carrying configuration.

2-9. ADVERSE WEATHER CONDITIONS

If the gunner can see the target, the Dragon can operate in all weather conditions. The Dragon and round stay the same temperature. For example, if the gunner removes a Dragon from a vehicle whose inside temperature is 45 degrees Fahrenheit, both the Dragon and the round will measure 45 degrees Fahrenheit, even if the outside temperature measures 32 degrees Fahrenheit.

a. **Cold Weather**. Before deployment, check all optical surfaces and all mechanical devices for proper operation. Ensure there is neither ice nor snow on the sight components. Never operate the Dragon in temperatures below minus 25 degrees Fahrenheit.

b. **Hot Weather**. To avoid subjecting the round and sight to prolonged, direct sunlight, provide as much shade or cover as possible. Never operate the Dragon in temperatures above 145 degrees Fahrenheit.

CHAPTER 3

TRAINING EQUIPMENT

Training equipment provides units with a low-cost means to develop and maintain gunnery skills. To train effectively, trainers must know the weapon and the training equipment. This chapter describes Dragon training equipment and provides a program to train instructors to use it. (Appendix A provides train-the-trainer tasks for the DGT and DFTT.)

3-1. PRECISION GUNNERY TRAINING SYSTEM

The Dragon PGTS includes one indoor training system and one outdoor training system. The Dragon gunnery trainer (DGT) is the indoor training system. It provides a means for precision gunnery training, practice, and qualification with the Dragon. The Dragon field tactical trainer (DFTT), the outdoor training system, provides a means for familiarization with the Dragon. (Appendix B and Appendix C discuss the DGT and DFTT, respectively.)

a. The realistic training offered by the Dragon PGTS allows units to achieve a high level of proficiency without firing live Dragons. Individual units are responsible for conducting sustainment training on the Dragon PGTS.

b. Soldiers can obtain instructional videotapes on the Dragon PGTS from training support centers (TSCs). Basic instructional fire (Firing Tables 1 through 4) is conducted on the DGT. These tables introduce new gunners to the basic gunnery skills required to engage targets with a Dragon. Firing Tables 5 and 6 are conducted in a simulated outdoor tactical situation using the DFTT. Firing Tables 7 and 8 are conducted on the indoor DGT. Firing Table 7 allows gunners to practice for qualification. If they score 16 out of 20 on Firing Table 8 (Table 3-1), gunners qualify on the Dragon. (Appendix D describes firing tables and provides examples of completed firing tables.)

FIRING TABLE NUMBER	PURPOSE OF FIRING TABLE	TRAINING DEVICE	PAGE IN MANUAL
1 2 3 4	Familiarization	DGT	D-1
5 6	Familiarization	DFTT	D-2
7	Practice for qualification	DGT	D-3
8	Qualification (Hit 16 out of 20)		

Table 3-1. Firing tables for Dragon training devices.

c. **Train-the-Trainer Program.** A successful Dragon training program requires well-trained NCOs who know and have confidence using both trainers. Each company should have at least one certified trainer. (Appendix A provides train-the-trainer tasks for both the DFTT and the DGT.)

3-2. DRAGON GUNNERY TRAINER

The DGT man-portable trainer simulates the Dragon guided missile system's sight(s), controls, switches, and indicators. (See Appendix B for a complete description of the DGT and its components.) Training programs selected by the training NCO display simulated battlefield scenes through the sight. These scenes include both enemy and friendly vehicle targets. The indoor, 110- or 220-volt AC Dragon gunnery trainer enables units to train personnel on the Dragon without a range or training area (Figure 3-1).

a. **Role**. Soldiers train on the DGT to—

- Assume the correct Dragon firing position.
- Identify targets.
- Determine engageability of targets.
- Engage targets (including tracking and firing).

b. **Missions**. Matched sets of videodiscs and floppy (data) disks provide a library of missions. Each disk set contains 30 missions of various lengths plus learning objectives. Each mission applies to both daysight and nightsight viewing.

c. **Target Options**. The DGT allows the trainer to adjust the difficulty of the scenario to the shooter's skill. To do this, trainers can—

- Change the target kill zone.
- Change the obscuration.
- Vary the amount of weight the system loses after a round is fired.
- Select night fire.
- Select multiple or moving targets.

Figure 3-1. Dragon gunnery trainer.

d. **Feedback**. The DGT enables the trainer to provide visual feedback to the firer in either of two ways. He can replay the mission on video, or he can print out two views (from the top and side) of the actual mission. Both show horizontal and vertical tracking errors and the firer's score. Figure 3-2 shows the instructor's console.

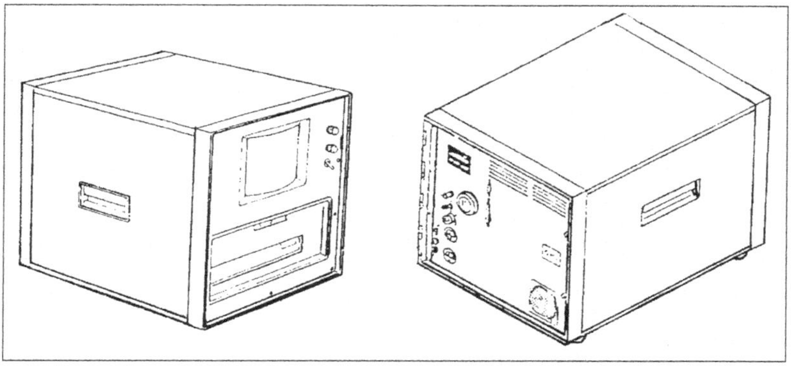

Figure 3-2. Instructor's console, Dragon gunnery trainer.

3-3. DRAGON FIELD TACTICAL TRAINER

The DFTT outdoor trainer realistically simulates the appearance, feel, launch, flight, and effects of the Dragon guided missile system. (Appendix C describes the DFTT and its components.) Training with the DFTT can occur on designated ranges, general outdoor areas, or representative tactical environments. The DFTT uses the Dragon daysight or nightsight and trainers can use it instead of the inert practice round. The DFTT is bipod-mounted, just like an actual Dragon.

a. **Role**. Trainers use the DFTT to familiarize gunners with firing the Dragon in an outdoor environment. Specifically, they use it to train soldiers to—

- Assume the correct Dragon firing position.
- Determine engageability of targets (whether targets can be engaged).
- Engage targets (including tracking and firing).

b. **Targets**. Each target has a retroreflector. This acts as a mirror and returns part of the laser beam sent by the trainer round. The DFTT transmits Dragon MILES code, so it can also kill targets equipped with a MILES (multiple integrated laser engagement system) harness, as long as the target also has a retroreflector.

CHAPTER 4
UNIT TRAINING

This chapter guides the unit leader in developing a successful unit training program.

4-1. TRAINING STRATEGY

A training strategy is the overall concept for integrating resources to train the skills—individual and collective—needed to perform a unit's wartime mission. Commanders at TRADOC institutions as well as in the units themselves implement the Dragon training strategy, whose four parts include *initial training*, *sustainment and advanced training*, *collective training*, and *leader training*. Commanders make sure that they include trainer certification as part of the overall training strategy. Using this manual, a unit commander can develop an effective Dragon training program that meets unit requirements.

a. **Initial Gunner Training**. Initial gunner training is a prescribed POI conducted in TRADOC institutions. Initial gunner training includes 15 blocks of instruction. To pass, the soldier must successfully achieve a score of 16 out of 20 engagements on the DGT and successfully complete the 11 tasks in the gunner's performance test. (Chapter 6 provides an example POI.)

b. **Sustainment and Advanced Training**. Sustainment and advanced training are conducted in the unit.

(1) ***Sustainment Training***. Sustainment training, conducted monthly, quarterly, and annually, ensures gunners retain the skills they learned in initial training. To retain their skills, gunners practice precision tracking on the DGT at least monthly. To practice, they fire selected engagements from the monthly sustainment table using the DGT and DFTT. They must also complete at least 50 percent of the gunner's skill test. (Chapter 7 discusses sustainment-training programs in detail.)

(2) ***Advanced Training***. The unit conducts advanced training. It consists of tracking exercises using the DGT, DFTT, or the Dragon MILES. The tracking engagements grow more difficult. Existing conditions, equipment, or tactical play determine difficulty. Examples of advanced training include—

- Night tracking exercises using artificial illumination or the AN/TAS-5 nightsight.
- MOPP tracking exercises.
- Other situational gunnery exercises.

c. **Live-Fire Training**. The number of live missiles the gunner may fire depends on the type of unit and on budget constraints. The unit may conduct annual live fire in an instructional setting or may integrate it into other live-fire exercises. Where possible, live missile firings should closely follow a scheduled qualification. Only currently qualified gunners should be allowed to fire live missiles.

d. **Collective Training**. The unit conducts collective training to fully integrate the Dragon into the unit's overall combat power. Collective training has two parts: force-on-force training and live fire training.

(1) The unit conducts *force-on-force training* with MILES during a squad and platoon FTX or STX. During semiannual external evaluations, the unit evaluates the platoons on Dragon employment.

(2) The unit conducts *live-fire training* using the laser target interface device (LTID) or actual missiles (live or inert) along with platoon LFXs. Dragon gunners or teams should participate in squad or platoon collective live-fire exercises at least twice a year.

e. **Leader Training**. Leader training occurs in the unit through NCO and officer development classes and personal initiative. Leaders selected as Dragon instructors must have already *demonstrated* proficiency in using the Dragon and should periodically recertify to maintain leader proficiency.

4-2. COMMANDER'S RESPONSIBILITY

The commander plans, executes, and supervises training, selecting instructors and gunners based on the criteria in this chapter. He makes sure that instructors and gunners receive training. The commander does the following:

a. Allows adequate time for effective training.

b. Ensures all training meets training standards.

c. Combines Dragon training with other unit tactical training.

d. Ensures the maintenance facility properly maintains all consolidated Dragon training equipment.

e. Conducts sustainment training.

f. Ensures subordinate leaders conduct operational readiness checks.

g. Periodically inspects all tactical and training equipment.

h. Keeps a formal record of—

- Inspections.
- Training results.
- Gunner qualifications.
- Gunner turnover.

i. Evaluates the Dragon training and corrects deficiencies in future training.

4-3. TRAINING PROGRAM GOALS

Training program goals provide leaders with the objectives necessary for success. For soldiers to achieve proficiency in Dragon gunnery, leaders must—

- Train them to successfully engage armored targets.
- Train them to maintain and operate the Dragon and related training devices.
- Provide clear sustainment training standards.
- Make sure gunners know how to meet the standards.
- Train gunners to those standards.
- Conduct sustainment training.
- Evaluate each gunner's performance.

4-4. INSTRUCTOR SELECTION

Dragon instructors have to meet the same standards that gunners do. The commander picks soldiers who want to and can instruct and who have a high level of tactical competence at squad and platoon levels. Experience as a Dragon gunner will serve an instructor well, but does not guarantee a soldier's success as an instructor. Finally, the overall training strategy must include trainer certification.

4-5. GUNNER SELECTION

First, the leader identifies soldiers who want to be Dragon gunners. Then, he makes sure each soldier closely approaches the physical standards for Dragon gunners shown in Table 4-1.

1. To track moving targets, the soldier must be able to flex his upper trunk to the left and right.
2. To hold the weapon steady during tracking and to carry it as required, a soldier must have considerable upper body strength and physical stamina
3. To use the Dragon, a soldier must stand between 5 feet tall and 6 feet 2 inches tall—the bipod only adjusts for this height range.
4. To acquire a target, fire the Dragon, and track the round, a soldier must show that he can hold his breath for at least 13 seconds.
5. To use the Dragon, a soldier must have unaided minimum vision of 20/100, correctable to 20/20. The focus (reticle adjustment) can only compensate for this amount of near-sightedness.
6. To use the Dragon, a soldier must have right-eye dominance. He must be able to close his left eye independently. That allows him to keep his left eye closed to protect it from distracting dust and debris.

Table 4-1. Physical standards for the Dragon gunner.

4-6. MATCHING THE TRAINING PROGRAM TO THE UNIT

Mandatory training programs seldom fit a unit's particular circumstances and needs. Only the unit's own commander can develop its training program. Table 4-2 shows the steps for developing a unit-training program. The key is to consolidate training at unit level. For this to succeed, the commander must—

- Assign a qualified NCO with additional duties as NCOIC of Dragon training.
- Include Dragon training on both short-range and long-range training calendars.
- Hold subordinate leaders accountable for execution of training.

4-7. LEADER TRAINING

Each commander must determine what specific, Dragon-related tasks he and other leaders must perform. Then he develops these tasks into training objectives. Next, instructors choose proper training methods, depending on the resources available and other training requirements. If incoming leaders do not know how to employ the Dragon, the commander makes sure they receive Dragon training—from the basics on up through the more difficult tasks. Initial Dragon training works well in the classroom, since it depends on the use of lectures, seminars, small discussion groups, and briefings. This cuts down on the use of expensive training resources such as ranges and transportation.

1. Identify all the unit's tactical and administrative missions.
2. Analyze the unit's missions. Determine what individual and unit tasks soldiers must accomplish in order to complete the mission.
3. Establish individual and unit training objectives to accomplish the unit's tasks.
4. Determine the level of individual and unit proficiency in the unit's tasks.
5. Determine individual and unit training needed to attain the training objectives.
6. Identify available training resources.
7. Program and schedule training based on the training resources available and on individual and unit training needs.
8. Conduct training.
9. Monitor and evaluate training, and revise the training program as required.

Table 4-2. Steps in developing a unit training program.

4-8. GUNNER TRAINING

Dragon gunners should be trained consistently and as repetitively as resources allow. All training includes gunnery and night fire.

a. **Training Methods**. Two main methods are used to train the Dragon gunner:

(1) ***Centralized Gunnery Training***. Centralized gunnery training requires the unit to train all Dragon gunners at once. Training type depends on the training resources available (equipment, facilities, personnel, and time) and the tasks to be taught.

(a) *Equipment and Facilities*. Limited distribution of training equipment suggests centralized control of training at the highest unit level possible. Also, the number of firing ranges suitable for the Dragon may dictate centralized control over these facilities.

(b) *Personnel*. A shortage of qualified Dragon instructors can hamper gunnery training.

(c) *Time*. Centralizing Dragon training can save training time since fewer instructors and classes are required. For example, if the training is centralized at brigade level or higher, battalion and company level instructors can focus on other training requirements. With centralized training, qualified Dragon instructors prepare and conduct training. They need less time to prepare than would unqualified instructors.

(d) *Tasks*. When considering centralized training, the commander also considers the task he wants to train. If the task relates to gunnery training or qualification, the number of training sets and ammunition available might limit the choice to centralized training. Dragon gunners first learn how to prepare a basic range card. Then they practice applying what they have learned using different pieces of terrain or terrain substitution (maps, sand tables, or 35-mm slides of the terrain). Gunners build these skills by working in small groups where they can ask questions, talk about the answers, and debate the advantages and disadvantages of the range cards.

(2) ***Round-Robin Training***. Battalions have a limited number of training sets, so soldiers cannot practice all at once. Rather than letting a few soldiers practice on the equipment while others watch, trainers can set up round-robin (multistation) training. Soldiers then rotate through the stations. This keeps everyone actively engaged in training the whole time. For example, soldiers might train on the equipment itself at Station Number 1, learn to prepare range cards at Station Number 2, and learn to identify enemy vehicles at Station Number 3 (Figure 4-1).

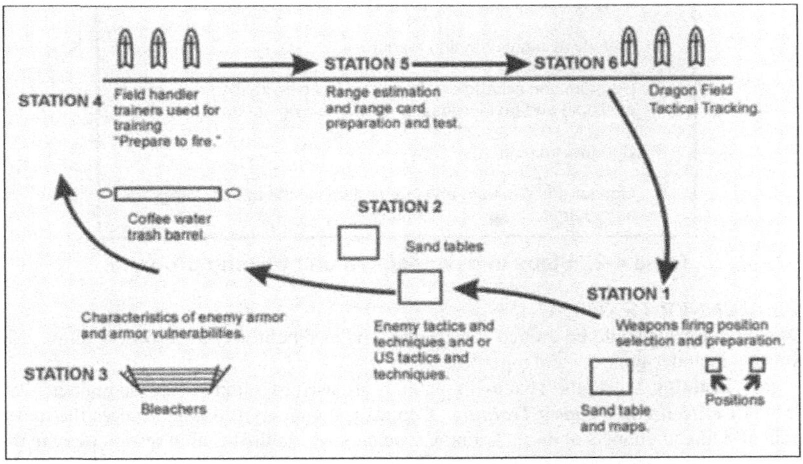

Figure 4-1. Example of round-robin training.

b. **Gunnery**. Gunnery qualifies gunners to fire the Dragon tactically and nontactically. To qualify, gunners must meet the following objectives:

(1) Detect vehicles at different ranges under varying field conditions, such as rolling hills and vegetation, with the vehicles moving and stationary.

(2) Determine whether a moving target, if engaged, will reach cover before impact.

(3) Prepare a firing position and range card.

(4) Know how to lessen the signature of the backblast.

(5) Know what suppressive fires the enemy can place on the firing position.

(6) Know how to use cover and concealment, deception, surprise, and movement.

(7) Know unit SOPs covering rules of engagement, including signals to lift or shift fires, priority of targets, and times to engage targets.

(8) Know where to obtain resupply of missiles.

(9) Know how to inspect the round before firing.

c. **Night Training**. Since threat doctrine stresses night operations, gunners should practice their skills at night. They can do this during FTXs, with the gunners using the FHTs with nightsights, or they can do it on a range, under controlled conditions, with artificial illumination.

4-9. TEAM TRAINING

Due to the frequent need for leaders to employ Dragons as part of a team, team training must become an integral part of Dragon training in the field. As a rule, Dragon gunners employed apart from their squad positions have at least one assistant gunner and ammunition bearer with them, or they may be employed in an antiarmor fire team. The other members of the team work with the gunner as a coordinated element. As such, their duties include providing security for the gunner, helping the gunner prepare his firing position, carrying ammunition (missiles), and locating targets.

a. Once gunners have qualified with the Dragon, and once leaders have trained in tactical employment of the Dragon, they can train together profitably in FTXs.

(1) During their training, leaders must decide how to employ the Dragon. They could use scenarios, simulated threat conditions, or OPFOR, for example. Based on these simulated conditions, platoon leaders practice choosing armor avenues of approach, selecting Dragon firing positions, and making related employment decisions.

(2) Leaders should have gunners prepare firing positions and range cards, and simulate engaging targets. Employing gunners with their squads helps train the other squad or fire team members to perform their duties (providing security, locating targets, and so forth). Then, leaders should evaluate and discuss the training exercises to decide if the techniques and tactics used were the best ones for the situation.

b. When qualifying on the range, Dragon gunners usually learn under near ideal conditions: They have excellent fields of fire and a target that moves conveniently back and forth on level ground directly in front of them. Except when other gunners fire, they have no distractions. So, after qualifying on the range, they are ready for more advanced training. This means using the Dragon under more realistic, and therefore more difficult, conditions, which the commander creates. The following paragraphs suggest a few ways to add realism to Dragon training:

(1) During both FTXs and practice firing exercises on the range, train the gunner to concentrate while tracking a target. Try to distract him with various combinations and amounts of smoke, haze, and harassing fires. Detonate explosives near him to simulate the noise of enemy artillery and tank fires. To simulate nearby explosions, detonate explosives under a bag of flour. Obscure the target with smoke in or near the gunner's position or target—this makes tracking more realistic.

(2) During MILES-Dragon tracking exercises, the target vehicle with the MILES receivers must travel over various terrain. This moves the target across the front or towards

the gunner's position over various terrain such as curving roads, or wooded or rolling terrain.

4-10. COLLECTIVE TRAINING

In addition to monthly or quarterly training and qualification (or both), Dragon gunners must demonstrate their proficiency during squad and platoon exercises. Each Dragon gunner takes part in live-fire exercises as part of a squad or larger unit. He does the same in a platoon external evaluation based on standards in ARTEP 7-8-MTP (RDL version). In live-fire exercises, the unit uses actual Dragon rounds or LTIDs with other small-arms targets. Platoon external evaluations should include medium antiarmor weapons tasks and should be conducted as part of a company FTX or STX.

4-11. EVALUATION

One of an instructor's major responsibilities is to evaluate training. The training to be evaluated may be training that the instructor conducted or training that was conducted by another instructor. The evaluation is concerned with the *effectiveness* and *efficiency* of training.

a. *Training effectiveness* refers to whether the soldiers, teams, or units met the standards established in the commander's training objective(s).

b. *Training efficiency* refers to how well instructors used the available training resources. Leaders should write all evaluations in the AAR format. This helps make sure that gunners remember what they learned from their mistakes. It also helps trainers improve future training.

CHAPTER 5

INSTRUCTOR'S GUIDE TO EFFECTIVE TRAINING

This chapter provides the instructor with the information needed to train and sustain Dragon gunners and to train soldiers as team members.

5-1. TRAINING PROGRAM SCHEDULE

The number of firing sites that can be staffed with instructors and equipped with DGTs and DFTTs dictates the number of gunners for each class. This allows for the effective use of time.

a. Instructors must allow sufficient time to orient students on the equipment. Negative results occur when the training schedule is shortened. The more information each soldier receives about how the weapon and training equipment operate and on the purpose of each, the better the soldier will perform as a gunner.

b. Initial marksmanship training, in the institution or unit, teaches essential skills and develops fixed and correct procedures in marksmanship before range practice begins. Thoroughly instructing and carefully supervising practice in the initial phase saves time and ammunition during range firing. It also allows students to develop techniques and procedures necessary for well-trained Dragon gunners and teams.

5-2. INDIVIDUAL INSTRUCTION

To prepare for live firing the Dragon, each gunner must learn various skills and habits. Although the Dragon training equipment closely simulates firing the Dragon weapon, gunners require personalized individual instruction to develop proper gunnery techniques and procedures. Instructors should key on the following points:

- Coach and stress the gunnery techniques shown in Table 5-1.
- Use and emphasize the sitting position and standing supported positions.
- Have gunners practice tracking in both directions—left to right and right to left.
- Have gunners clean the training equipment at the end of the day.

• Tight eye contact with the eyepiece.
• Proper position.
• Steady hold at the launch.
• Steady tracking at all times, especially through smoke.
• Slow and steady aim point correction to the target.

Table 5-1. Gunnery techniques.

5-3. TRAINING SEQUENCE

The Dragon gunner should receive training in the following sequence to control the missile launch and flight:

- Positions.
- Sighting, aiming, and firing.

- Breathing.
- Tracking exercises.
- Qualification and verification.

5-4. FIRING POSITIONS

The gunner must acquire and maintain a stable body position relative to the weapon and be able to move smoothly when tracking a moving target. The round must be solidly anchored on the muscle of the gunner's shoulder. His arms and hands must be properly placed to squeeze the trigger and to maintain the stability of the round. The position of the eye against the eyepiece is critical. Keeping the eye firmly pressed into (against) the eyepiece reduces launch-induced movement and prevents obscuration. The three basic firing positions for the Dragon are the sitting, standing supported, and kneeling. The gunner uses a modified sitting position to fire the Dragon from the M175 mount. The M175 mount fits the M3 or M122 machine gun tripod. (For more information about use of the tripod, see Appendix E.)

a. **Sitting Position.** Demonstrate the sitting position, then have the soldiers assume the position also (Figure 5-1). Instruct them as follows:

(1) The sitting position is the most stable. Sit with your legs extended as far as possible. Place the notch of your boot heels on the bipod and push outward.

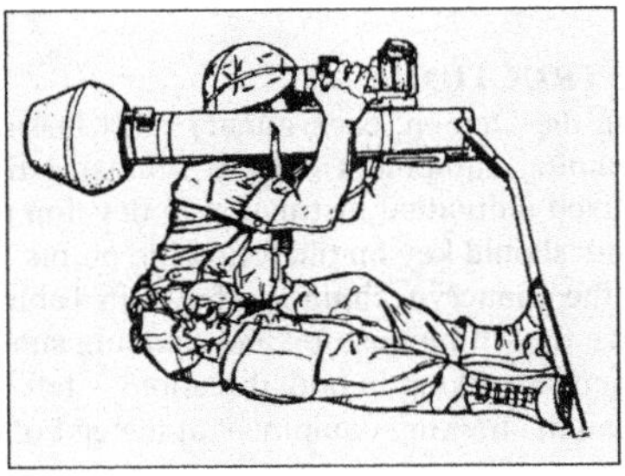

Figure 5-1. Sitting position.

(2) Lean forward from your waist as far as possible. Pick up the round and place it on the muscle portion of your shoulder, keeping it tight against the curve of your neck (Figure 5-2).

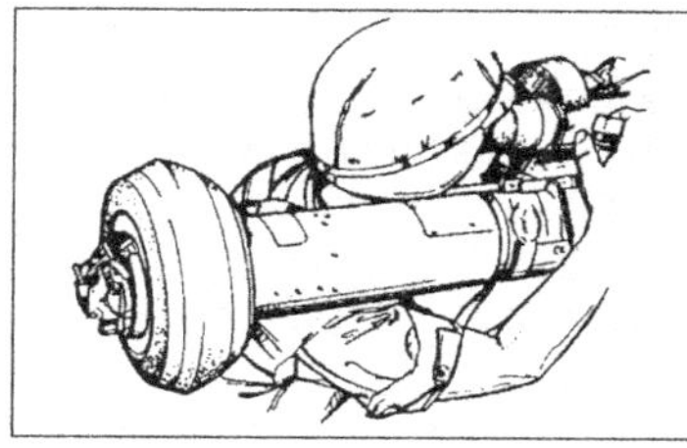

Figure 5-2. Position of round on shoulder.

(3) Grasp the barrel of the sight with your left hand, curling your thumb under the tube. Grasp the firing mechanism with your right hand, with your thumb on the safety, three fingers on the firing lever, and your little finger on the front of the firing mechanism. Place the heel of your hand on the base of the firing mechanism to provide a firm grip and reduce slippage. When firing, *hold* the trigger in the depressed position; if you release your hand, you will experience an involuntary muscle reaction that will affect the sight and in turn the path of the round.

(4) Lift your head to align your right eye with the telescopic sight. Press your head forward and press your eye firmly against the eye guard. This forces your eye to stay open. Close your left eye, and keep it closed. If necessary, focus the sight.

(5) Pull down and back with your hands while pushing out with your feet. Try to touch your elbows together and to your chest at the same time (Table 5-2).

(6) Keep your back as straight as possible while leaning forward for better breath control. This limits discomfort and increases your ability to move your upper body.

(7) Maintain arm, back, and leg muscle tension. Use enough force that you do not experience an involuntary muscle reaction when the weight of the missile is removed from your shoulder.

1. Stress the need for keeping the pull-down force on the sight and the eye tight in the eye guard.
2. Body position and breath control are the two key elements to effectively engage targets.
3. Ensure the gunner keeps his body and limbs clear of the backblast area. He must keep the round at least 6 inches off the ground in order for the missile fins to clear the ground.

Table 5-2. Training notes.

b. **Standing Supported Position**. Demonstrate the standing supported position, then have the soldiers assume it also (Figure 5-3, page 5-4). Instruct them as follows:

(1) While standing in an individual fighting position or behind a support, place the bipod legs to your front. Place it at such a distance that you must reach for the round.

(2) Spread your legs a comfortable distance apart, keeping them straight. Place the round on your shoulder muscle.

(3) Lean forward against the wall of the fighting position to support your body from the waist down, so that you are in a stable firing position.

(4) Grip the sight as you did in the sitting position. Pull backward and down, while straightening your upper body slightly; this removes any slack in the bipod.

(5) Place your upper body, arms, hands, head, and eyes in the same position as you did when you assumed the sitting position.

c. **Kneeling Position**. Kneel, and spread your knees a comfortable distance apart. Position the bipod so that you have to lean forward to position your eye in the eye guard (Figure 5-4). Grasp the sight as previously described. Assume the same upper body position you did for the other firing positions. Place the round on your shoulder muscle, keeping it tight against your neck. As you lower your buttocks to your heels, take the slack out of the bipod. Try to sit on your heels so you will have a stable firing platform.

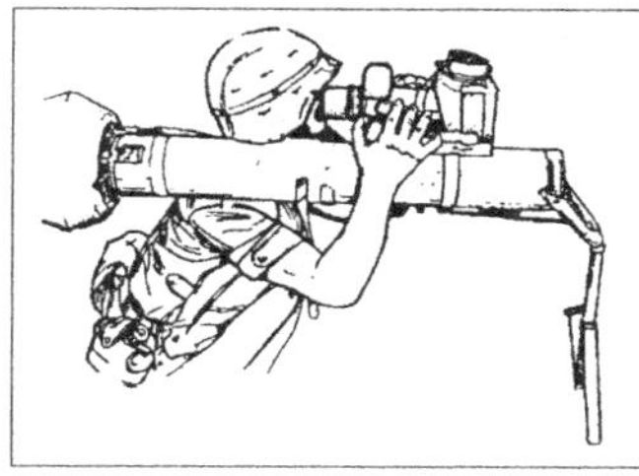

Figure 5-3. Standing supported position (fighting position).

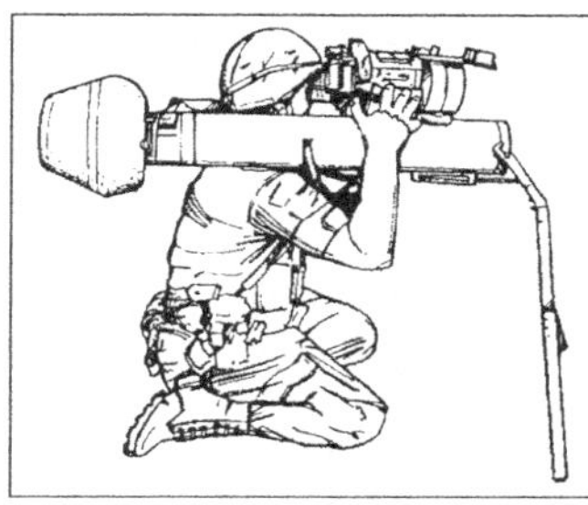

Figure 5-4. Kneeling position.

5-5. M113 INSTRUCTIONAL TECHNIQUES

Certain instructional techniques apply only to M113-equipped units. (See Appendix E for more information.)

5-6. SIGHTING, AIMING, AND FIRING

Once the gunner has mastered the firing positions, the instructor can teach him how to sight, aim, and fire properly.

a. **Sighting**. Place the eye in the eyecup and pull the weapon tight enough against your eye that you cannot blink. Wrap your small finger around the front of the firing mechanism to add pulling force and to help keep the weapon tight against your eye. Visually select a target, and then acquire it through the daysight or nightsight by adjusting the upper portion of your body.

(1) Keep the weapon tight to keep the sight picture.

(2) Do not move your eye in the eyecup. If you move your eye, you will see the side of the telescope prism, which will blur your sight picture.

b. **Aiming**. Confirm that the target is within range by using the stadia lines. Place the cross hairs of the sight on the center of the target's visible mass. Regardless of the target's range or speed of movement, keep the cross hairs on the firing point until after you have fired and observed the impact of the round. To maintain the proper sight picture on a moving target, you have to move the upper portion of your body laterally (sideways). When you fire from a seated position, never rest your elbows on your knees. This transfers any movement of your leg directly to the sight.

c. **Firing**. Fully depress the safety before you try to squeeze the trigger. Then, squeeze—do not pull—the trigger. Since the Dragon has little recoil, many gunners move more when they pull the trigger than when they experience the missile's launch effects (recoil and backblast).

5-7. BREATH CONTROL

Aiming the Dragon is similar to aiming a rifle, except that the cross hairs must be kept on the desired impact point for 1 to 12 seconds following missile launch, depending on range to the target. Training must stress the importance of starting to hold the breath two seconds before squeezing the trigger and continuing to hold it while acquiring the target, firing, and tracking. That is, to prevent breathing from interfering with tracking, the gunner takes a breath and holds it while pressing the trigger. He must not breathe while tracking a target, because body movements cause the launcher to move. To check for breath control, the instructor watches the gunner's back.

5-8. TRACKING EXERCISES

Reacting properly to temporary obscuration is an important gunner skill. Occasionally, the target may be obscured by launch gases, dust, and so forth. The gunner's instinctive reaction is to look for a target. This can either cause erratic missile flight, or it can terminate the flight. Training prepares the gunner to "freeze" on a stationary target or to continue tracking at the established rate on a moving target until the target reappears. Instructors should simulate obscuration during DGT and DFTT exercises. They should show soldiers how an improper reaction causes the LOS to move outside the established aiming error limits. Most training should be conducted at moving targets to promote gunner concentration. A gunner who can hit a moving target consistently can hit a stationary target easily. Most *misses* occur on moving targets. Training in the tracking of a moving target should begin when a gunner achieves proficiency in assuming positions, sighting and aiming, and breathing.

5-9. RANGE PROCEDURES

Leaders normally consolidate qualification, verification, and sustainment range firing for gunners and assistant gunners at battalion or higher.

a. **Officer or NCOIC**. Table 5-3 (page 5-6) shows officer or NCOIC duties. (See also Appendix F.)

• Organizes the range. • Assigns, coordinates, and supervises the firing lines. • Issues fire commands and general instructions to the firing line. • During all firing exercises, enforces safety precautions as prescribed in AR 385-62, local SOPs, and applicable range regulations. • Stresses precision, a steady tracking rate, point training, and use of a firm firing position posture in all instructional firing. • Fires exercises in the order listed in the firing tables. Ensures that instructors control the exercises using appropriate fire commands. • Ensures that a qualified instructor inspects all DGTs, DFTTs, and sights before and after each firing day for cleanliness, serviceability, and operation. • Instructs gunners and assistant gunners on duty assignments and range operating procedures before training them with the equipment. Divides them into teams and assigns each of them a position.

Table 5-3. Officer or NCOIC duties.

b. **Qualification Training**. For qualification, the gunner must fire the DGT.

c. **Verification Training**. A gunner must verify quarterly to meet qualification standards.

d. **Sustainment Training**. For monthly sustainment training, see Table 5-4 (page 5-7). The commander can select any target speed and monitor *DFTT* operation for gunner proficiency.

e. **Coach and DFTT Operator**. During instructional firing, a coach or DFTT operator is at each DFTT to instruct and assist the gunner. The coach or DFTT operator—

- Requires each gunner to observe safety precautions.
- Supervises each gunner's actions at the DFTT.
- Ensures gunners execute all the commands.
- Repeats orders and instructions to ensure understanding and timely execution.

- Reports misfires, malfunctions, and discrepancies to the OIC or NCOIC.
- Critiques the tracking runs.
- Sets target size on the computer.
- Sets obscuration on the computer.

VEHICLE SPEEDS		NUMBER OF ROUNDS	RANGE
KPH	MPH		
0	0	2	1,000
0	0	2	500
0	0	1	300
10	6	5	1,000
18	11	5	500
29	18	5	300

Table 5-4. Sustainment speeds.

f. **Training Team**. (Figure 5-5).

- DGT or DFTT gunner.
- Loader.
- Scorekeeper (optional).
- Safety monitor.
- One target vehicle operator, who does not participate in training.

NOTE: Rotate duty assignments between students.

Figure 5-5. Training team.

g. **Safety Precautions**. Implementing the precautions shown in Figure 5-6 ensures safety for all personnel.

- Observe the DFTT backblast area.
- Wear properly fitted earplugs.
- Reset the dummy weight before loading or reloading.

Figure 5-6. Safety precautions.

h. **Duty Assignments and Tasks of Training Team Personnel**. Figure 5-7 (page 5-8) shows training team personnel duty assignments and tasks.

Duty Assignment	Tasks
DFTT Gunner	☐ Avoids pointing either end of the DFTT at personnel.
	☐ Fires only when cleared by the loader.
	☐ Does not look at the sun or at searchlights when sighting.
	☐ Considers all misfires to be hangfires and proceeds accordingly.
	☐ Does not leave a loaded DFTT unattended.
	☐ Prepares the DFTT for operation.
	☐ Fires the DFTT.
	☐ Reviews each tracking run with the DFTT operator.
Loader	☐ Considers all misfires to be hangfires and proceeds accordingly.
	☐ Helps perform misfire procedures.
	☐ Loads or reloads the DFTT.
	☐ Has all soldiers move their hands clear of the launcher before giving permission to fire.
	☐ Informs the gunner when to fire.
Scorekeeper	☐ Prepares the scorecard.
	☐ Informs the instructor of the gunner's progress.
Safety Monitor	☐ Ensures all participating team members use earplugs during the firing exercises.
	☐ Keeps personnel clear of the DFTT's backblast area during the firing exercises.
	☐ Reports all safety violations to the instructor for corrective action.
Target Vehicle Operator	☐ Complies with all safety precaution warning signs.
	☐ Drives the target vehicle.

EXAMPLE

Figure 5-7. Example format for checklist of duty assignments and tasks.

DANGER

THE DFTT USES HIGH VOLTAGES AND HIGH-EXPLOSIVE BLAST SIMULATORS.

YOU MUST FOLLOW ALL SAFETY PRECAUTIONS WHEN USING THIS EQUIPMENT. FAILURE TO DO SO COULD CAUSE YOU OR SOMEONE ELSE TO LOSE YOUR HEARING, SUFFER TRAUMATIC INJURY, OR EVEN DIE.

FOR INFORMATION ABOUT ARTIFICIAL RESPIRATION, SEE FM 21-11.

1. ***HIGH VOLTAGES.***
   - **AVOID TOUCHING THE METAL PINS WHEN CONNECTING OR DISCONNECTING POWER CORDS OR SYSTEM CABLES.**
   - **NEVER TRY TO DISASSEMBLE THE COMPUTER.**
   - **NEVER USE THE EQUIPMENT IF THE CATHODE RAY TUBE (CRT) OR CABLE IS DAMAGED.**
   - **FOLLOW THESE SAFETY PRECAUTIONS, OR YOU COULD DIE.**
2. ***HIGH-EXPLOSIVE BLAST SIMULATORS.***
   - **WITHIN 70 METERS OF A LOADED BLAST SIMULATOR, WEAR HEARING PROTECTION.**
   - **HANDLE BLAST SIMULATORS CAREFULLY. IF POSSIBLE, WEAR FULL HELMET AND GLOVES.**
   - **KEEP BLAST SIMULATORS AWAY FROM FIRE.**
   - **NEVER HANDLE A DAMAGED BLAST SIMULATOR.**
   - **STAY CLEAR OF THE DANGER ZONE AT ALL TIMES.**

CHAPTER 6
GUNNER QUALIFICATION PROGRAM

This chapter guides the unit through training soldiers to qualify as Dragon gunners. AR 614-200, paragraph 3-13, authorizes unit commanders to award the additional skill identifier (C2) to soldiers who successfully complete the qualification program. This chapter includes the qualification requirements and provides an example POI. (See Appendix G for gunner's performance test.)

6-1. REQUIREMENTS

The gunner qualification program consists of three categories of instruction: informational classes (four of these), Dragon tasks (eight of these), and tracking qualifications.

a. **Informational Classes.**

- Introduction to the DGT.
- Introduction to the DFTT.
- Emergency decontamination procedures for the Dragon (Appendix H).
- Emergency destruction procedures for the Dragon (Appendix H).

b. **Tasks.** Soldiers must obtain a GO on each of the following tasks (the order is not important):

071-052-0001 — Maintain an M47 medium antitank weapon.
071-052-0003 — Construct a fighting position for an M47 medium antitank weapon.
071-052-0004 — Restore an M47 medium antitank weapon to carrying configuration.
071-052-0005 — Operate a night vision sight AN/TAS-5.
071-052-0006 — Engage targets with an M47 medium antitank weapon.
071-317-0000 — Prepare an antiarmor range card.
071-317-3302 — Prepare an M47 medium antitank weapon for firing.
071-317-3306 — Perform misfire procedures on an M47 medium antitank weapon.
071-317-3324 — Select a fighting position for an M47 medium antitank weapon.

c. **Tracking Qualification.** Soldiers must successfully engage 16 out of 20 targets on the DGT. Using the DGT requires an indoor training area.

6-2. PROGRAM OF INSTRUCTION

Initial gunnery trainers at Fort Benning, GA use the example POI shown in Figure 6-1 (page 6-2) to qualify *initial entry soldiers* on the Dragon. Trainers in each unit can adapt this POI to the resources available.

DEPARTMENT OF THE ARMY
BRAVO COMPANY 2ND BATTALION 29TH INFANTRY REGIMENT
FORT BENNING, GEORGIA 31905

ATSH-INB-B 9 AUG 99

MEMORANDUM FOR RECORD

SUBJECT: FIVE-DAY POI for Dragon Gunnery

1. **Schedule.** The following is a five-day POI training schedule for the Dragon Gunnery Course. It includes uniform, location and times for each class:

DAY 1-LEE FIELD	SUBJECT	UNIFORM
0900-1000	Introduction to the M47 Dragon (Characteristics, Safety, and Limitations)	C
1000-1100	Prepare the M47 for Firing	B
1100-1200	Prepare and Restore the M47 for Firing	B
1200-1300	Eat Lunch	C
1300-1330	Demonstrate Correct M47 Firing Position	B
1330-1430	Demonstrate Target Engagement Techniques	B
1430-1530	Perform Misfire Procedures on the M47 Dragon	B
DAY 2-BLDG 4, RM 26	SUBJECT	
0900-1100	Identify Combat Vehicles	C
1100-1200	Prepare an Antiarmor Range Card	C
1200-1300	Eat Lunch	C
1300-1600	DGT Firing Tables 1-4	B
DAY 3-LEE FIELD	SUBJECT	
0900-1000	Review/Test Prepare an Antiarmor Range Card	C
1000-1100	Identify the six major components of an M47 Fighting Position	B
1100-1200	Maintain the M47 Dragon	B
1200-1300	Lunch	C
1300-1400	Operate the AN/TAS-5 nightsight	B
1400-1600	DFTT daysight firing. Table 5 (10 rounds)	A
1600-1700	Completion of DFTT nightsight firing during daylight conditions. Table 6 (5 rounds)	A

Figure 6-1. Suggested POI for Dragon gunnery qualification.

DAY 3—LEE FIELD	SUBJECT	UNIFORM
1700-1800	Dinner	C
1800-1830	Safety Briefing	A
1830-UTC	Conduct DFTT Night Sight firing during darkness. Table 6 (5 rounds)	A
DAY 4—BLDG 4, RM 26	**SUBJECT**	
0900-1000	DGT Table 7, Practice Qualification	B
1000-1200	DGT Table 8, Qualification	B
1200-1300	Lunch	C
1300-1400	Completion of Tables 7 and 8	B
1400-UTC	Retest of Table 8	B
DAY 5—BLDG 4, RM 26/ COOLIDGE RANGE	**SUBJECT**	
0900-1030	Maintenance	C
1100-1130	End of Course Critique	C
1130-1230	Lunch	C
1230-1300	Movement to Coolidge Range	B
1300-UTC	Dragon Live Fire	B

2. **Uniform code:**

A = LCE and BDU Cap

B = LCE and helmet

C = BDUs

NOTE: Instructors and students will wear the same uniform.

3. **Questions:** If you have questions about this memorandum, contact the undersigned.

PHIL S. CLIFFORD
SSG, USA
Master Trainer
CML 706-545-9918
DSN 835-9918

Figure 6-1. Suggested POI for Dragon gunnery qualification (continued).

CHAPTER 7
SUSTAINMENT TRAINING PROGRAM

This chapter recommends a sustainment program for units to conduct for qualified Dragon gunners. Training should progress into collective training for applicable MTPs and field manuals. Unit commanders can get qualified Dragon gunners either from the United States Army Infantry School or from a unit-run gunner qualification program (Chapter 6). After completing either of these training programs, each gunner must take part in sustainment training to maintain his tracking skills.

7-1. GUNNER SUSTAINMENT TRAINING

Units conduct gunner sustainment training to maintain a high level of gunner proficiency. Training is divided into monthly, quarterly, and annual training phases.

a. **Monthly Sustainment Training**. Units conduct monthly training eight months of the year. The other four months, they conduct quarterly qualification or verification (on the DGT).

- One month, gunners perform half the tasks in Table 7-1 (page 7-2) using Firing Tables 1 and 2.
- In alternate months, gunners perform the other half of the tasks using Firing Tables 3 and 4.
- Each month, each gunner must score 16 out of 20 hits.

b. **Quarterly Training**. Each quarter, gunners fire practice qualification (Firing Table 7) and qualification (Firing Table 8). Each table includes 20 target engagements. Gunners must complete successfully all the tasks shown in Table 7-1. Anyone who fails to verify his qualification must retrain and retest. The qualification standards follow:

19 or 20 hits	—	Expert
17 or 18 hits	—	Sharpshooter
16 hits	—	Marksman

c. **Annual Training**. The commander follows one quarterly training period a year with an annual live-fire exercise. This exercise really tests the gunners' training. Commanders must ensure that only qualified Dragon gunners fire the live missiles. They must exclude from the live-fire exercise any gunner who fails to qualify on DFTT Table 6 or DGT Table 8. Allowing unqualified gunners to fire jeopardizes everyone else present.

7-2. COLLECTIVE TRAINING

Though the Dragon MILES offers an excellent way for gunners to keep their tracking skills honed (Table 7-2, page 7-2), leaders must still provide sustainment training.

TASK NO	TASK	TNG FREQ	TIME	REMARKS
—	Explain Emergency Decontamination Procedures for an M47 Medium Antitank Weapon	M	10 min	Oral Presentation
—	Explain Emergency Destruction Procedures for an M47 Medium Antitank Weapon	M	10 min	Oral Presentation
—	Tracking Sustainment Training	M	4 hrs	Practical Exercise
071-052-0001	Maintain an M47 Medium Antitank Weapon	M^1	10 min	Practical Exercise
071-052-0004	Restore an M47 Medium Antitank Weapon to Carrying Configuration	M^1	5 min	Practical Exercise
071-317-3306	Perform Misfire Procedures on an M47 Medium Antitank Weapon	M^1	15 min	Practical Exercise
GTA 17-2-11 GTA 17-2-13	Recognize Friendly and Threat Armored Vehicles	M^2	15 min	Practical Exercise
071-052-0003	Construct a Fighting Position for an M47 Medium Antitank Weapon	M^2	—	Perform as part of an ARTEP[3] or FTX
071-317-3302	Prepare an M47 Medium Antitank Weapon for Firing	M^2	30 sec	Practical Exercise
071-317-3324	Prepare an Antiarmor Range Card	M^2	15 min	Practical Exercise
071-052-0006	Engage Targets with an M47 Medium Antitank Weapon	A	8 to 16 hrs	Practical Exercise

M = Every month
M^1 = First month
M^2 = Second month
A = Annually

[3] ARTEPs are available at Reimer Digital Library.

Table 7-1. Training tasks for Dragon gunners and assistant gunners.

TRAINING SEQUENCE	RECOMMENDED FREQUENCY: TIMES PER YEAR	EQUIPMENT USED
Squad, platoon, or company FTX	4	Dragon/ MILES
Battalion FTX	2	Dragon/ MILES
Platoon or company MOUT exercise	1	Dragon/ MILES
Squad, platoon LFX	2	Dragon/ MILES or LTIDs (laser-target interface devices)
Externally evaluated battalion MTP	1	Dragon/ MILES

Table 7-2. MILES training program.

CHAPTER 8

EMPLOYMENT

This chapter discusses roles and duties, employment considerations, and target-engagement techniques for employing the Dragon in defensive operations. Though this chapter discusses the Dragon as an infantry weapon, the techniques described apply to any situation in which when a soldier uses a Dragon, regardless of the type of unit. Specific discussions include selection and preparation of firing positions, target-engagement techniques, and fire-control procedures. To fully understand and properly integrate the Dragon into unit TTP, leaders must know FM 7-7, FM 7-7J, FM 7-8, FM 7-10, *or* FM 71-1, *whichever applies.*

Section I. EMPLOYMENT IN THE DEFENSE

This section provides guidance for employing the Dragon in defensive operations. This information applies to all types of infantry and other types of units. In infantry units, the platoon has the mission in the defense to repel the enemy's assault-by-fire and close combat. The Dragon has combat characteristics that are important to the defense. (FM 7-8, FM 7-7J, and FM 7-10 cover defensive techniques thoroughly.) The Dragon(s) can—

- Destroy or immobilize armored vehicles, depending on type.
- Deliver accurate fire, day or night.

8-1. FIRING POSITIONS

Figure 8-1 shows the three types of firing positions: primary, alternate, and supplementary.

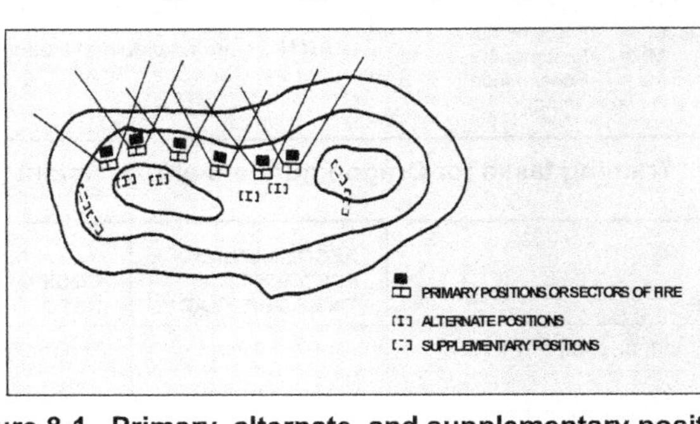

Figure 8-1. Primary, alternate, and supplementary positions.

a. **Primary Position**. From this position, the gunner or team can cover the assigned sector of fire. The position should offer good observation, good cover and concealment, and a good field of fire.

b. **Alternate Position**. The *alternate position* is either to the flank or slightly to the rear of the primary position. The gunner or team must be able to cover the same sector of fire as they could from the primary position. He (or they) occupies the alternate position when they must leave or cannot occupy the primary position.

c. **Supplementary Position**. The *supplementary position* covers avenues of approach and any TRPs not covered by the primary and alternate positions. This position usually falls close enough to the primary position to share mutual support with other positions.

8-2. MOVEMENT

Whether mounted or dismounted, the gunner should observe certain basics when moving to and from firing positions.

a. Avoid disturbing natural foliage.

b. Move in and around positions as little as possible.

c. Move into the position from the rear. In the defense, the only movement forward of the position should be the gunner and the other soldiers clearing the fields of fire. The leader checks for cover and concealment and paces the distances for the range card.

d. Select good covered and concealed routes to and from positions. Gullies and reverse slopes are excellent options for protection and ease of movement.

8-3. SELECTION OF FIGHTING POSITIONS

The two main factors in positioning the Dragon include gunner protection and effective use of the weapon's capabilities. The gunner must remain exposed while tracking his targets. When firing at a target at maximum range, he remains susceptible to counterfires for as long as 12 seconds. The gunner avoids positions that would force him to fire into the sun, which could affect his ability to track the target. Many of the steps for protecting gunners also make the most of the Dragon's capabilities. Gunners can enhance mission accomplishment by following some basic rules when selecting positions:

a. Use natural cover and concealment. Choose positions where the terrain provides *cover* from enemy fire and *concealment* from enemy ground and aerial observation (Figure 8-2).

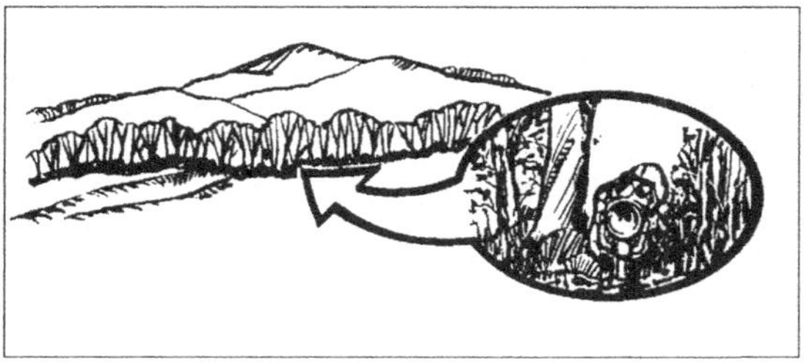

Figure 8-2. Use of natural cover and concealment.

b. Try to choose positions where the gunner can engage the enemy's flank or rear from behind frontal cover. A gunner firing obliquely should do so from beneath the protection of a parapet or natural cover. Dragon missiles probably will not defeat a tank hit in the frontal 60-degree arc. You will also find other armored vehicles easier to kill from the flanks and rear. The enemy has a tougher time tracing the origin of a Dragon round fired from his flank than one fired head-on.

(1) From an oblique weapon position, a gunner can provide frontal cover to protect against direct suppressive fires. An oblique position also helps to conceal his location from the view of any person or vehicle approaching from the front (A, Figure 8-3).

(2) In the attack, a tank orients firepower and observation mainly to the front. This helps keep the tank from detecting the launch site (B, Figure 8-3).

c. Avoid using Dragon positions that must engage targets from the front (especially tank targets).

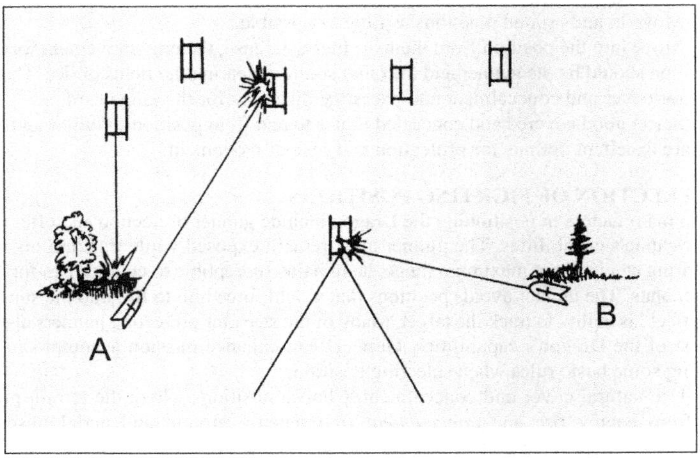

Figure 8-3. Engagement of the enemy with flank or rear shots.

d. Position Dragons so they can provide mutual support. Ideally, position them 300 meters apart. This helps protect the gunners by ensuring continuous coverage of enemy armored vehicles. To do this—

(1) Employ Dragons so their fires interlock with and support other Dragons, TOWs, or tanks. Ensure that sectors of fire overlap. Cover each sector with more than one antiarmor weapon.

(2) Position Dragons to engage any enemy armored vehicles that assault another Dragon, TOW, or tank position.

e. For security, integrate Dragon gunners with nearby infantry. Leaders should provide local security for any Dragon gunner employed away from the squad or platoon such as a Dragon team or armor-killer team.

8-4. PREPARATION OF FIGHTING POSITIONS

The enemy must not learn the true location of the Dragon. Clearing away loose debris behind the launcher, wetting down the backblast area, and covering the ground with shelter halves reduces the Dragon's launch signature (backblast). To prevent detection, soldiers move only in and around the position. As long it can see the target clearly, the unit can use indirect fires (HE, smoke, and WP) and small-arms weapons to distract the enemy. Other deception measures include preparing partly visible dummy positions to draw enemy fire away from the actual positions, then positioning Dragons on less obvious firing positions. The Dragon fighting position offer have unobstructed fields of fire, mask clearance (minimum dead space that could hide targets in the sector), and a clear backblast area. Just as it can do with other weapons organic to the platoon, the unit can employ the Dragon either from hasty or improved positions. Soldiers situate and orient a fighting position to cover a sector of fire.

a. After receiving a sector of fire and firing location from the leader, the gunner constructs the Dragon position to cover the sector. He clears only what he must clear for effective fields of fire. He camouflages the position using available materials and improves the position as time permits.

b. Leaders must consider the backblast and the muzzle blast when employing the weapon. To prepare a Dragon fighting position, the gunner follows these guidelines:

(1) When the gunner fires from an improved position, the muzzle end of the launcher must extend 15 centimeters (6 inches) beyond the front of the hole. The rear of the launcher must extend out over the back of the hole. As the missile leaves the launcher, the unfolding stabilizing fins require at least 15 centimeters (6 inches) of clearance above ground. The position should offer protection to the front (a parapet) or other natural or man-made cover.

(2) Gunners clear the ground in front of and behind the position. They remove rocks, sand, and debris. This prevents a dust cloud from forming when the gunner fires. Dust would obscures a gunner's vision and marks the location for enemy observers. When the gunner must fire in only one direction, a one-person fighting position works best (Figure 8-4, following; and Figure 8-5, page 8-5).

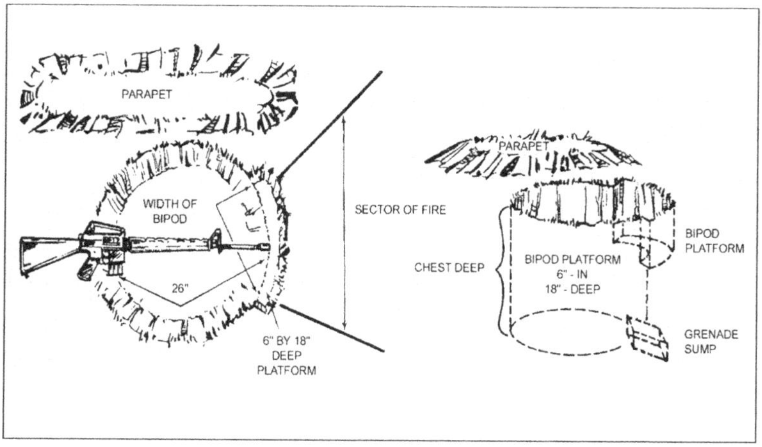

Figure 8-4. Construction of a one-person fighting position.

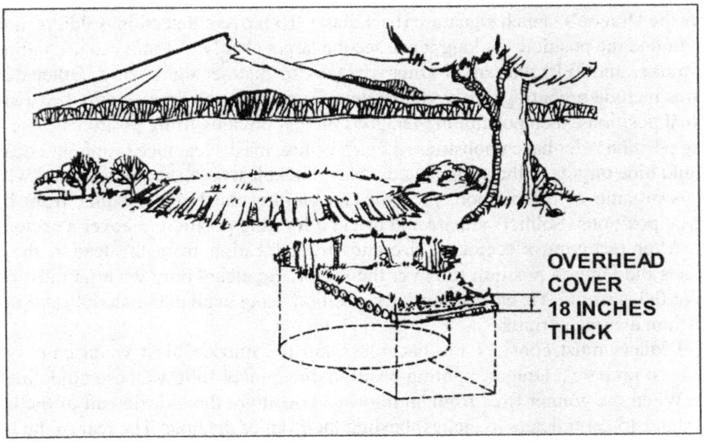

Figure 8-5. Construction of overhead cover.

(3) The gunner should be positioned to fire obliquely. This protects the gunner from frontal fire while he engages the target from the flank. If necessary, the gunner can fire to the front as well as to the oblique from a one-person fighting position.

(4) The wedge shape of the two-person fighting position gives the gunner frontal protection. It also allows the gunners to engage their targets from the oblique or flank. The team constructs the position as follows:

(a) *Trench.* Construct the trench position the length of three M16s, in an inverted "V." Dig it waist-deep. Make the trench waist-wide plus about 15 centimeters (6 inches) (A, Figure 8-6, page 8-6).

(b) *Front Parapet.* Construct the front parapet as long and as wide as the length of an M16. Build it up until it measures two helmets high (B, Figure 8-6, page 8-6). Build the front parapet in front of the trench.

(c) *Grenade Sump.* Make a grenade sump as long as an entrenching tool, and as wide as its blade. Dig the floor of the main trench such that it slopes gently downward from each end toward the center of the position, and so that it slopes gently downward from the rear to the front (C, Figure 8-6, page 8-6).

(d) *Overhead Cover.* Construct overhead cover at each end of the trench large enough to protect one soldier and extra rounds. It should measure 31 centimeters deep by 1 meter wide (12 inches deep and 3 feet wide). It should extend 46 centimeters (18 inches) over each side (D, Figure 8-6, page 8-6).

(e) *Flank Parapet.* Construct a flank parapet at each end of the trench. The width of each should measure the same as the length of an M16, as high as two helmets, and a length sufficient to provide good flank protection. To increase overhead protection, build flank parapets are built on top of the overhead cover (E, Figure 8-6, page 8-6).

(f) *Bipod Trench.* Dig a bipod trench for each sector of fire. The back of the bipod trench should measure 10 to 15 centimeters (4 to 6 inches) forward of the main trench. Make the bipod trench two helmets long, one helmet wide, and 15 centimeters (6 inches) deep (F, Figure 8-6).

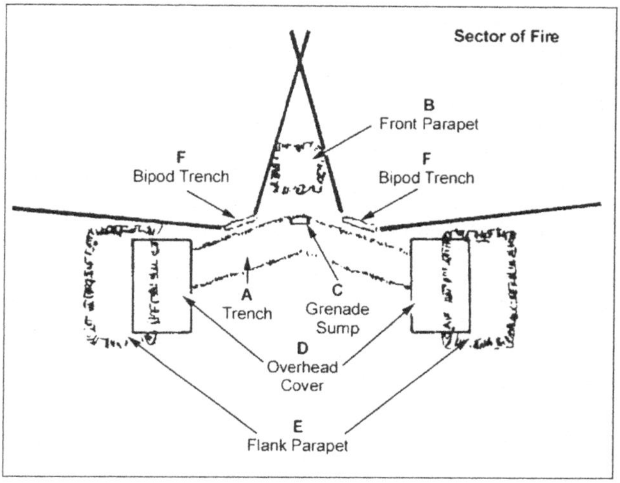

Figure 8-6. Construction of a two-man fighting position.

(g) *Front Cover*. Sometimes the gunner can fire only in one direction. If so, construct front cover so the gunner should engage targets from the flank (G, Figure 8-6). Also construct cover and concealment from other directions (Figure 8-7).

Figure 8-7. Fire in one direction.

8-5. PREPARATION OF RANGE CARDS

A range card consists of a sketch of the terrain that a specific weapon system covers. The range card contains information that helps in planning and controlling fires, in quickly detecting and engaging targets, and in orienting replacement personnel or units. Using a range card, a gunner can quickly find the correct information he needs to engage targets. In order to engage targets rapidly in all visibility conditions, gunners need range cards. They also need them so another soldier could continue the mission if the gunner can no longer fire. For this reason, after he prepares the Dragon for firing, the gunner prepares a range card in duplicate for each position. That is, for each position, he makes one to keep at the position and another for the leader. The two types of range cards are standard (DA Form 5517-R) and field-expedient.

a. **Information Provided on All Range Cards**. All range cards must show the following:

- Weapon symbol, position, or both.
- Sector of fire.
- Maximum engagement line.
- Range and azimuth TRPs.
- Dead space.
- Distance and azimuth from a known point (gunner reference point).
- Magnetic north arrow.
- Data section.

b. **Standard Range Card**. Once the leader provides the necessary information, the gunner prepares a standard range card in duplicate (Figure 8-8, page 8-8).

c. **Field-Expedient Range Card**. Because he may not be able to find any standard range cards in a combat situation, the gunner can draw one on anything available. He prepares a field-expedient range card the same as he would any other range card, except that he uses the weapon symbol to show only the location of the weapon system (Figure 8-9, page 8-9).

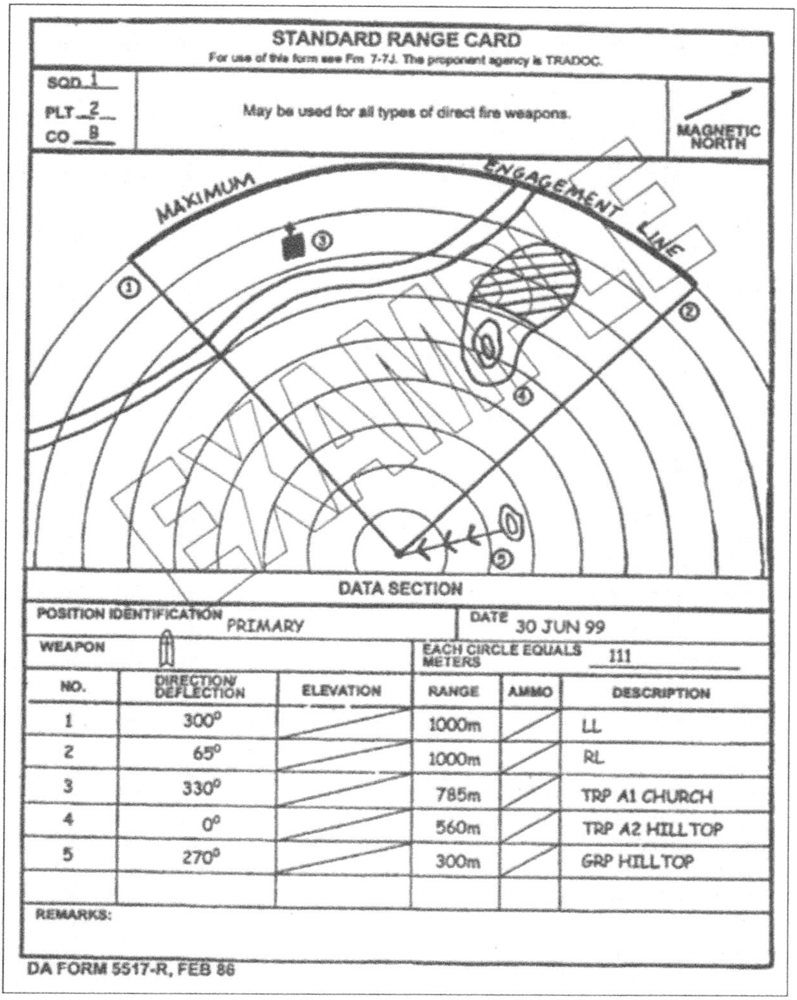

STANDARD RANGE CARD

For use of this form see FM 7-7J. The proponent agency is TRADOC.

SQD 1
PLT 2
CO B

May be used for all types of direct fire weapons.

MAGNETIC NORTH

DATA SECTION

POSITION IDENTIFICATION PRIMARY

DATE 30 JUN 99

WEAPON

EACH CIRCLE EQUALS 111 METERS

NO.	DIRECTION/ DEFLECTION	ELEVATION	RANGE	AMMO	DESCRIPTION
1	300°		1000m		LL
2	65°		1000m		RL
3	330°		785m		TRP A1 CHURCH
4	0°		560m		TRP A2 HILLTOP
5	270°		300m		GRP HILLTOP

REMARKS:

DA FORM 5517-R, FEB 86

Figure 8-8. Example completed DA Form 5517-R, Standard Range Card.

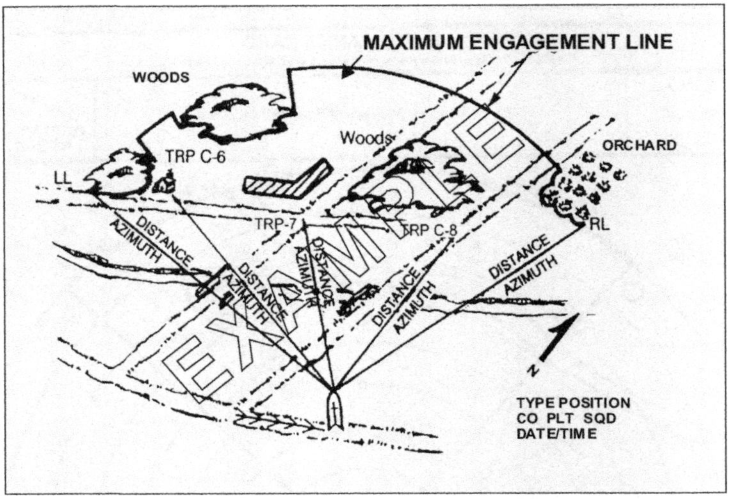

Figure 8-9. Example field-expedient range card.

8-6. TARGET IDENTIFICATION

The activities, locations, or signatures (visual or otherwise) of potential targets identify them as enemy. Dragon gunners must receive sufficient training to recognize the sizes, shapes, and thermal images of all types of targets. Turrets and main guns offer the most recognizable identifiers.

a. Friendly foreign units may operate with or near US forces. This complicates the task of identifying friendly vehicles. To reduce the chance of engaging an allied vehicle, the commander can establish target priorities. Antiarmor gunners then engage only specific *types* of enemy vehicles. Which type the commander tells them to engage depends on the enemy situation. Dragon gunners must be able to perform the following in sequence:

(1) Determine whether a vehicle is tracked or wheeled.

(2) Determine whether a vehicle is friendly or enemy.

(3) If an enemy vehicle, use Table 8-1, page 8-10 to determine its type.

(4) From the type of vehicle, identify the type of enemy unit. This aids the gunner, because each type unit has a unique organization and target value to the gunner, S2, and intelligence community.

(5) State the nomenclature of the vehicle.

b. Most weapons and vehicles produce telltale signatures. For example, most tracked vehicles use diesel fuel, which emits a large amount of black smoke. Tracked vehicles make more noise than wheeled vehicles. Antiarmor units can use these and other signatures to help them locate and identify enemy targets.

RECON	APC	IFV	TANK
• AML-90 • BRDM-1 • BRDM-2 • EE-9 • M-93 Scorpion	• BTR-50 • BTR-60 • BTR-70 • BTR-80 • LAV-25 • M-113 • MT-LB • PRC TYPE 63/YW 531 • V-150	• BMP-1 • BMP-2 • BMP-3 • BRM • BMD • M2A2 Bradley • VTT-323 (M1973)	• Centurion • Challenger • Chieftain • M1A1 Abrams • M60 • M1985 (NK) • PT-76 • T-54 • T-55 • T-62 • T-64 • T-72 • T-80

Table 8-1. Vehicles categorized by function.

c. Dragon gunners mainly detect target signatures by sight, sound, and smell. If they detect anything suspicious or unusual, they should thoroughly investigate it. Sun shining off a flat surface, such as off a windshield, sounds of diesel or turbine engines, or the clanking or squeaking of end connectors can indicate the locations of targets.

(1) ***Soldier Signatures.***

- Fighting positions.
- Trash.
- Cut or missing vegetation (cleared for fields of fire or camouflage).
- Freshly dug earth (may indicate a fighting position).
- Noise from equipment or talking.
- Light from a match, cigarette, or fire.

(2) ***Tracked Vehicle Signatures.***

- Large dust clouds.
- Diesel smoke.
- Noise made by tracks and engine.
- Vehicle tracks on the ground.
- Distinctive silhouette or shape.

(3) ***Antitank Weapon Signatures.***

- "Swish" of missile launch.
- Long, thin wires in brush, trees, or along the ground.
- Dismounted soldier looking through a periscope-type device. (Launcher could be up to 100 meters away from him.)

(4) ***Aircraft Signatures.***

- Reflection of the sun from aircraft canopies and rotor blades.
- Vapor trails.

- Dust and movement of foliage caused by a hovering helicopter.
- Sound of a turbine engine (high-pitched whirring sound).

(5) ***Obstacles and Mines.***

- Loose dirt or dirt that has been disturbed in a regular pattern.
- Areas where large trees have been removed.

8-7. TARGET RECOGNITION BY TYPE

Learning to recognize targets by type presents little challenge. However, identifying them as friendly or enemy requires careful study and attention to detail. Tanks are most difficult, because many friendly and enemy tanks share many design similarities. When camouflaged and moving at a distance of 1,500 to 2,000 meters, the gunner may not detect the differences. Soldiers must know which friendly and threat armored vehicles most likely will appear on the battlefield (STP 21-1-SMCT, available in Reimer's Digital Library). Soldiers can use training aids, such as GTA 17-2-13, to study the armored vehicles of other nations (Figure 8-10, page 8-12). To identify most armored vehicles or tanks, the gunner considers the type, location, and absence or presence of certain equipment. Specifically, he looks at the suspension system, turret, and main gun. However, he should remember that, just like friendly forces, the enemy also uses camouflage and deception.

a. **Suspension System**. Vegetation and terrain often conceal a vehicle's suspension system, which makes it the least identifiable part of the vehicle. If the gunner can see it, he can distinguish a suspension system by its—

- Road wheels and support rollers.
- Road wheels only.
- Number of road wheels.
- Spacing between road wheels.
- Armored skirt.

b. **Turret**. The gunner distinguishes turrets by these characteristics—

- Position on the hull: well forward, center, or to the rear.
- Presence, absence, or location of a searchlight.
- Shape of the turret: rounded, elongated, or boxy.
- Externally mounted storage racks and other equipment.

c. **Main Gun**. The gunner distinguishes tank main guns by looking for—

- A bore evacuator on the gun tube.
- A muzzle brake or blast deflector.
- The presence or absence of a thermal jacket.

d. **Commander's Station (Some Tanks*)***. Normally a simple hatch or cupola, the commander's station projects from the top left or right side of the commander's turret.

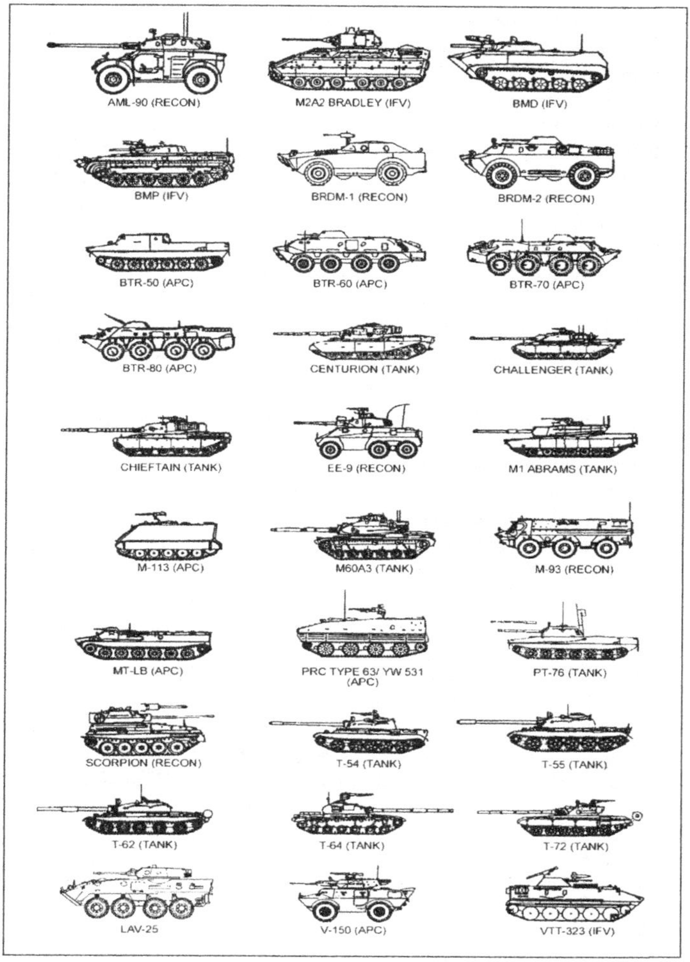

Figure 8-10. Differences between armored vehicles.

8-8. THERMAL IDENTIFICATION

Using the AN/TAS-5 to identify targets by their thermal signature presents a challenge. Doing it successfully requires extensive training. Appendix I discusses in detail how to identify targets by their thermal signatures. Gunners use the same steps to identify vehicles thermally as they would to identify them through the daysight.

Section II. TARGET ENGAGEMENT

The Dragon gunner should engage the enemy within his own capabilities and the capabilities of the weapon.

8-9. ENGAGEMENT RANGE

At 1,000 meters, targets viewed through the daysight look about the size of a postage stamp. Engaging a target while viewing its thermal image in the nightsight presents more difficulty.

a. The Dragon's best engagement range for moving targets falls between 200 and 800 meters, for two reasons:

(1) The shorter flight time reduces the gunner's vulnerability to enemy counterfires.

(2) At closer ranges, the target appears larger in the sights.

b. The gunner's skill in tracking and hitting the target reflects his level of proficiency. Before he fires a Dragon missile at an enemy target, the gunner must determine whether he can engage the target. He can do so if—

- It falls within engagement range.
- Its exposure allows the gunner to identify and track it.
- Its continued exposure gives the missile time to reach the target.

8-10. RANGE DETERMINATION

The Dragon gunners use the stadia lines in the daysights and nightsights to determine if a target falls within range. Moving and stationary vehicles may present flank, oblique, and frontal or rear targets (Figure 8-11). At maximum range (1,000 meters), A 6-meter (20-foot) long target completely fills the area between the stadia lines and exceeds the stadia lines at a closer range.

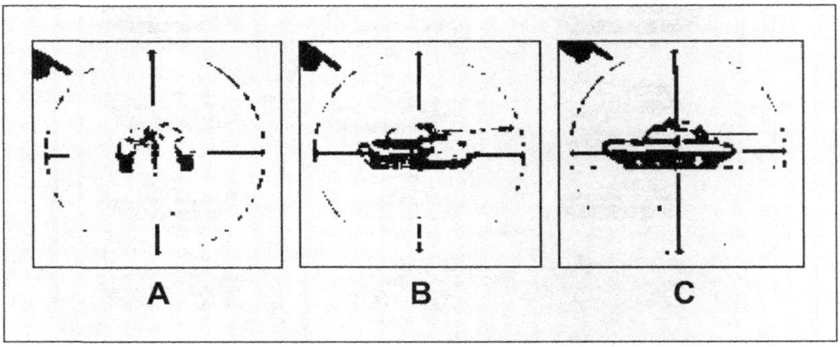

Figure 8-11. Frontal, oblique, and flank targets.

a. **Flanking Targets (Full Stadia)**. Adjust the sight picture by moving the launcher so the target centers between the stadia lines (Figure 8-12, page 8-13).

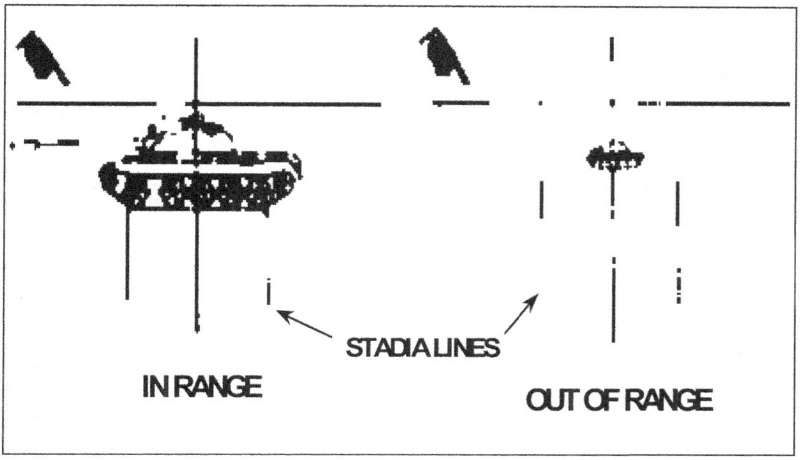

Figure 8-12. Range determination for flank target.

b. **Oblique Targets**. If you can see more of the flank, use the full-stadia method (Figure 8-13). If you see more of the front or rear, use the half-stadia method (Figure 8-14, page 8-15).

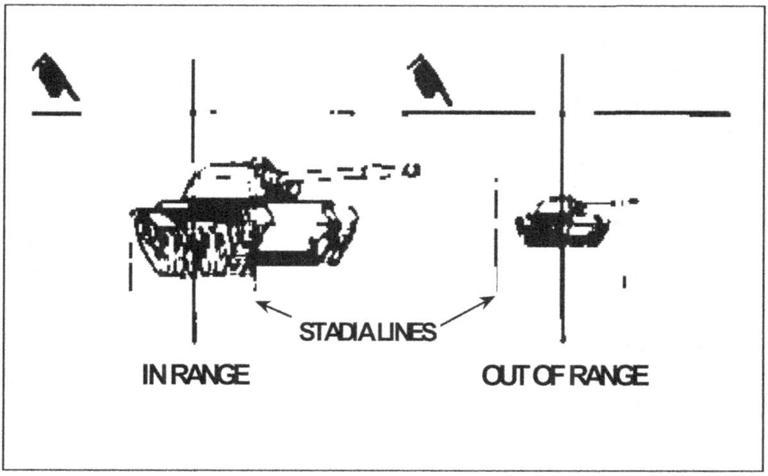

Figure 8-13. Range determination for oblique target, more flank visible.

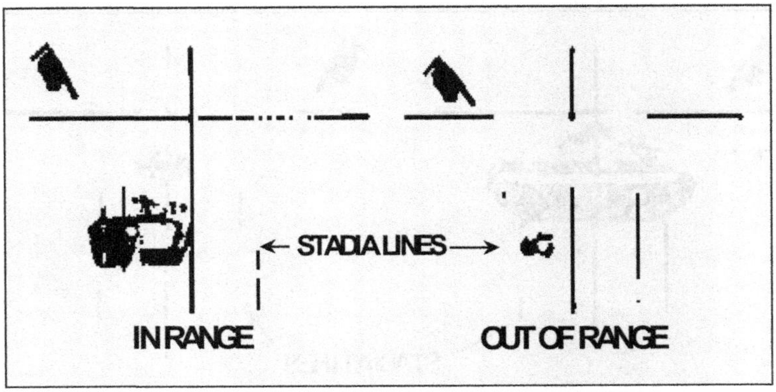

Figure 8-14. Range determination for oblique target, more front or rear visible.

c. **Frontal (Head-On) or Rear (Going Away, Half-Stadia) Targets**. Adjust the sight picture by moving the launcher to align the vertical cross hair and one of the stadia lines on the target (Figure 8-15).

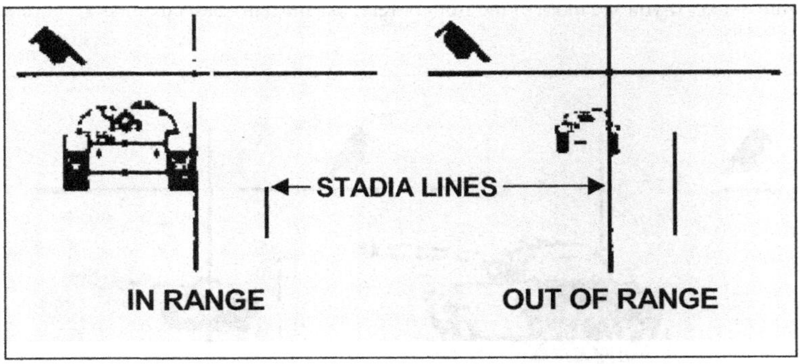

Figure 8-15. Range determination for frontal or rear target.

8-11. TIME AND SPACE

The gunner uses the time or space factor to determine whether he can engage the target.

a. The gunner looks through his sight to determine whether the missile can hit a moving target before it can find a covered position. He aims the sight at a point directly in front of the moving target. With a target moving at a speed of 35 KPH or less, if the gunner sees no obstructions or covered areas on the apparent path of the target, he can destroy the target (Figure 8-16, page 8-16).

Figure 8-16. Target able or unable to reach cover in time.

b. If the gunner places the cross hairs center of mass and fires at once, the target will not reach the protection of the hill in time (Figure 8-17). If the gunner does fire—the target will reach the protection of the hill before the missile hits (Figure 8-18).

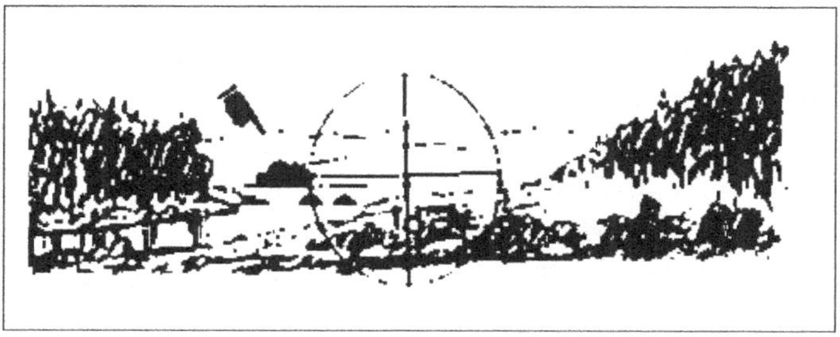

Figure 8-17. Target unable to reach cover in time.

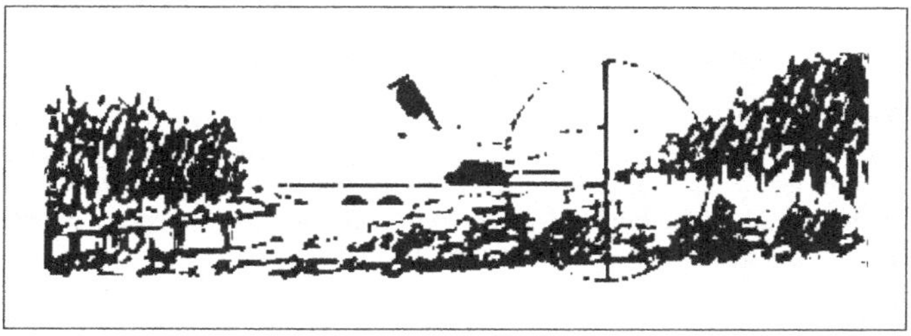

Figure 8-18. Target able to reach cover in time.

8-12. VULNERABLE POINTS OF ARMORED VEHICLES

After the gunner decides that he can engage a target, he should try to hit it at its weakest points. Leaders can help by positioning Dragons where the gunners can take advantage of those weak points.

a. To do so, the gunners must *know* the weak points of each type of enemy armored vehicle. An enemy armored vehicle usually has the most armor protection on its front glacis (slope). It has less on its flanks, and still less on its top, back, and belly. Any armored vehicle's weakest areas include its internal fuel tanks, ammunition storage areas, and engine. Destroying the engine not only immobilizes the vehicle, but may ignite ruptured fuel lines, causing a fire or explosion.

b. Gunners have a better chance to get a kill with oblique or flanking shots. They must keep themselves ready for opportunities. For example, a potential target may expose its flank when it tries to bypass an obstacle or evade an oncoming ATGM. Gunners produce more mobility kills with flank and oblique shots. However, they can do so even if they only hit the vehicle's wheels, track, or suspension system (Figure 8-19). Flank shots deflect off armor less often than oblique shots. The top or bottom (belly) of an enemy armored vehicle may show briefly, while the vehicle breaches an obstacle or antitank ditch, fords a river with steep banks, or traverses a shallow valley. By carefully analyzing the terrain in the assigned sector of fire, the gunner can determine where approaching armor units will expose their weaker armor.

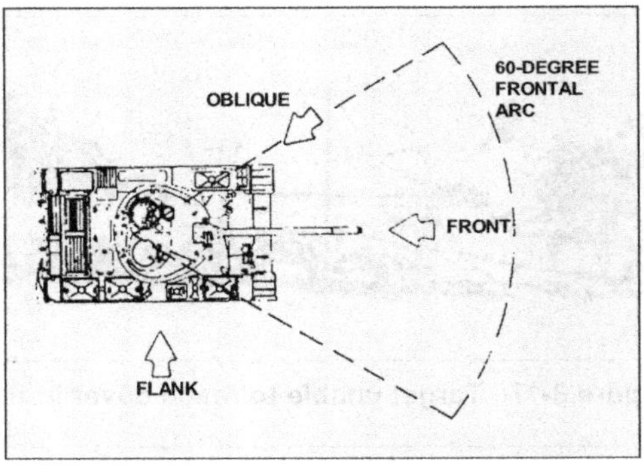

Figure 8-19. Flank and oblique target shots.

8-13. FIRE CONTROL

Effective Dragon fire requires well-planned and executed fire-control measures. Proper fire control ensures that Dragons engage targets at the best possible times. Leaders should never endanger Dragon gunners by ordering them to fire prematurely. Firing at a target traveling out of range offers little chance of success. So would firing when, for any reason, a target offers little chance of a first-round kill. Leaders should employ only the Dragon best sited to hit the target. Lack of fire control reduces a unit's antiarmor capability. Under ideal conditions, Dragons dispersed as far apart as 1,600 to 2,000 meters can concentrate their fires on the same group of targets (Figure 8-20).

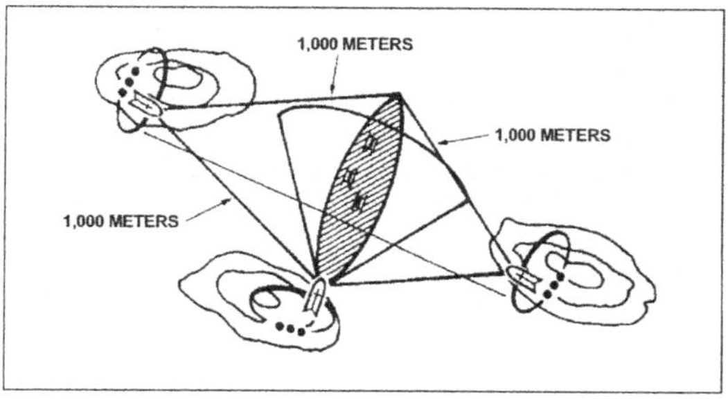

Figure 8-20. Dragon dispersion.

a. **Fire Control Methods**. Effective fire control methods prevent the wasteful firing of more than one Dragon missile at the same target, and they prevent premature firings. Unnecessarily firing a Dragon may disclose the location of the defensive position to the enemy. Dragon fire control methods include sectors of fire, TRPs, engagement priorities, fire patterns, and fire commands.

b. **Sector of Fire**. A sector of fire refers to an area limited by boundaries and assigned to a unit or weapon to cover by fire (Figure 8-21). The gunner or unit assigned a sector of fire may fire only within that sector. Leaders make sure that sectors overlap in order to cover all areas and so that the Dragon gunners do not hesitate when enemy vehicles come into range.

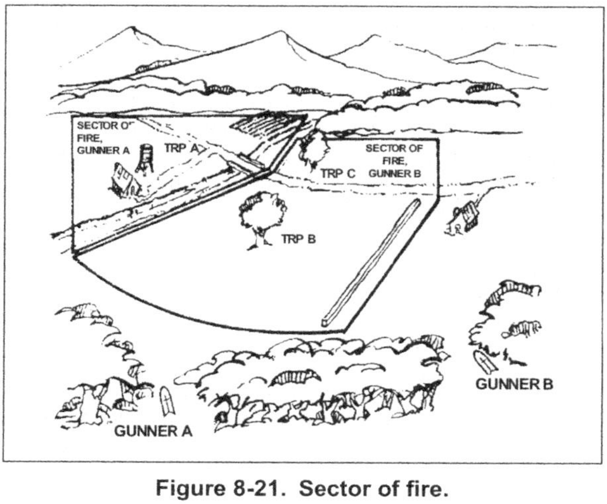

Figure 8-21. Sector of fire.

c. **Target Reference Point**. By definition, a TRP identifies an easily recognizable, natural or man-made point on the ground. Each gunner in the unit's sector must know how to identify all TRPs visually or by locating them on a map. TRPs offer a great way to define visually a gunner's sector of fire.

d. **Engagement Priorities**. Engagement priority means the order in which Dragons engage the various types of vehicles in an enemy formation. Because many enemy tanks have improved armor, Dragons have the best chance of killing command and control vehicles and APCs. If the unit has TOWs and tanks, it should use those weapons to kill enemy tanks. The warning or OPORD lists the engagement priority, which includes as its main priority the completion of the mission. The engagement priority (Figure 8-22) lists C2 vehicles first, because the enemy relies heavily on its leaders for combat operations. Eliminating leadership will cause confusion and disrupt the enemy's mission.

1. Command and control vehicles.
2. Armored personnel carriers.
3. Antiaircraft vehicles.

Figure 8-22. Example of engagement priority.

e. **Fire Patterns**. Fire patterns describe the relationship between Dragons and their targets. Firing patterns help leaders control gunners. They also give gunners specific targets to focus on. Two basic patterns include cross fire and depth fire. Leaders should not fixate on one, but should remain flexible and change the pattern as needed. Within the unit's sector, the leader could use both patterns at the same time.

(1) ***Cross Fire***. Leaders use cross fire with targets dispersed laterally or when obstructions prevent the Dragon from firing to the front. Gunners must stagger their cross fire. That is, when the first missile hits, only then may the next gunner fire. Cross fire prevents Dragon gunners from picking up the infrared flare from, and thus controlling, each other's missiles instead of their own. Cross fire increases the chance of a kill.

(a) *Flanking Targets*. With flanking targets, leaders have each gunner engage the target at a diagonal to his position.

(b) *Frontal Targets*. With a frontal target, that is, a target moving straight at the Dragon, cross fire helps prevent ***detection***.

(c) *Follow-Up*. As they destroy their targets, Dragon gunners shift their fire to the center of the enemy formation (A, Figure 8-23, page 8-20).

(2) ***Depth Fire***. Leaders use depth fire when targets are exposed in depth. Dragons on one side engage the nearest targets, while Dragons on the other side engage the farthest targets. The gunners then shift their fire toward the center of the formation. This can be specified by unit SOP or in the section leader's order (B, Figure 8-23, page 8-20).

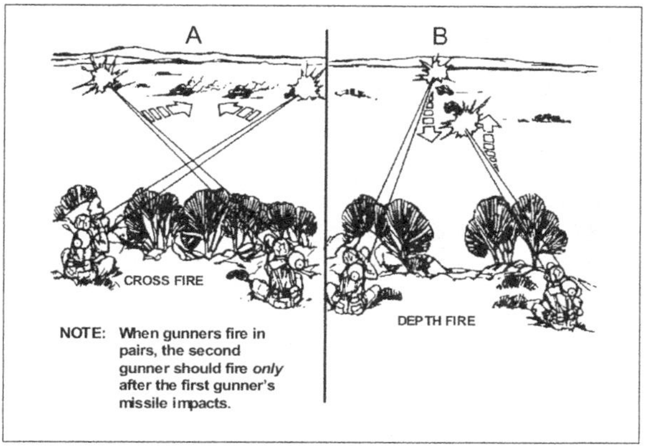

Figure 8-23. Fire patterns.

f. **Fire Commands**. The leader can use fire commands to control the gunner's rate of fire, time of fire, and point of fire. However, when distance dictates radio delivery of fire commands, the gunner will have to divide his concentration between listening to the radio and firing the Dragon. His assistant, if he has one, can listen to the radio and relay the radio commands (Table 8-2).

ELEMENT	DEFINITION	EXAMPLE
ALERT	Warns the gunner(s) of a fire mission.	"Enemy in sector, prepare to fire."
TARGET DESCRIPTION/ LOCATION	Briefly describes the target location relative to a TRP or the gunner	"BMP, 200 meters right of TRP 1."
METHOD OF ENGAGEMENT	Provides instructions for engaging a formation — fire cross or depth fire, fire front to rear, fire left to right or right to left, and so forth.	"Team Alpha, fire from rear to center." "Team Bravo, fire from front to center."
EXECUTION	Gives the command to fire.	"Team Alpha, fire." "Team Bravo, stand by."

Table 8-2. Fire commands and examples.

g. **Emergency Fire Signals**. How would the leader control fires at a distance without radio? The next quickest option includes using sectors of fire and alternate signals with pyrotechnics. Units must establish SOPs and practice what communications procedures to use when they lose communications.

h. **Range Cards**. In some situations, such as when moving from position to position to cover the movement of an advancing force or during retrograde operations, preparing and using range cards does not work. Then, TRPs offer the best way to control the distribution of fire. However, if time and circumstances permit, gunners should prepare range cards for primary, alternate, and supplementary positions. These valuable tools help the gunner engage targets successfully.

8-14. LIMITED VISIBILITY

The gunner has little opportunity to engage targets in darkness, haze, smoke, or fog. The platoon or company must have a simple and quick procedure or SOP that will help him conduct quick limited visibility engagements and that includes provisions for—

a. **Target Acquisition**. The gunner can find targets using his sight, binoculars, NVDs, or a combination of these.

b. **Communications**. The platoon leader ensures he has the communications means to issue target acquisition data and fire commands rapidly. These means must allow him to disseminate SOPs, signals, and fire commands. In a static position, he can "hot loop" all essential personnel on wire communications.

c. **Illumination**. When natural and battlefield light sources, such as burning vehicles, provide too little light to illuminate the target, the gunner uses the AN/TAS-5 nightsight. Leaders may have responsive artificial sources of illumination they can dedicate to supporting antiarmor fires. Two examples of illumination available to platoon leaders for marking TRPs at night include artillery and mortar illumination. (Chapter 1 discusses AN/TAS-5 operational capabilities.)

APPENDIX A

TRAIN-THE-TRAINER TASKS FOR THE DGT AND DFTT

Table A-1 lists train-the-trainer tasks for the Dragon gunnery trainer (DGT) and for the Dragon field tactical trainer (DFTT).

TASK	CONDITIONS	STANDARD
Prepare the DGT for Operation	Given a DGT and a 110-volt AC power supply.	Assemble the DGT.
Program the DGT for Operation	Given a DGT, the correct program disks, and a 110-volt AC power supply.	Program the DGT for the correct firing table.
Operate the DGT	Given a DGT, the correct program disks, and a 110-volt AC power supply.	Operate the DGT using the correct firing table.
Evaluate Gunner Performance	Given a DGT, the correct program disks, and a 110-volt AC power supply.	Operate the DGT using the correct firing table. Provide the gunner with an after-action review of his tracking and score.
Conduct Shutdown Procedures for DGT	Given a DGT, the correct program disks, and a 110-volt AC power supply.	Shut down the DGT.
Maintain the DGT	Given a DGT, a 110-volt AC power supply, and the proper cleaning materials.	Perform maintenance on the DGT.
Prepare the DFTT for Operation	Given a DFTT and either a 24-volt DC or a 120- or 240-volt AC power supply.	Assemble the DFTT.
Operate the DFTT	Given a DFTT and either a 24-volt DC or a 120- or 240-volt AC power supply.	Operate the DFTT using the correct firing table.
Evaluate Gunner Performance	Given a DFTT and either a 24-volt DC or a 120- or 240-volt AC power supply.	Operate the DFTT using the correct firing table. Provide gunner with an after-action review of his tracking and score.
Conduct Shutdown Procedures for DFTT	Given a DFTT and either a 24-volt DC or a 120- or 240-volt AC power supply.	Shut down the DFTT.
Maintain the DFTT	Given a DFTT and either a 24-volt DC or a 120- or 240-volt AC power supply.	Perform maintenance on the DFTT.

Table A-1. Train-the-trainer tasks for the DGT and the DFTT.

APPENDIX B

DRAGON GUNNERY TRAINER

This appendix provides technical data for the Dragon gunnery trainer (DGT) and discusses its components and limitations. The DGT and related accessories are shipped and stored in two shipping containers, one for the instructor station and one for the student station (Figure B-1).

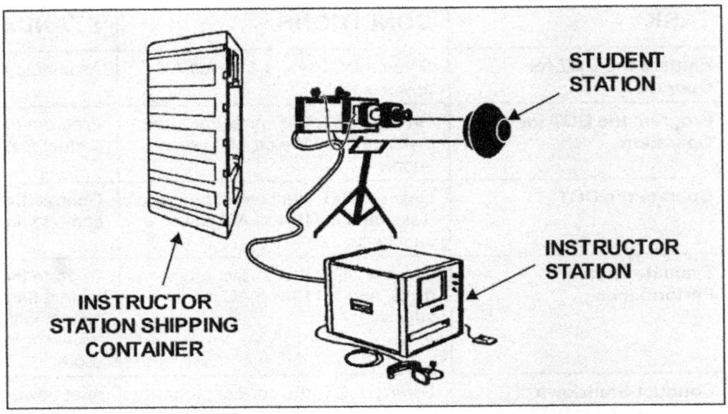

Figure B-1. Dragon gunnery trainer.

Section I. INSTRUCTOR STATION

The instructor station consists of a console, a headset with microphone, a mouse, two power cables, and a shipping container (Figure B-2).

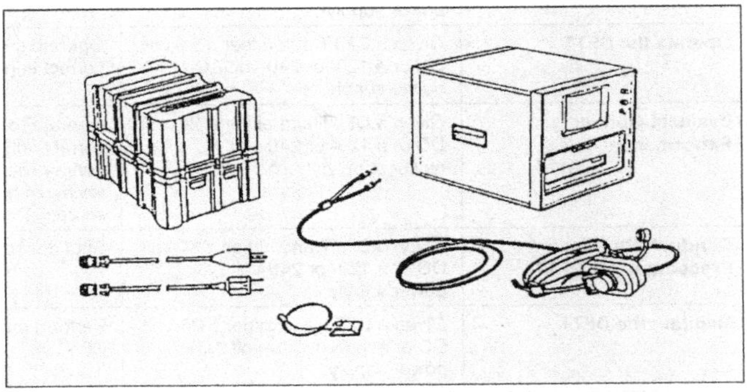

Figure B-2. Components of instructor station.

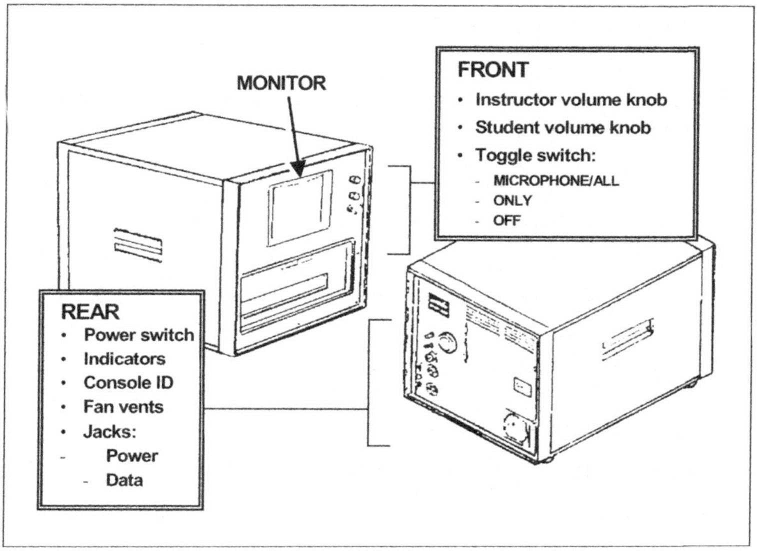

Figure B-3. Instructor console.

B-1. INSTRUCTOR CONSOLE

The trainer programs the DGT instructor console (Figure B-3), which has two carrying handles.

a. **Monitor**. The instructor console displays the mission scenario on its own monitor as well as on the student viewer. This allows the trainer to watch the student's actions during mission execution. Afterwards, the trainer can play back and evaluate the student's performance.

b. **Front Panel**. From the front panel of the instructor console, the trainer controls the volume of his own and the student's headsets (INSTRUCTOR VOLUME and STUDENT VOLUME). The toggle switch on the front panel has three settings:

(1) ***MICROPHONE/ALL***. This setting allows both trainer and student to hear the DGT mission.

(2) ***ONLY***. This setting allows only the trainer to hear the mission.

(3) ***OFF***. This setting is self-explanatory.

c. **Rear Panel**. The rear panel of the instructor console contains the power switch, indicators, console identification data, fan vents, and power and electronic data transmission receptacles (jacks).

(1) ***ON/OFF Switch and Indicator***. The ON/OFF switch controls power to the instructor console; when the switch is set to ON, the indicator light comes on.

(2) ***Hour Meter Indicator***. The hour meter shows how long the trainer has been in use.

(3) ***Name Plate***. The nameplate identifies equipment.

(4) ***Fan Assembly***. The fan assembly filters and circulates air through the instructor console.

(5) ***Receptacles***. The receptacles (jacks) shown in Table B-1 are located on the rear panel of the instructor console:

J01	Data output to the printer.
J02	Signal interface with the student station sight assembly.
J03	Power input from a 110VAC or 220VAC source.
J04	Power output to the student station sight assembly.
J05	Data input from the mouse.
J06 **J07**	Audio signal input from and output to the headset/microphone (phone jack).
J08	Power output to the printer.

Table B-1. Receptacles (jacks) on rear panel of instructor console.

d. **Media**. Each DGT comes with its own set of four floppy disks (3.5 inch) and two videodisks (12 inch) (Figure B-4).

(1) ***Floppy Disks***. These are standard 3.5-inch floppy disks.

(a) *Program Disk*. This disk contains the DGT simulation software.

(b) *Diagnostic Disk*. This disk contains the self-test for the DGT.

(c) *Mission Data Disks*. Each of these two numbered floppy disks corresponds to one of the similarly marked 12-inch videodisks.

(2) ***Videodisks***. Two 12-inch videodisks store the DGT's audiovisuals. Each is numbered and color-labeled like one of the two mission data disks. Each videodisk comes in a protective jacket.

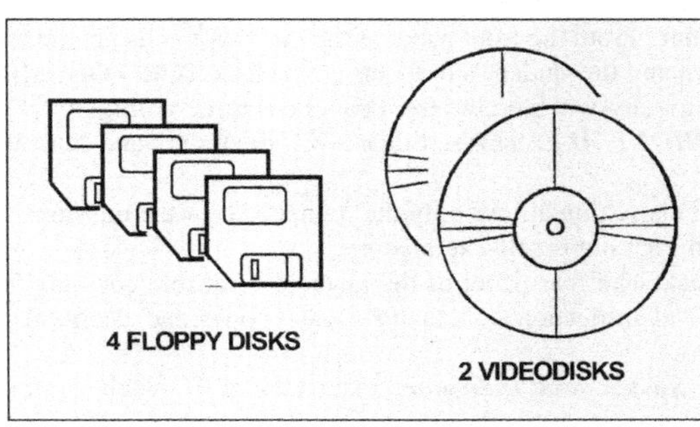

Figure B-4. DGT media.

B-2. HEADSET-MICROPHONE

The instructor uses the standard communication headset with boom microphone (Figure B-5). This text will simply refer to "the headset." The headset allows the trainer to hear the audio portion of the simulation and to feed back instructions to the student. The headset's 6-foot long cable has two phone plugs that plug into jacks on the instructor console.

B-3. MOUSE

The mouse (Figure B-5) is a hand-controlled device the trainer uses to move the cursor on the video monitor. Using the mouse, the trainer controls the cursor, selects options, and controls exercises. To select an option, the instructor positions the cursor on it and clicks the left mouse button once (the right button does not work with this program). The connector on the mouse's 4-foot long cable plugs into a jack on the instructor console.

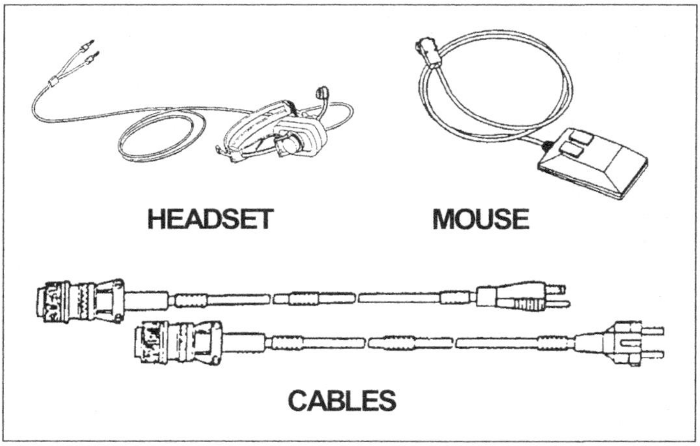

Figure B-5. Headset, mouse, and cables.

B-4. POWER CABLES

The DGT comes with two power cables (Figure B-5): one for use with 110VAC power (cable W2) and the other for use with 220VAC power (cable W1). Each is 8 feet long.

B-5. INSTRUCTOR STATION SHIPPING CONTAINER

Instructor station shipping containers (Figure B-6, page B-5) are made from olive drab-colored polyethylene. Soldiers can stack these containers four deep. Each has specially shaped inserts to prevent damage to the components. The shipping container resists dust and water. The base and cover each have two carrying handles, one on each end. The container has four latches on each side and three latches along each end.

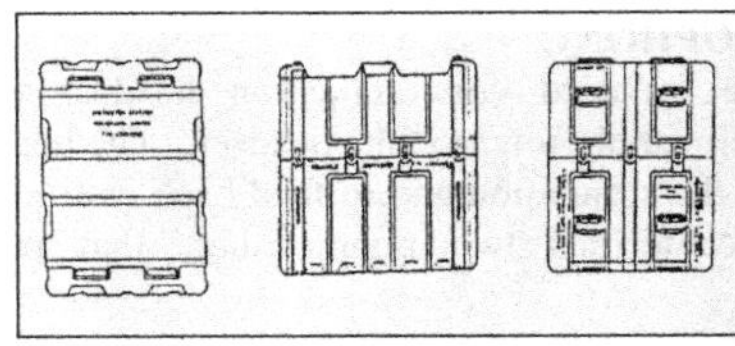

Figure B-6. Instructor console shipping container.

a. The container bears the stenciled message on each end in two places: CAUTION, 198 POUNDS, 4-MAN LIFT (MALE ONLY).

b. The shipping container has an identification number and a part number stenciled on the right side of the base along with an arrow icon: REUSABLE CONTAINER P/N 13363801.

c. The instructor station identification and part number are stenciled on the top of the container cover: INSTRUCTOR STATION DRAGON/DRAGON INDOOR P/N 13363800. This part number includes the instructor station and shipping container.

d. The caution HANDLE WITH CARE, REUSABLE CONTAINER is stenciled on both ends of the base.

Section II. STUDENT STATION

The student station consists of the Dragon unit, the weight assembly, the student headset, a special-purpose cable, the student station's shipping container, disks, and an operator's manual (Figure B-7).

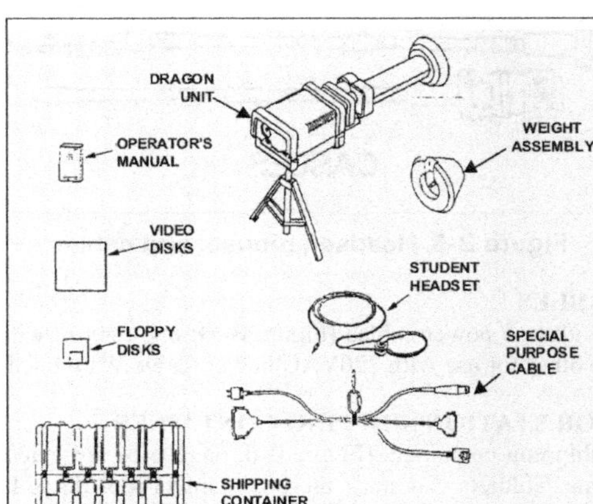

Figure B-7. Student station components.

B-6. DRAGON UNIT

The Dragon unit (Figure B-8) consists of a sight assembly, bipod assembly and retaining strap, and launcher assembly. The student views the mission through the sight assembly. The retaining strap secures the bipod assembly during storage. The bipod assembly supports the launcher and sight assemblies. The launcher assembly simulates the Dragon launch tube.

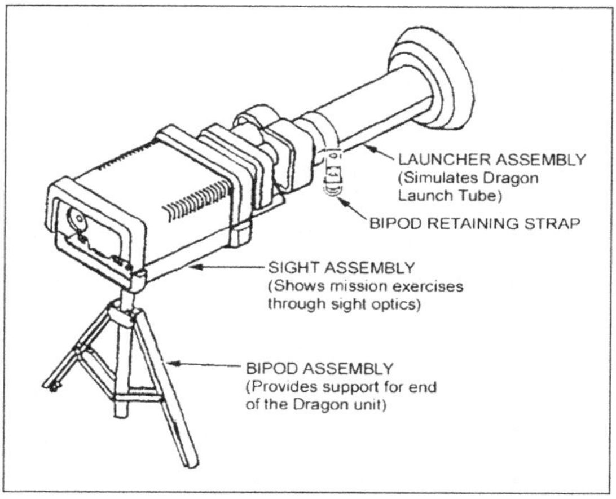

Figure B-8. Dragon unit.

a. **Sight Assembly**. The DGT sight assembly (Figure B-9) operates as either the daysight or nightsight, depending on the mission loaded at the instructor console. The sight displays the mission visual signals. The sight assembly attaches to the Dragon launch assembly.

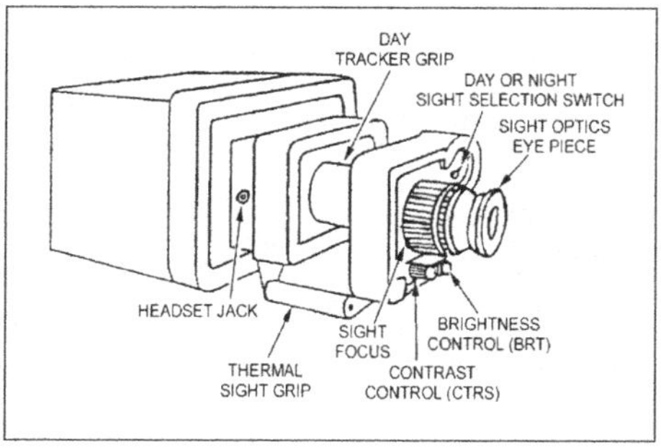

Figure B-9. Sight assembly.

(1) The sight assembly has a daysight grip, a switch to select the daysight or nightsight, an eyepiece, a brightness control (BRT), a contrast control (CTRS), a sight focus ring, a thermal sight grip, and a headset jack. The eyepiece simulates both the daysight and nightsight eyepieces.

(2) On its right side (Figure B-10), the sight assembly has a safety and a right-hand trigger grip. Before the student can pull the trigger, he must first depress and hold the safety with his thumb.

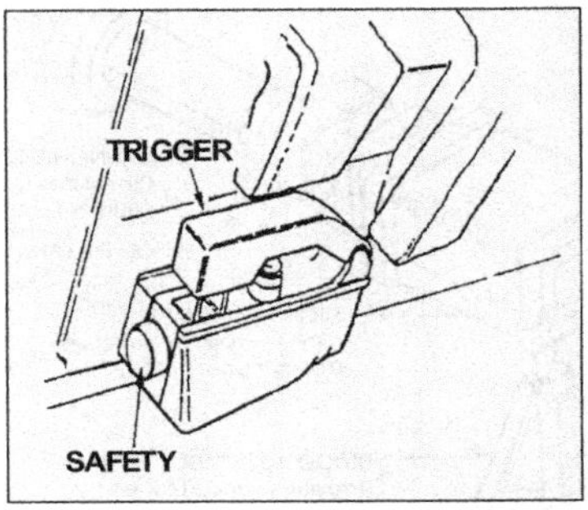

Figure B-10. Sight assembly—right side.

(3) The sensor port, power connectors, video receptacle, and recessed analog, digital, and audio jacks are located on the rear panel of the sight assembly (Figure B-11). The trainer may record launch tube movements from the reflector assembly via a sensor port. A CAUTION label designates the sight assembly as a sensitive electronic device.

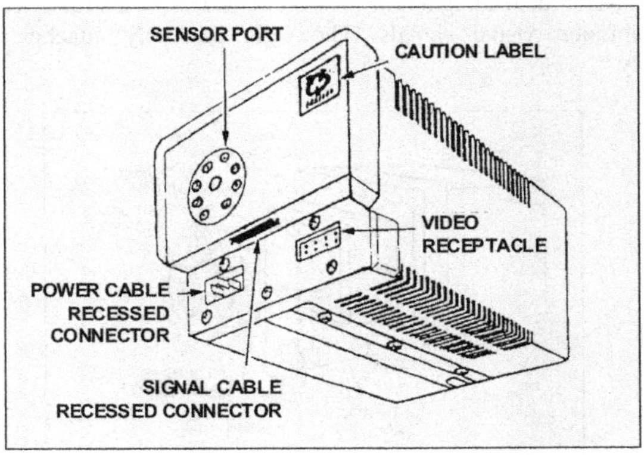

Figure B-11. Sight assembly—rear panel.

b. **Bipod Assembly**. The bipod assembly (Figure B-12), which supports the launcher and sight assemblies, attaches to the front end of the launcher assembly.

c. **Launcher Assembly**. This assembly houses the pull-down sensor (Figure B-12). The sensor measures the force with which the student is pulling down on the launch tube. The weight-drop assembly and bipod attach to the launcher assembly.

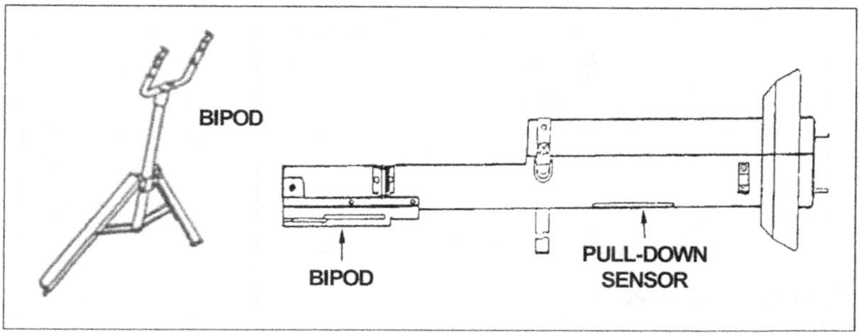

Figure B-12. Bipod and launcher.

B-7. WEIGHT ASSEMBLY

The weight assembly (Figure B-13) drops to simulate the effect that occurs when the Dragon missile leaves the launch tube. A pin in the center of the weight assembly attaches it to the rear of the launch tube. The weight assembly drops 0.6 seconds after the soldier depresses the trigger.

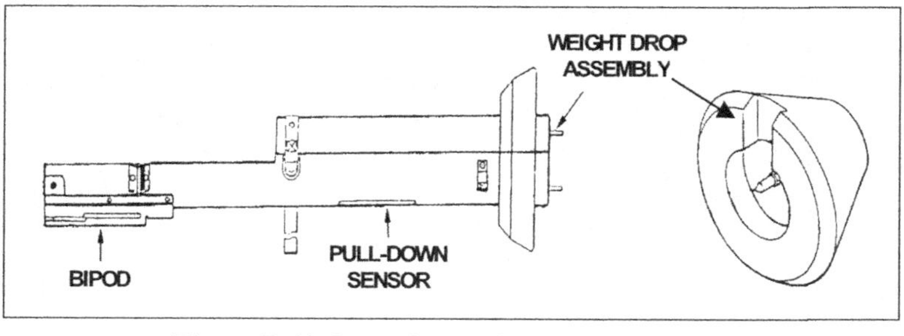

Figure B-13. Launcher and weight assembly.

B-8. HEADSET

The headset (Figure B-14, page B-5) allows the student to hear the instructor's commands as well as the audio portion of the mission. A 3-foot long cable and a headphone plug connect the headset to the sight assembly.

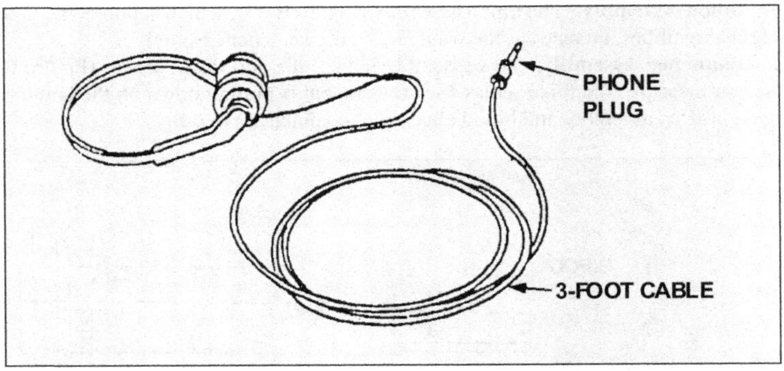

Figure B-14. Student headset.

B-9. SPECIAL-PURPOSE CABLE

The molded yellow 12-foot (W3) special-purpose cable (Figure B-15) provides signal and power connections between the instructor console and the sight assembly.

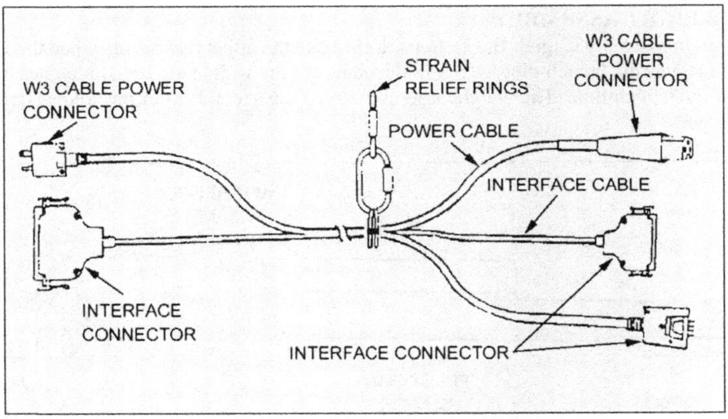

Figure B-15. Special-purpose cable.

B-10. STUDENT STATION SHIPPING CONTAINER

The olive drab-colored, polyethylene, student station shipping containers (Figure B-16, page B-10) have specially shaped inserts to protect the components from damage. The shipping containers resist dust and water, and they stack four deep. The container base has two carrying handles and three latches on each end and seven latches along each side (TM 9-1425-484-10).

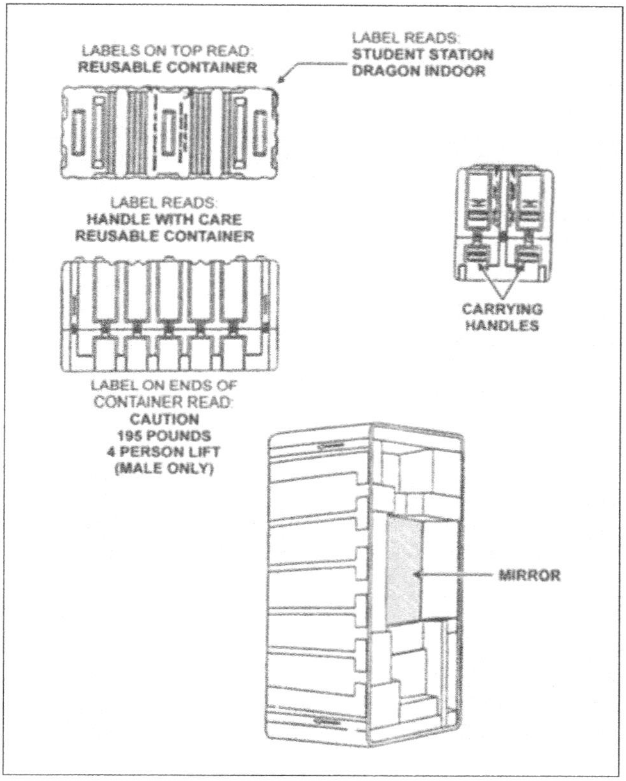

Figure B-16. Student station shipping container.

a. **Ends**. The lifting requirement, CAUTION: 195 POUNDS 4 PERSON LIFT (MALE ONLY), is stenciled on each end of the container.

b. **Cover**. The cover has the following labels:

(1) The part number for entire student station (including the shipping container): STUDENT STATION DRAGON INDOOR P/N 13364400.

(2) Part number for shipping container cover only: REUSABLE CONTAINER P/N 13364390.

(3) The warning HANDLE WITH CARE; REUSABLE CONTAINER.

(4) An arrow icon that shows which side goes upward.

c. **Inside Cover**. A mirror is located inside the cover of the student station shipping container.

.B-11. LIMITATIONS AND RESOURCE REQUIREMENTS

The DGT has certain limitations and requirements.

a. **Limitations**. The DGT—

(1) Requires an indoor location to protect it from humidity, blowing dust and sand, and extreme temperatures.

(2) Requires a trainer, a student, another person. The other person adjusts the bipod legs, replaces the weight assembly, and supports the DGT between firings.

(3) Requires an X-feet wide by Y-feet tall mirror. (If the one in the cover of the student station shipping container breaks or is missing, any mirror this size may be substituted.)

b. **Electrical Requirements**. The DGT—

(1) Requires grounded (three-pronged) extension cords, because its power cables, which are 96 inches (8 feet) long, may be too short to reach the appropriate outlet.

(2) Requires one of the following electrical outlets:

- 110VAC 1 Phase 60 Hertz.
- 220VAC 1 Phase 50 Hertz.

c. **Environmental Requirements**. Table B-2 shows the environmental requirements for the Dragon.

	LOWEST	HIGHEST
Operating Temperature	5.0 degrees Celsius	35.0 degrees Celsius
	41.0 degrees Fahrenheit	98.0 degrees Fahrenheit
Storage Temperature	-34.4 degrees Celsius	62.8 degrees Celsius
	-30.0 degrees Fahrenheit	145.0 degrees Fahrenheit
Humidity	20 percent	80 percent

Table B-2. Environmental requirements for the Dragon.

B-12. OPERATION AND FUNCTION

The trainer programs the target. Using the DGT mouse and controls, he selects a target and its missions. He also chooses from several options that affect the target's level of difficulty. The target will appear in both the trainer's and the firer's viewers. The firer's viewer has crosshairs. He moves the DGT launcher, placing the crosshairs on the target. He keeps the crosshairs on the target until the missile explodes. Through the headset, he will hear the simulated sounds of the Dragon: the firing of thrusters (when he adjusts his aim) and the blast itself. He will see the impact and know at once whether he hit or missed the target. Throughout this training, he can hear the trainer's commands as well as the battlefield sounds of small arms and artillery fire. At the end of each mission, the trainer's screen displays the firer's score. He can replay or save the mission for future use. If the trainer sets up a printer to use with the DGT, he can print a copy of the mission performance results.

DANGER

EVEN "LOW VOLTAGES" PRESENT A HAZARD. IN SOME CONDITIONS, VOLTAGE AS LOW AS 50 VOLTS COULD KILL YOU.

B-13. SAFETY

Wherever the DGT is used, commanders and leaders must develop local directives and SOPs to supplement existing safety precautions. For the protection of soldiers, these documents should spell out locations; safety requirements; individual responsibilities and soldier's limits; and sequence of operations. Everyone must follow the safety precautions. The DGT uses high voltage. Failure to observe safety precautions could cause death. Before using this equipment, you must know the following rules:

a. Never try to operate electronic equipment alone. Be sure someone else is nearby who knows the operation and hazards of the equipment and can give you first aid.

b. Never touch the metal pins when connecting or disconnecting the power cords.

c. On each piece of equipment, learn which areas carry high voltages. Also, be careful not to contact high-voltage connections when installing or operating this equipment.

d. Before working inside the equipment, turn off the power. Before touching anything, ground anything that might carry a high voltage.

APPENDIX C

DRAGON FIELD TACTICAL TRAINER

Leaders use the outdoor DFTT system to train advanced field tracking. This appendix provides technical information about the DFTT and discusses its components and limitations.

C-1. RETROREFLECTOR

The DFTT tracks a small retroreflector mounted on the target vehicle. This retroreflector can be placed on virtually any vehicle or item. Only the number of retroreflectors available restricts the number of targets.

C-2. FEEDBACK

During the simulated flight of the missile, the DFTT provides feedback to the firer. As the missile travels downrange, the DFTT displays the missile's alignment with the target. At the end of the mission, the DFTT displays whether the missile hit or missed; if a hit, it shows where the missile struck the target.

C-3. MULTIPLE INTEGRATED LASER ENGAGEMENT SYSTEM

Although trainers can use MILES with the DFTT, doing so is impractical. If trainers use this method of training, they will need a retroreflector and MILES belt. The MILES laser requires a clean LOS to the retroreflector. This means that trainers must not hide the retroreflector behind any obstruction such as a tree or brush pile. Leaders should not count MILES training as force-on-force training.

C-4. SCENARIO

The DFTT also allows the trainer to adjust the difficulty of the scenario to the firer's proficiency by changing—

- The target kill zone.
- The amount of target obscuration caused by the rocket blast.
- The weight lost after the gunner fires a round.

C-5. CAPABILITIES AND LIMITATIONS

The DFTT can be used either during daylight hours or periods of limited visibility.

a. **Capabilities.**

(1) DFTT provides feedback to the firer during the simulated flight of the missile. It shows how the missile aligns with the target as it travels downrange.

(2) After each mission, the DFTT displays whether or not the missile hit the target and, if so, where on the target it hit.

b. **Limitations.**

(1) Only the number of retroreflectors available to the unit restricts the number of targets presented to the firer.

(2) The afternoon or evening before scheduled training, trainers must recharge the DFTT computer from a 110-volt outlet for at least five hours.

C-6. SAFETY

Local commanders and leaders should work together to write local directives and SOPs to supplement safety precautions. The SOPs should state individual responsibilities, safety requirements, distance limits for soldiers and explosives, training locations, sequence of operations, equipment needed for handling munitions, and protection of soldiers. Directives or SOP must clearly designate each person's responsibilities in any operation involving explosives. The unit must observe safety precautions.

C-7. FIRING TABLES

Basic instructional Firing Tables 5 and 6 help new gunners learn about the DFTT.

WARNING

This equipment uses laser light. Always follow laser range safety procedures in AR 385-63 and TB MED 524. Remember—*under the following conditions,* looking directly at a laser light can blind you permanently:

- **For long periods of time.**
- **At close range.**
- **Through binoculars, telescope, or periscope at a range of less than 154 meters.**

DANGER

THIS EQUIPMENT USES HIGH VOLTAGES AND HIGH-EXPLOSIVE BLAST SIMULATORS.

YOU MUST FOLLOW ALL SAFETY PRECAUTIONS WHEN USING THIS EQUIPMENT. FAILURE TO DO SO COULD CAUSE YOU OR SOMEONE ELSE TO LOSE YOUR HEARING, SUFFER TRAUMATIC INJURY, OR DIE.

FOR INFORMATION ABOUT ARTIFICIAL RESPIRATION, SEE FM 21-11.

1. ***HIGH VOLTAGES.***
   - **FAILURE TO FOLLOW SAFETY PRECAUTIONS COULD RESULT IN DEATH ON CONTACT.**
   - ***DO NOT*** **TOUCH METAL PINS WHEN CONNECTING OR DISCONNECTING POWER CORDS OR SYSTEM CABLES.**
   - ***DO NOT*** **TRY TO DISASSEMBLE THE COMPUTER.**
   - ***DO NOT*** **USE THE EQUIPMENT IF THE CATHODE RAY TUBE (CRT) OR CABLE IS DAMAGED.**
2. ***HIGH-EXPLOSIVE BLAST SIMULATORS.***
   - **WITHIN 70 METERS OF A LOADED BLAST SIMULATOR, WEAR HEARING PROTECTION.**
   - **HANDLE BLAST SIMULATORS CAREFULLY. IF POSSIBLE, WEAR FULL HELMET AND GLOVES.**
   - **KEEP BLAST SIMULATORS AWAY FROM FIRE.**
   - **NEVER HANDLE A DAMAGED BLAST SIMULATOR.**
   - **STAY CLEAR OF THE DANGER ZONE AT ALL TIMES.**

APPENDIX D

FIRING TABLES FOR DRAGON PRECISION GUNNERY TRAINING DEVICES

The Dragon PGTS replaces the LES/LET trainers. The PGTS includes the Dragon gunnery trainer (DGT) and the Dragon field tactical trainer (DFTT). Up to ten gunners can fire simultaneously at any vehicle or stationary target equipped with a mounted retroreflector. The system provides extensive individual feedback after each engagement.

D-1. DGT BASIC INSTRUCTIONAL FIRING TABLES

The basic instructional firing tables for the DGT include Firing Tables 1 through 4.

a. **Firing Table 1.** This firing table introduces the new gunner to the DGT. The gunner will engage stationary targets using the daysight. The hit zone will be set at 100 percent. Students will be allowed to rest for three to five minutes between events. The student will learn—

(1) To use the proper force when pulling down on the Dragon.

(2) To control his breathing.

(3) To aim the weapon properly under different obscuration settings from the sitting position.

TASK	Engage stationary and moving targets.
CONDITIONS	Given a DGT, DGT disk number 14, LCE, protective mask, and helmet.
STANDARD	Breathe and aim properly while in the sitting position.

b. **Firing Table 2.** The students must be able to track the moving targets with a smooth transition between targets. Trainers will set the hit zone to 100 percent. Students will be allowed to rest for three to five minutes between events.

TASK	Engage stationary and moving targets.
CONDITIONS	Given a DGT, DGT disk number 14, LCE, protective mask, and helmet.
STANDARD	Apply proper breath control and aiming procedures while firing from the sitting position.

c. **Firing Table 3.** Trainers will set the hit zone to 100 percent. Students will be allowed to rest for three to five minutes between events.

TASK	Engage moving targets.
CONDITIONS	Given a DGT, DGT disk number 14, LCE, protective mask, and helmet.
STANDARD	Apply proper breath control and aiming procedures while in the sitting position.

d. **Firing Table 4.** The table introduces the new gunner to using the DGT with multiple targets and to using both the daysight and nightsight with different obscuration settings. Students will be required to track while wearing a protective mask. Trainers will set the hit zone to 100 percent. Students will be allowed to rest for three to five minutes between events.

TASK	Engage multiple targets.
CONDITIONS	Given a DGT, DGT disk number 14, LCE, protective mask, and helmet.
STANDARD	Apply proper breath control and aiming procedures while in the sitting position.

e. **Scorecard**. Use DA Form 7445-R, Dragon Gunnery Trainer (DGT) Scorecard, Moving and Stationary Targets, to score Firing Tables 1 through 4.

(1) This blank, reproducible form may be copied from the back of this manual onto 8 1/2 by 11-inch paper. It is also available on the Army Electronic Library (AEL) CD-ROM (EM0001) and at the USAPA website located at (http://www.usapa.army.mil).

(2) Write the soldier's personal data, the unit and point number, and the date as indicated. Complete the remainder of the form as follows. If you need help, see the example completed scorecard in Figure D-1, page D-5:

- ***ENG.*** Engagement numbers are preprinted on the scorecard.
- ***MSN.*** Mission numbers are preprinted on the scorecard.
- ***OBSC.*** Obscuration level is preprinted on scorecard.
- ***HIT.*** Scorer marks this block if student hits the target.
- ***MISS.*** Scorer marks this block if student misses the target.
- ***TRACKING SCORE.*** Scorer marks the student's tracking score.
- ***THRUSTERS.*** Scorer marks how many thrusters fired.

D-2. DFTT BASIC INSTRUCTIONAL FIRING TABLES

The basic instructional firing tables for the DFTT include Firing Tables 5 and 6.

a. **Firing Table 5.** The gunner will engage multiple targets using both the daysight and nightsight under varying obscuration settings. Students will be allowed to rest for three to five minutes between events.

TASK	Engage multiple targets.
CONDITIONS	Given a DFTT, flak vest, LCE, protective mask, helmet, and hearing protection.
STANDARD	Fire for familiarization only.

b. **Firing Table 6.** Students will be allowed to rest for three to five minutes between events.

TASK	Engage stationary, moving, and multiple targets.

CONDITIONS	Given a DFTT, flak vest, LCE, protective mask, helmet, and hearing protection.
STANDARD	Fire for familiarization only.

c. **Scorecard**. Use DA Form 7446-R, Dragon Field Tactical Trainer (DFTT) Scorecard, Multiple Targets, to score Firing Tables 5 and 6.

(1) This blank, reproducible form may be copied from the back of this manual onto 8 1/2 by 11-inch paper. It is also available on the Army Electronic Library (AEL) CD-ROM (EM0001) and at the USAPA website located at (http://www.usapa.army.mil).

(2) Write the soldier's personal data, the unit and point number, and the date as indicated. Complete the remainder of the form as follows. If you need help, see the example completed scorecard in Figure D-2, page D-6:

- ***TGT***. This box is already completed.
- ***RANGE***. For each target fired, enter range to target.
- ***OBS***. For each target fired, enter level of obscuration.
- ***HIT KILL***. Check this box if student hit and killed the target.
- ***HIT NO KILL***. Check this box if student hit but did not kill the target.
- ***MISS***. Check this box if student missed the target.
- ***MISSILE GROUNDED***. Check this box if the missile grounded.
- ***EXCEEDED RANGE***. Check this box if the missile exceeded range.
- ***THRUSTERS***. Write the number of thrusters that fired.
- ***SCORER'S SIGNATURE AND DATE.*** Sign and date the form.

D-3. DGT QUALIFICATION FIRING TABLES

The qualification firing tables for the DGT include Firing Tables 7 and 8.

a. **Firing Table 7.** Soldier practices firing the DGT. This table is for practice qualification only. It evaluates the gunner's ability to engage stationary and moving targets using both the daysight and nightsight. Some targets are more difficult to hit than others.

TASK	Engage stationary, moving, and multiple targets.
CONDITIONS	Given a DGT, DGT disk number 14, LCE, a protective mask, and helmet.
STANDARD	Kill 16 out of 20 target vehicles.

b. **Firing Table 8.** This table is for qualification. It evaluates the gunner's ability to engage stationary and moving targets using both the daysight and nightsight. Some targets are more difficult to hit than others.

TASK	Engage stationary, moving, and multiple targets to qualify.
CONDITIONS	Given a DGT, DGT disk number 15, LCE, protective mask, and helmet.
STANDARD	Kill 16 out of 20 target vehicles.

c. **Scorecard**. Use DA Form 7447-R, Dragon Gunnery Trainer (DGT) Scorecard, Multiple Targets, to score Firing Tables 7 and 8.

(1) This blank, reproducible form may be copied from the back of this manual onto 8 1/2 by 11-inch paper. It is also available on the Army Electronic Library (AEL) CD-ROM (EM0001) and at the USAPA website located at (http://www.usapa.army.mil).

(2) Write the soldier's personal data, the unit and point number, and the date as indicated. Complete the remainder of the form as follows. If you need help, see the example completed scorecard in Figure D-3, page D-7:

(a) *ENG*. Engagement number is preprinted on the scorecard.

(b) *MSN*. Mission number is preprinted on the scorecard.

(c) *OBSC*. Level of obscuration is preprinted on the scorecard.

(d) *TEST*. The soldier must score at least 16 in order to qualify on the Dragon. He gets three chances to do so. If the soldier misses 5 engagements, he will not be able to hit the required sixteen targets. He must move on to the first retest, which another scorer oversees.

— *GO*. Check in this column if student scores a hit.

— *NO GO*. Check in this column if student misses.

(e) *RETEST (1)*. The same conditions apply to this test. If the soldier again fails to hit sixteen targets, he moves on to his last retest.

— *GO*. Check in this column if student scores a hit on the retested target.

— *NO GO*. Check in this column if student misses the retested target.

(f) *RETEST (2)*. A third scorer oversees this test. If the soldier misses five (fails to hit sixteen) on this last test, he fails to qualify on the trainer.

— *GO*. Check in this column if student scores a hit on the retested target.

— *NO GO*. Check in this column if student misses the retested target.

(g) *TS*. Enter tracking score.

(g) *TH*. Enter number of thrusters fired.

d. **Scorecard.** On the scoring scale shown to the left of the legend, mark the box next to the scoring category the soldier achieved.

DRAGON GUNNERY TRAINER (DGT) SCORECARD
MOVING AND STATIONARY TARGETS

For use of this form see FM 3-23.24 (FM 23-24); the proponent agency is TRADOC

DATA REQUIRED BY PRIVACY ACT OF 1974. AUTHORITY: 10 USC 3012(g)/Executive Order 9397. PRINCIPLE PURPOSE(S): Aids in individual training on targets at various ranges. ROUTINE USE(S): To evaluate individual proficiency. SSN used for positive identification purposes only. MANDATORY OR VOLUNTARY DISCLOSURE AND EFFECT ON INDIVIDUAL NOT PROVIDING INFORMATION: Voluntary. Individuals not providing information cannot be rated or scored on a mass basis.

LAST NAME	FIRST NAME	MI	SSN	UNIT NO	POINT NO	DATE (YYYYMMDD)
Frost	Martin	W	431-19-4824	2-29 IN	1	20010801

FIRING TABLE 1
STATIONARY AND MOVING TARGETS
DISK: DGT #14
PERCENT HIT ZONE SETTING 50 75 100 %

EVENT A. OPTICAL SIGHT. WEIGHT LOSS OR PULL-DOWN, AIMING POINT AND TARGET ASPECT.

ENG	MSN	OBSC	HIT	MISS	TRACKING SCORE	THRUSTERS
1	1	0.0	X		66	11
2	2	0.0		X	9	14
3	3	0.5	X		89	15
4	3	1.0	X		90	23
5	4	2.0	X		100	16

EVENT B. OPTICAL SIGHT. INTRODUCTION TO OBSCURATION AND MOVING.

ENG	MSN	OBSC	HIT	MISS	TRACKING SCORE	THRUSTERS
1	8	0.5	X		63	23
2	9	1.0	X		91	28
3	10	0.5		X	25	16
4	11	0.5	X		65	[illegible]
5	12	0.5		X	[illegible]	28

FIRING TABLE 3
MOVING TARGETS
DISK: DGT #14
PERCENT HIT ZONE SETTING 50 75 100 %

EVENT A. OPTICAL SIGHT.

ENG	MSN	OBSC	HIT	MISS	TRACKING SCORE	THRUSTERS
1	8	1.0	X		50	27
2	9	1.0	X		22	28
3	10	1.0	X		0	14
4	11	1.0	X		41	9
5	12	1.0		X	54	30

EVENT B. THERMAL SIGHT.

ENG	MSN	OBSC	HIT	MISS	TRACKING SCORE	THRUSTERS
[illegible]	[illegible]	[illegible]	X		25	7
[illegible]	27	0.0		X	0	14
3	28	0.0	X		0	20
[illegible]	[illegible]	0.0	X		0	14
[illegible]	[illegible]	0.0	X		45	16

FIRING TABLE 2
STATIONARY AND MOVING TARGETS
DISK: DGT #14
PERCENT HIT ZONE SETTING 50 75 100 %

EVENT A. OPTICAL SIGHT. STATIONARY AND MOVING TARGETS.

ENG	MSN	OBSC	HIT	MISS	TRACKING SCORE	THRUSTERS
1	8	0.5	X		[illegible]	23
2	9	0.5	X		[illegible]	28
3	18	0.0	X		36	28
4	19	0.0	X		49	16
5	20	0.0	X		77	12

EVENT B. BRIGHTNESS AND CONTRAST.

ENG	MSN	OBSC	HIT	MISS	TRACKING SCORE	THRUSTERS
1	26	0.0	X		1	7
2	27	0.0		X	34	15
3	28	0.0		X	0	22
4	29	0.0	X		10	20
5	30	0.0	X		0	15

LEGEND: ENG - ENGAGEMENT
MSN - MISSION
OBSC - OBSCURATION

REMARKS:

FIRING TABLE 4
INTERMEDIATE TARGETS
DISK: DGT #14
PERCENT HIT ZONE SETTING 50 75 100 %

EVENT A. OPTICAL SIGHT.

ENG	MSN	OBSC	HIT	MISS	TRACKING SCORE	THRUSTERS
1	7	0.5	X		88	19
2	9	1.0	X		0	27
3	10	1.5	X		0	15
4	11	0.5	X		78	9
5	12	1.0	X		62	30

EVENT B. OPTICAL AND THERMAL SIGHT.

ENG	MSN	OBSC	HIT	MISS	TRACKING SCORE	THRUSTERS
1	13	THERMAL	X		20	18
2	26	THERMAL	X		56	7
3	28	THERMAL		X	0	22
4	29	THERMAL	X		56	17
5	30	THERMAL	X		6	15

EVENT C. OPTICAL SIGHTS. MOPP, PROTECTIVE MASK FAMILIARIZATION.

ENG	MSN	OBSC	HIT	MISS	TRACKING SCORE	THRUSTERS
1	7	0.0	X		84	19
2	9	0.0	X		0	28
3	10	0.0	X		76	15
4	11	0.0	X		60	9
5	12	0.0	X		61	29

SCORER'S NAME (PRINT): SGT Jon Flores
SCORER'S SIGNATURE: SSG Jon Flores

DA FORM 7445-R, DEC 2001

EXAMPLE

Figure D-1. Example completed DA Form 7445-R, Dragon Gunnery Trainer (DGT) Scorecard (Moving and Stationary Targets), Firing Tables 1 through 4.

DRAGON FIELD TACTICAL TRAINER (DFTT) SCORECARD

MULTIPLE TARGETS

TARGET SIZE SETTING: 1 2 3

For use of this form see FM 3-23.24 (FM 23-24); the proponent agency is TRADOC.

DATA REQUIRED BY PRIVACY ACT OF 1974. AUTHORITY: 10 USC 3012(g)/Executive Order 9397. PRINCIPLE PURPOSE(S): Aids in individual training on targets at various ranges. ROUTINE USE(S): To evaluate individual proficiency. SSN used for positive identification purposes only. MANDATORY OR VOLUNTARY DISCLOSURE AND EFFECT ON INDIVIDUAL NOT PROVIDING INFORMATION: Voluntary. Individuals not providing information cannot be rated or scored on a mass basis.

LAST NAME	FIRST NAME	MI	SSN	UNIT NO	POINT NO	DATE (YYYYMMDD)
Frost	Martin	W	431-19-4824	2-29 IN	1	20011206

EXAMPLE

FIRING TABLE 5

MULTIPLE TARGETS, FAMILIARIZATION ONLY

EVENT A. OPTICAL SIGHT, SITTING.

TGT	RANGE	*OBS	HIT KILL	HIT NO KILL	MISS	MISSILE GROUNDED	EXCEEDED RANGE	THRUSTERS
1	450	2		X				12
2	507	2			X			5
3	852	2	X					16
4	757	2	X					8
5	510	2		X				19

EVENT B. OPTICAL SIGHT, STANDING.

TGT	RANGE	*OBS	HIT KILL	HIT NO KILL	MISS	MISSILE GROUNDED	EXCEEDED RANGE	THRUSTERS
1	912	2					X	30
2	805	2		X				26
3	650	2			X			18
4	450	2	X					7
5	375	2	X					6

EVENT C. THERMAL SIGHT, SITTING.

TGT	RANGE	*OBS	HIT KILL	HIT NO KILL	MISS	MISSILE GROUNDED	EXCEEDED RANGE	THRUSTERS
1	450	0	X					11
2	523	0		X				14
3	648	0	X					17
4	719	0			X			20
5	863	0		X				21

EVENT D. THERMAL SIGHT, STANDING.

TGT	RANGE	*OBS	HIT KILL	HIT NO KILL	MISS	MISSILE GROUNDED	EXCEEDED RANGE	THRUSTERS
1	920	0					X	30
2	412	0				X		14
3	557	0		X				17
4	650	0		X				20
5	787	0	X					25

FIRING TABLE 6

STATIONARY, MOVING, AND MULTIPLE TARGETS, FAMILIARIZATION ONLY

EVENT A. OPTICAL SIGHT, SITTING.

TGT	RANGE	*OBS	HIT KILL	HIT NO KILL	MISS	MISSILE GROUNDED	EXCEEDED RANGE	THRUSTERS
1	445	2	X					10
2	574	2			X			15
3	63[illegible]	2		X				18
4	[illegible]	2					X	30
[illegible]	85[illegible]	2				X		24

EVENT B. OPTICAL SIGHT, STANDING.

TGT	RANGE	*OBS	HIT KILL	HIT NO KILL	MISS	MISSILE GROUNDED	EXCEEDED RANGE	THRUSTERS
1	560	2		X				17
[illegible]	39[illegible]	2	X					8
[illegible]	7[illegible]7	2	X					21
4	869	2	X					24
[illegible]	910	2		X				29

EVENT C. THERMAL SIGHT, SITTING.

TGT	RANGE	*OBS	HIT KILL	HIT NO KILL	MISS	MISSILE GROUNDED	EXCEEDED RANGE	THRUSTERS
1	854	0	X					23
2	570	0	X					12
3	451	0	X					9
4	912	0	X					28
5	787	0	X					21

EVENT D. THERMAL SIGHT, STANDING.

TGT	RANGE	*OBS	HIT KILL	HIT NO KILL	MISS	MISSILE GROUNDED	EXCEEDED RANGE	THRUSTERS
1	633	0	X					19
2	557	0	X					17
3	480	0		X				14
4	857	0	X					23
5	920	0					X	30

REMARKS:

SCORER'S NAME (PRINT): SSG Jonathan Flores

SCORER'S SIGNATURE: SSG Jonathan Flores

* OBSCURATION

DA FORM 7446-R, DEC 2001 USAPA V1.00

Figure D-2. Example completed DA Form 7446-R, Dragon Field Tactical Trainer (DFTT) Scorecard (Multiple Targets), Firing Tables 5 and 6.

DRAGON GUNNERY TRAINER (DGT) SCORECARD

MULTIPLE TARGETS

For use of this form see FM 3-23.24 (FM 23-24); the proponent agency is TRADOC

DATA REQUIRED BY PRIVACY ACT OF 1974. AUTHORITY: 10 USC 3012(g)/Executive Order 9397. PRINCIPLE PURPOSE(S): Aids in individual training on targets at various ranges. ROUTINE USE(S): To evaluate individual proficiency. SSN used for positive identification purposes only. MANDATORY OR VOLUNTARY DISCLOSURE AND EFFECT ON INDIVIDUAL NOT PROVIDING INFORMATION: Voluntary; individuals not providing information cannot be rated or scored on a mass basis.

LAST NAME	FIRST NAME	MI	SSN	UNIT NO	POINT NO	DATE (YYYYMMDD)
Frost	Martin	W	431-19-4824	2-29 IN	1	20011206

FIRING TABLE 7

DGT PRACTICE QUALIFICATION
DISK: DGT #14
PERCENT HIT ZONE SETTING
50 % 75 % 100 %

EVENT A. OPTICAL SIGHT

ENG	MSN	OBSC	TEST GO	TEST NO GO	TEST TS	TEST TH
1	1	0.5	X		97	11
2	2	1.5	X		0	11
3	7	0.5	X		86	18
4	8	1.5	X		76	23
5	9	0.5	X		40	28
6	10	1.5	X		0	15
7	11	0.5	X		57	9
8	12	1.5	X		35	29
9	13	0.5		X	0	17
10	14	1.5	X		[illegible]	14

EVENT B. THERMAL SIGHT

ENG	MSN	OBSC	TEST GO	TEST NO GO	TEST TS	TEST TH
1	26	0.0	X		49	11
2	27	0.0	X		0	10
3	28	0.0		X	1	12
4	29	0.0	X		40	21
5	30	0.0	X		19	16
6	26	0.0	X		0	8
7	27	0.0		X	0	12
8	28	0.0	X		23	22
9	29	0.0	X		0	10
10	30	0.0	X		22	15

FIRING TABLE 8

DGT QUALIFICATION
DISK: DGT #15
PERCENT HIT ZONE SETTING
50 % 75 % 100 %

EVENT A. OPTICAL SIGHT

ENG	MSN	OBSC	TEST GO	TEST NO GO	TEST TS	TEST TH	RETEST GO	RETEST NO GO	RETEST TS	RETEST TH	RETEST GO	RETEST NO GO	RETEST TS	RETEST TH
1	1	0.5	X		42	9								
2	6	1.5	X		77	20								
3	7	0.5	X		[illegible]	[illegible]								
4	4	[illegible]	X		83	19								
5	[illegible]	0.5	X		78	12								
6	11	[illegible]		X	0	28								
7	12	0.5	X		82	23								
[illegible]	13	[illegible]	X		33	24								
9	14	0.5	X		0	26								
10	15	[illegible]	X		65	23								

EVENT B. THERMAL SIGHT

ENG	MSN	OBSC	TEST GO	TEST NO GO	TEST TS	TEST TH	RETEST GO	RETEST NO GO	RETEST TS	RETEST TH	RETEST GO	RETEST NO GO	RETEST TS	RETEST TH
1	17	0.0	X		63	17								
2	18	0.0	X		48	22								
3	19	0.0	X		64	24								
4	20	0.0	X		24	30								
5	21	0.0	X		71	13								
6	22	0.0	X		55	20								
7	23	0.0	X		2	11								
8	24	0.0	X		56	22								
9	25	0.0	X		65	16								
10	26	0.0	X		69	13								

LEGEND: ENG - ENGAGEMENT; MSN - MISSION; OBSC - OBSCURATION; TH - THRUSTERS (NARROW COLUMN); TS - TRACKING SCORE (NARROW COLUMN)

REMARKS:

SCORING (CHECK ONE):

EXPERT	X	19 to 20
SHARPSHOOTER		17 to 18
MARKSMAN		16
UNQUALIFIED		Less than 16

SCORER'S NAME (PRINT): SSG Jonathan Flores

SCORER'S SIGNATURE: SSG Jonathan Flores

DA FORM 7447-R, DEC 2001

USAPA V1.00

Figure D-3. Example completed DA Form 7447-R, Dragon Gunnery Trainer (DGT) Scorecard (Multiple Targets), Firing Tables 7 and 8.

APPENDIX E

M113 ARMORED PERSONNEL CARRIER

This appendix is included for all the units that still have M113 APCs.

E-1. FIRING FROM THE M113 APC

You can fire the Dragon from the M113 in three ways.

a. The best way to fire the Dragon from a stationary M113 APC in a hull-defilade position is to use the M175 mount (Figure E-1), though field-expedient methods can be used.

Figure E-1. Firing the Dragon from an M113 APC.

(1) Adjust the commander's chair in the APC so that, when you assume the firing position, both feet are flat on the commander's chair and your knees and body are straight.

(2) Keep your feet apart at a comfortable angle.

(3) Ensure that your left arm and hand exert maximum rearward pressure on the elevation damper. This action pulls the cradle down against your shoulder (Figure E-2).

Figure E-2. M175 firing position.

(4) Position your right hand on the M175 mount firing mechanism the same as you would on the sight's firing mechanism.

- Right thumb on safety.
- Index, middle, and third fingers on the trigger lever.
- Little finger forward of the trigger mechanism.

WARNING

To prevent a missile fired from an M175-mounted Dragon from dipping and flying into the ground, press heavily back and down on the elevation damper.

b. If not using the M175 mount, the gunner stands in the track commander's hatch, places the Dragon's bipod legs in the grill on top of the APC, and fires the weapon (Figure E-3). He could also sit on the closed cargo hatch and fire the Dragon as if he were on the ground (Figure E-4, page E-3).

c. The gunner can also fire the Dragon from a dismounted position, using either the M3 or M122 machine gun tripod (Figure E-5, page E-3).

(1) ***M3 Tripod.*** Sit along the left side of the tripod to ensure that no portion of your body is forward of the muzzle. Sight, aim, and fire the Dragon the same as when it is mounted on an APC.

(2) ***M122 Tripod.*** With the M122 tripod, put the right leg over the top of the two rear tripod legs. Put the left leg over the top of the front tripod leg. Sight, aim, and fire the same as when mounted on an APC.

NOTE: Body position is the most important factor in successful firing and tracking. Providing soldiers with continual feedback during the early phase of practical training will help them form good habits.

WARNING

TO ENSURE THE SAFETY OF THE PERSONNEL IN THE APC, CLOSE THE DRIVER'S AND CARGO HATCHES BEFORE FIRING.

Figure E-3. Firing the Dragon from the commander's hatch.

Figure E-4. Firing from a closed hatch.

Figure E-5. M3 position (left) and M122 tripod position (right).

E-2. SUPPORT EQUIPMENT (M113-EQUIPPED UNITS ONLY)

Units equipped with the M113 APC are authorized the following Dragon support equipment: M175 guided missile launcher mounts, M3 or M122 machine gun tripods, and vehicle storage kits.

a. **M175 Mounting Assembly**. The guided missile launcher mount (Figure E-6, page E-4) provides a stable platform for firing the Dragon missile from the M113 APC. The unit track mechanic installs the M175 mounting assembly on the APC. Before turning in the APC for overhaul or repair, the unit track mechanic must remove all support equipment.

(1) The M175 mount is an improved, stable firing and tracking base for the gunner that increases the effectiveness of the weapon system.

(2) The gunner uses the M175 mount to fire the M222 tactical round or the M223 practice round.

(3) The round and sight mate electrically through two connectors on the M175 mount. The remote firing mechanism is located on the right rear of the cradle.

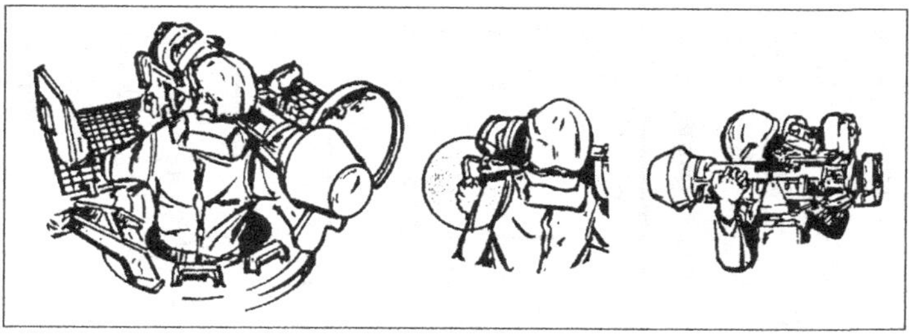

Figure E-6. M175 guided missile launcher mount.

(4) Azimuth and elevation dampers on the M175 mount reduce vibrations and help the gunner track more smoothly.

b. **M3/M122 Machine Gun Tripods**. The gunner can use either the M3 or the M122 machine gun tripod (Figure E-7) to fire the Dragon from a ground support position.

c. **Vehicle Storage Kit**. Each M113 APC assigned to a mechanized infantry squad has one of these kits installed (Figure E-8, page E-5). It consists of the following nightsight support equipment:

(1) A nightsight storage rack, located just below the daysight storage case.

(2) A vehicle power conditioner (VPC) to "step down" the power of the 24-volt battery to the required level.

(3) A battery charger rack and a PP-7382/TAS battery charger also operate from the APC power source.

(4) Three storage racks for coolant cartridge containers and one storage rack for a battery container are located under the personnel seats.

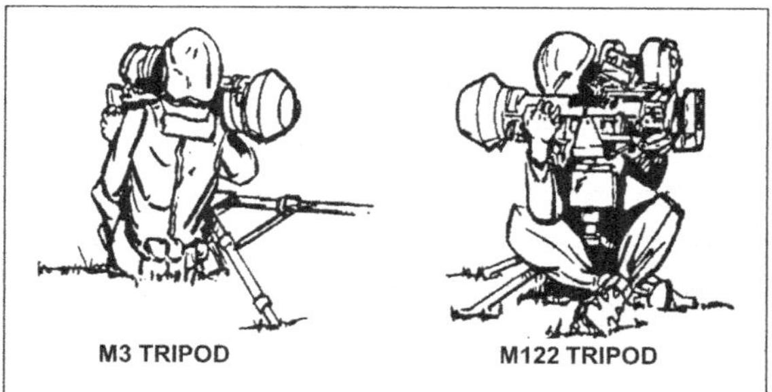

Figure E-7. M3/M122 machine gun tripods.

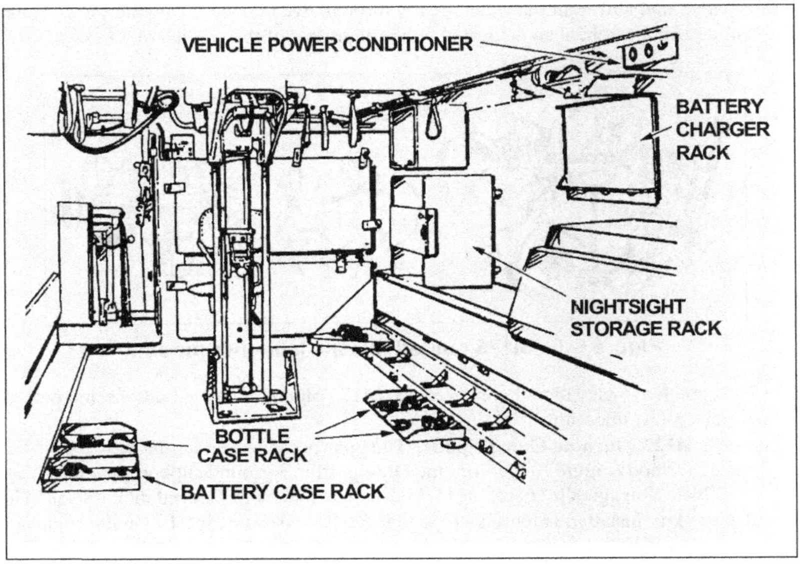

Figure E-8. Vehicle storage kit.

APPENDIX F
SAFETY

Commanders and leaders supplement this appendix and the other published materials referenced in it with local directives and SOPs. They can require soldiers to obtain and use this appendix and the published materials referenced in it, or they can incorporate this appendix and the materials it references.

In addition, local directives and SOPs must supplement this appendix with the following information:

- *Individual responsibilities.*
- *Additional or more stringent local safety requirements.*
- *Distance limitations for soldiers and explosives.*
- *Locations and sequence of operations.*
- *Equipment needed for handling munitions.*
- *Soldier protection.*

Finally, in any operation involving explosives, directives and SOPs should clearly designate and explain each soldier's responsibilities.

F-1. SAFETY PRECAUTIONS

After removing the Dragon's front shock absorber and desiccant package, all soldiers must strictly observe *additional* safety precautions to prevent personnel from tampering with the encased missile.

a. Before they may fire the Dragon, leaders and soldiers must know and follow these guidelines:

(1) Avoid looking at the sun, flares, lasers, or searchlights while looking through the sight. Doing so magnifies their effect and could severely burn your eyes.

(2) The warhead will not arm within 65 meters, so try to avoid firing it within that range.

(3) Due to the risk of an accidental discharge, use extreme care when mating the sight to the round.

(4) Avoid going or placing any equipment or other items forward of the firing position.

(5) Never perform PMCS when the sight is mated to the round.

(6) If you are within 50 meters of a Dragon when it is fired, you must wear your earplugs.

(7) Never fire the Dragon over friendly soldiers or vehicles.

b. The antireflective coating on the AN/TAS-5 infrared lens contains thorium fluoride, which is slightly radioactive. However, this coating material is only hazardous if swallowed or inhaled.

c. The Dragon backblast area extends 50 meters to the rear and 30 meters to the flanks of the launcher (Figure F-1, page F-2). This area is divided into two zones: a danger zone and a caution zone. Trainers must keep both zones clear during training.

(1) The *danger zone* extends 30 meters to the rear of the launcher in a 90-degree cone. Within this area, the blast, flame, and flying debris could cause deaths or serious casualties, so personnel must avoid this zone.

(2) The *caution zone* extends an additional 20 meters to the rear and 30 meters to the flanks of the danger zone. Personnel should try to avoid this area as well. To protect their eyes from flying debris, they *must* face away from the rear of the Dragon. To protect their ears from the blast and overpressure, they must wear earplugs.

d. Due to the danger of the backblast, everyone must take care during all phases of instruction. Team drills, position exercises, and tracking exercises must be conducted with the DGT and DFTT just as though a tactical missile were being fired.

e. After flight, a practice missile can still contain live rocket thrusters. Though the warhead on this type of missile is inert, the live rocket thrusters present a hazard to personnel. The locations of all practice missiles must be reported to EOD so they can dispose of the missiles.

f. During live-fire exercises, safety personnel must stand on either side of the gunner. However, due to the gases, flame, and debris that escape from the front and rear of the Dragon, safety personnel must stand at least 1 meter away from the Dragon.

WARNING

Missile guidance wire, which can hang up on trees or bushes, is thin, strong, and hard to see. Avoid leaving any of it around. Control access to any areas where you think some might remain. Remove any wire that you find, then store or dispose of it properly.

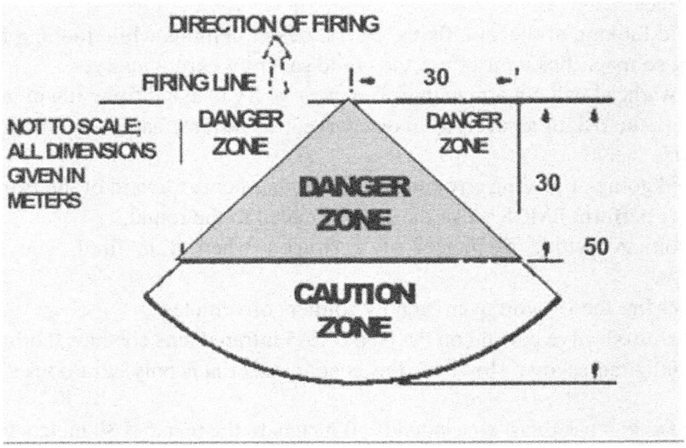

Figure F-1. Dragon backblast area.

WARNINGS

1. **Wear ear protection devices to prevent ear damage.**
2. **Keep personnel clear during training exercises and nontactical firings. Noise levels and flying debris could cause serious ear or eye damage.**

F-2. ROUND HANDLING

Improperly or carelessly handling the Dragon round can damage its components and cause the missile to malfunction when it is launched. If the gunner notices any sign that a missile may have been dropped, or if the launcher is deformed or fractured, the gunner should return it to the responsible ammunition personnel for inspection and disposition.

F-3. FIRING LIMITATIONS

The gunner must not fire the Dragon from within buildings, bunkers, or other enclosures, or within 15 meters (50 feet) of a vertical or nearly vertical backstop. Trainers may be granted waivers from this limitation under the provisions of AR 385-62. In combat, the gunner may have to risk firing the Dragon from an enclosure. If so, he must ensure the enclosure is at least 3 meters by 4.6 meters (10 feet by 15 feet). Also, he must ensure all debris and loose objects are cleared from behind the launch site. When possible, he should ensure all doors and windows are opened. The team should make holes at least 0.6 meter (2 feet) square in the walls and ceilings to allow the backblast and overpressure to escape. Even when the enclosure meets these requirements, the gunner must wear double hearing protection. Also, firing a Dragon from a confined space will almost certainly cause structural damage and create falling debris. It also concentrates the escaping toxic gases and can cause building fires. Any one of these hazards can injure the gunner or team. A Dragon should never be positioned so the gunner has to fire over power lines, through brush or brush fires, or through limbs or other obstructions. Any of these could damage the Dragon's command-link wire and interfere with missile guidance.

F-4. FIGHTING POSITION

When firing from a fighting position, the gunner should ensure there are no obstructions to his front or rear. These could deflect the launcher's backblast onto him. When firing from a downhill or uphill slope, he must be especially careful to ensure that the angle of the launcher relative to the ground is no more than 20 degrees. When firing from either a hasty or prepared fighting position, the gunner must allow at least 15 centimeters (6 inches) above and below the muzzle of the launcher. This gives the missile fins the room they need to unfold and lock into place.

F-5. FIRING OVER WATER

When firing the Dragon over salt water, the gunner must avoid firing at targets beyond 300 meters. Salt water can short-circuit the command link wire. Raising the launcher 0.3 meter

(1 foot) increases the distance the Dragon can be fired over water by 100 meters. Fresh water does not affect command-link wire, so the missile normally can be fired over it.

F-6. TRAINING EQUIPMENT

Safety precautions apply in training just as they do during live firing. For example, during training, soldiers must practice and observe the same backblast precautions they have to observe during live fire (Figure F-1). Even the trainers (DFTT and MILES) generate sound and pressure levels that can damage hearing. Consequently, anyone in or near the firing area must wear proper hearing protection.

APPENDIX G

GUNNER'S PERFORMANCE TEST

The trainer sets up the equipment mentioned in the conditions statement, and provides the student with information IAW the evaluation preparation. This appendix lists tasks, conditions, and standards for the gunner's performance test.

G-1. EVALUATION PREPARATION

The trainer tells each student that he will be required to perform without error all the actions to complete each task being tested.

G-2. TEST ADMINISTRATION

The trainer administers the tests and completes a scorecard for each student.

a. Destroy or recycle any copies of the 1990 edition of DA Form 4242-1-R, Dragon Performance Scorecard--they are obsolete. Copy the new, blank, reproducible DA Form 4242-1-R from the back of this manual onto 8 1/2 by 11-inch paper. Soon after the publication of this manual, those with Internet access should be able to download an electronic copy of the revised blank DA Form 4242-1-R from the USAPA web site. Figure G-1 shows an example of the completed form. Instructions for completing it follow:

(1) **LAST NAME**. Write student's last name.

(2) **FIRST NAME**. Write student's first name.

(3) **MI**. Write student's middle initial.

(4) **SSN**. Write student's social security number.

(5) **RANK**. Write student's rank.

(6) **UNIT**. Write student's unit.

(7) **TEST DATE**. For student's first test, mark the GO or NO-GO column for the appropriate training objective.

(8) **RETEST DATES**. For student's second and third tests, mark the GO or NO-GO column for the appropriate training objective.

(9) **REMARKS**. Write your comments about the student's performance.

(10) **QUALIFIED**. Check this column when the student qualifies.

(11) **UNQUALIFIED**. Check this column if the student fails to qualify on the first test or either retest. If the student gets three checks in this column, he fails to qualify on the weapon.

(12) **INSTRUCTOR'S SIGNATURE**. Sign here for each test that you score.

b. Record results on the Dragon scorecard.

c. Provide added training on each task failed, and retest each failed task twice.

d. Eliminate students who fail any task after three tries.

e. Substitute an FHT wherever a task states to use a round.

G-3. TRAINING OBJECTIVES

Training objectives begin on the page that follows Figure G-1.

a. Training Objective 1, Maintain the M47 Medium Antitank Weapon, page G-3.

b. Training Objective 2, Prepare the M47 for Firing, page G-5.

c. Training Objective 3, Restore the M47 to Carrying Configuration, page G-6.

d. Training Objective 4, Demonstrate Correct Firing Positions, page G-7.

e. Training Objective 5, Determine If Targets Are Engageable, page G-8.

f. Training Objective 6, Perform Immediate-Action Procedures, page G-9.

g. Training Objective 7, Prepare an Antiarmor Range Card, page G-11.

h. Training Objective 8, Identify the Six Major Components of an M47 Fighting Position, page G-12.

I. Training Objective 9, Identify Armored Vehicles, page G-13.

j. Training Objective 10, Explain Emergency Destruction Procedures, page G-14.

k. Training Objective 11, Explain Emergency Decontamination Procedures, page G-15.

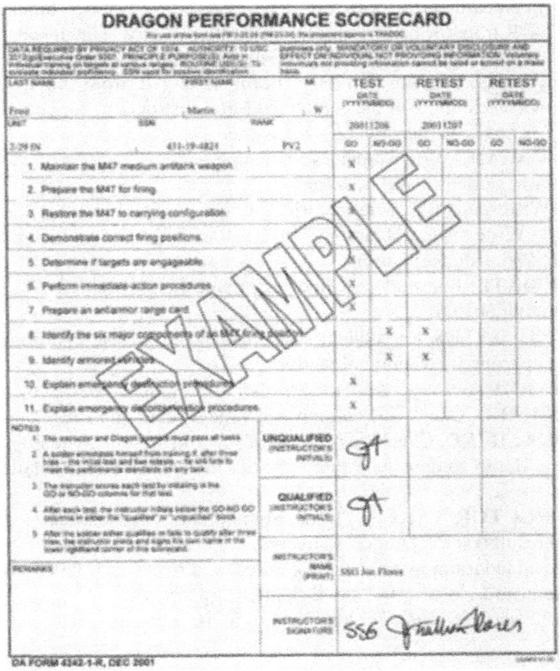

DRAGON PERFORMANCE SCORECARD

TEST RETEST RETEST

1. Maintain the M47 medium antitank weapon
2. Prepare the M47 for firing
3. Restore the M47 to carrying configuration
4. Demonstrate correct firing positions
5. Determine if targets are engageable
6. Perform immediate-action procedures
7. Prepare an antiarmor range card
8. Identify the six major components of an M47 firing position
9. Identify armored vehicles
10. Explain emergency destruction procedures
11. Explain emergency decontamination procedures

UNQUALIFIED

QUALIFIED

DA FORM 4242-1-R, DEC 2001

Figure G-1. Example completed DA Form 4242-1-R, Dragon Performance Scorecard.

TRAINING OBJECTIVE 1
MAINTAIN THE M47 MEDIUM ANTITANK WEAPON

Conditions. Given Dragon sights and a round of ammunition.

Standards. Conduct a complete preoperational check of the Dragon sights and a round of ammunition.

Evaluation Preparation.

Brief student: "When I say 'Begin,' you are to conduct a preoperational check of the Dragon sights and a round of ammunition. Do you have any questions?"

Answer the student's questions. If the student still does not understand, say, "Do the best you can."

Instructor's Note. Because the daysight and nightsight have several parts in common, the student may mention these parts on only one or the other. If he mentions it on either, give him a "GO."

> **WARNING**
> **Never perform a preoperational check on a Dragon sight while you still have it mated to a round of ammunition.**

Performance Measures	GO	NO-GO	NO-GO	NO-GO
1. **Daysight**. a. Pushed safety plunger. b. Squeezed trigger lever bar. c. Ensured safety was functional and trigger clicked.				
2. **Guide pins**. Looked for the guide pins, ensured there were four of them present, and ensured they were tight.				
3. **Connector**. a. Checked connector cover for rubber seals. b. Ensured connector was clean. c. Ensured connector was undamaged.				
4. **Sight housing**. a. If connector sustained damage, reported this to supervisor. b. Spot-painted as needed.				
5. **Eye guard**. Rotated the eye guard.				
6. **Nightsight**. a. Spot-painted as needed. b. Checked range focus lever. c. Checked for freedom of movement.				

Performance Measures	GO	NO-GO	NO-GO	NO-GO
7. **Daysight**. Cleaned the lenses.				
8. **Actuator switch**. a. Rotated the switch to AIR-BATT check. b. Checked the coolant gauge (2,500 PSI). c. Checked the battery monitor on the coolant cartridge. d. Ensured the reticles were illuminated. e. Rotated the actuator switch to the ON position. f. Waited 15 seconds. g. Removed the lens cover.				
9. **Infrared contrast**. a. Selected an object with infrared contrast. b. Adjusted the reticle focus. c. Adjusted brightness. d. Adjusted contrast. e. Adjusted range focus. f. Rotated the actuator switch to the OFF-LOCK position. g. Did not turn the actuator switch to the RELEASE position. h. Replaced the lens cover.				
10. **Ammunition round**. a. Checked the exterior for dirt, oil, and grease. b. Checked the exterior for dents, holes, and punctures.				
11. **Raceway**. Ensured raceway was not damaged.				

Signature Block

INSTRUCTOR'S SIGNATURE: ____________________

STUDENT'S INITIALS: GO ____________

NO-GO ____________

NO-GO ____________

NO-GO ____________

TRAINING OBJECTIVE 2
PREPARE THE M47 FOR FIRING

Conditions. Given a Dragon sight in the carrying bag and a round of ammunition in the carrying configuration.

Standards. Mate the sight to the round of ammunition IAW the performance measures described herein.

Evaluation Preparation.

Brief student. "When I say 'Begin,' you are to prepare the Dragon for firing by mating the sight to the round of ammunition. Do you have any questions?"

Answer the student's questions. If the student still does not understand, say, "Do the best you can."

Performance Measures	**GO**	**NO-GO**	**NO-GO**	**NO-GO**
1. **Prepared the round.** a. Unsnapped bipod-retaining strap. b. Lowered bipod to vertical-lock position. (Shook bipod and ensured it was locked.) c. Depressed friction lock and extended legs about halfway down. d. Removed the electrical connector cover from the round. 2. **Prepared the daysight.** a. Removed the sight properly. b. Removed electrical connector cover. c. Checked lenses. 3. **Mated the sight to the round.** a. Aligned the guide pins with the sight support assembly and locked the sight in place. (Shook sight to ensure it was locked.) b. Leveled sight picture. (Adjusted the bipod friction lock and the adjusted the foot.)				

Signature Block

INSTRUCTOR'S SIGNATURE: ______________________

STUDENT'S INITIALS: GO ____________

NO-GO ____________

NO-GO ____________

NO-GO ____________

TRAINING OBJECTIVE 3
RESTORE THE M47 TO CARRYING CONFIGURATION

Conditions. Given a Dragon sight, carrying bag, and round of ammunition.

Standards. Properly return the sight and round to the carrying configuration without damaging the equipment.

Evaluation Preparation.

Brief student. "When I say 'begin,' you will restore the Dragon to the carrying configuration. Do you have any questions?"

Answer the student's questions. If the student still does not understand, say, "Do the best you can."

Performance Measures	GO	NO-GO	NO-GO	NO-GO
1. **Removed and stored the sight.** a. Released the spring clip. b. Removed from support assembly. c. Removed the sight. d. Replaced the lens cover. e. Replaced electrical cover. f. Placed sight in the carrying bag. 2. **Restored the round of ammunition.** a. Replaced the connector cover. b. Collapsed the bipod legs. c. Replaced the shock absorber. d. Secured legs to the round.				

Signature Block

INSTRUCTOR'S SIGNATURE: ______________________

STUDENT'S INITIALS: GO ________

NO-GO ________

NO-GO ________

NO-GO ________

TRAINING OBJECTIVE 4
DEMONSTRATE CORRECT FIRING POSITIONS

Conditions. Given a Dragon sight already mated to a Dragon round.

Standards. Correctly assume the three correct Dragon firing positions.

Evaluation Preparation.

Brief student. "When I say 'Begin,' you will assume the three firing positions for the Dragon. We will test you on all three. After you assume each correct position, tell me which position it is, and I will inform you if you are a GO or NO-GO for that position. Do you have any questions?"

Answer the student's questions. If the student still does not understand, say, "Do the best you can."

Performance Measures	GO	NO-GO	NO-GO	NO-GO
1. **Sitting position**.				
a. Sat down, feet on bipod. Pushed legs out.				
b. Leaned forward and placed round on the meaty portion of shoulder, keeping the upper body straight.				
c. Properly grasped the sight with both hands.				
d. Rotated head and aligned right eye in the eye guard. Pushed head forward.				
e. Pulled elbows in and backward, and tried to touch them together on the chest. Removed slack from the bipod.				
2. **Standing supported**.				
a. Stood with legs spread comfortably apart and straight, but not locked.				
b. Leaned forward. Rested the abdomen against the edge of the fighting position.				
c. Placed Dragon on the meaty part of the shoulder. (Placed bipod legs far forward.)				
d. Placed hands as for the sitting position.				
e. Rotated head and right eye as in the sitting position.				
f. Removed slack from the bipod. Pulled down and back. Tried to touch the elbows and drive them into the chest. If there was slack in the bipod, straightened the back slightly.				
3. **Kneeling position**.				
a. Kneeled with knees spread comfortably apart.				

Performance Measures	GO	NO-GO	NO-GO	NO-GO
b. Placed the bipod well out in front and removed slack from the bipod by rocking backward. Tried to sit on heels.				
c. Rotated head and right eye as in the sitting position.				
d. Placed hand and elbows the same as in the sitting position.				

Signature Block

INSTRUCTOR'S SIGNATURE: ______________________

STUDENT'S INITIALS: GO ________

NO-GO ________

NO-GO ________

NO-GO ________

TRAINING OBJECTIVE 5
DETERMINE IF TARGETS ARE ENGAGEABLE

Conditions. Given a Dragon sight mated to a round of ammunition and targets (scale or actual targets).

Standards. Correctly determine if three targets are in range and engageable.

Evaluation Preparation.

Brief student. "To your front are armored vehicles at varying (simulated) ranges. When I say 'Begin,' determine whether the targets are in range. Then you must accurately determine the engageability of all three targets. Do you have any questions?"

Answer the student's questions. If the student still does not understand, say, "Do the best you can."

Performance Measures	GO	NO-GO	NO-GO	NO-GO
Determine if the target is—				
1. In range or out of range, engageable or not. **Target Number 1, ___, ___.**				
2. In range or out of range, engageable or not. **Target Number 2, ___, ___.**				
3. In range or out of range, engageable or not. **Target Number 3, ___, ___.**				

Signature Block

INSTRUCTOR'S SIGNATURE: ______________________

STUDENT'S INITIALS: GO __________

NO-GO __________

NO-GO __________

NO-GO __________

TRAINING OBJECTIVE 6
PERFORM IMMEDIATE-ACTION PROCEDURES

Conditions. Given a malfunctioning Dragon sight mated to a Dragon round.

Standards. Perform the proper immediate-action procedures for hot and cold sight batteries.

Evaluation Preparation.

Brief student. "When I say 'Begin,' perform immediate-action procedures for the Dragon. Do you have any questions?"

Answer the student's questions. If the student still does not understand, say, "Do the best you can."

Performance Measures	GO	NO-GO	NO-GO	NO-GO
1. **Hot sight battery**.				
a. Tried and failed to fire.				
b. Immediately resqueezed the trigger for 15 seconds.				
c. Announced "Misfire" if round still did not fire.				
d. Turned, cautiously felt near the sight battery, and found it to be hot.				
e. Removed the sight and placed the round a safe distance away.				
f. Obtained a new round and mated the sight to it.				
g. Acquired the target, said, "I am continuing my mission" or squeezed the trigger.				
h. Squeezed the trigger.				
2. **Cold sight battery**.				
a. Tried and failed to fire.				
b. Immediately resqueezed the trigger for 15 seconds.				
c. Announced "Misfire" if round still did not fire.				

Performance Measures	GO	NO-GO	NO-GO	NO-GO
c. Turned, cautiously felt near the sight battery, and found it to be cold.				
d. Released the sight and moved it forward; reseated the sight on the round.				
e. Acquired the target, said, "I am continuing my mission" or squeezed the trigger.				
f. When sight again failed to fire, rechecked battery, which was still cold.				
g. Removed sight; placed round a safe distance away.				
h. Got a new round and said, "I will continue the mission."				
i. When round again failed to fire, rechecked battery, which was still cold.				
j. Replaced the sight.				

Signature Block

INSTRUCTOR'S SIGNATURE: ______________________

STUDENT'S INITIALS: GO ________

NO-GO ________

NO-GO ________

NO-GO ________

TRAINING OBJECTIVE 7
PREPARE AN ANTIARMOR RANGE CARD

Conditions. Given the standard range card and squad leader's briefing containing all pertinent data.

Standards. Within 15 minutes, complete the antiarmor range card. Include all data from the squad leader's briefing, marginal data, a scaled-down sketch of the terrain, and the data section.

Evaluation Preparation.

Brief student. "When I say 'Begin,' you will have 15 minutes to complete the antiarmor range card, including the data section and terrain sketch. Do you have any questions?"

Answer the student's questions. If the student still does not understand, say, "Do the best you can."

Performance Measures	GO	NO-GO	NO-GO	NO-GO
1. **Marginal information.** a. Unit designation. b. Magnetic north. 2. **Terrain sketch.** a. Left limit. b. Right limit. c. Maximum engagement line. d. Terrain features. e. Dead space. f. Target reference point(s). g. Gunner reference point. 3. **Data section.** a. Position. b. Date. c. Weapon symbol. d. Meter increments. e. Item numbers. f. Direction. g. Range. h. Description. i. Remarks. 4. **Two copies of the range card.**				

Signature Block

INSTRUCTOR'S SIGNATURE: ____________________

STUDENT'S INITIALS: GO ________

NO-GO ________

NO-GO ________

NO-GO ________

TRAINING OBJECTIVE 8

IDENTIFY THE SIX MAJOR COMPONENTS OF AN M47 FIGHTING POSITION

Conditions. Given a predrawn fighting position.

Standards. Within 10 minutes, correctly label the six main components of a fighting position.

Evaluation Preparation.

Brief student. "When I say 'Begin,' you will have 10 minutes to draw and label the six main components of a Dragon fighting position, IAW the instructions given in the respective performance measure. Do you have any questions?"

Answer the student's questions. If the student still does not understand, say, "Do the best you can."

Performance Measures	**GO**	**NO-GO**	**NO-GO**	**NO-GO**
1. **Trench.** a. Three M16s long, shaped like an inverted "V." b. Waist deep. c. Waist wide plus 6 inches.				
2. **Front parapet.** a. One M16 long. b. One M16 wide. c. Two helmets high.				
3. **Flank parapet.** a. One M16 wide. b. Two helmets high. c. Long enough to provide good flank protection.				
4. **Bipod trench.** a. Two helmets long. b. One helmet wide. c. Six inches deep (one bayonet blade).				
5. **Grenade sump.** a. One entrenching tool-length long. b. One entrenching tool-length deep. c. One entrenching tool-blade wide.				
6. **Overhead cover.** a. Each end of the trench. b. Dug 12 inches deep. c. Eighteen inches longer than the side of the trench. d. Three feet wide. e. The hole under the overhead cover is large enough for one student and extra ammunition.				

Signature Block

INSTRUCTOR'S SIGNATURE: ______________________

STUDENT'S INITIALS: GO ________

NO-GO ________

NO-GO ________

NO-GO ________

TRAINING OBJECTIVE 9
IDENTIFY ARMORED VEHICLES

Conditions. In a garrison or field environment, given armored vehicles visible in a tactical or simulated tactical setting, a Dragon daysight, and a simulated round of ammunition (FHT), prepared for firing.

Standards. Within two minutes, recognize 10 out of 10 vehicles as friendly or enemy.

Evaluation Preparation.

Setup. The basic combat vehicle identification training aid, GTA 17-2-13, shows multiple views of 30 friendly and enemy armored vehicles. These are the vehicles that every student should know. Show each student the pictures of any six enemy and any four friendly armored vehicles. Show each picture for 10 seconds.

Brief student. "I will show you each picture for 10 seconds. During that time, you must state whether the vehicle is friendly or enemy."

Answer the student's questions. If the student still does not understand, say, "Do the best you can."

Performance Measures	GO	NO-GO	NO-GO	NO-GO
Recognized 10 out of 10 armored vehicles as friendly or enemy.				

Signature Block

INSTRUCTOR'S SIGNATURE: ______________________

STUDENT'S INITIALS: GO ________

NO-GO ________

NO-GO ________

NO-GO ________

TRAINING OBJECTIVE 10
EXPLAIN EMERGENCY DESTRUCTION PROCEDURES

Conditions. Given a simulated situation in which capture is imminent, and the order to destroy the sight and rounds of ammunition.

Standards. Explain how to destroy Dragon components and ammunition.

Evaluation Preparation.

Brief student. "When I say 'begin,' you must explain emergency destruction procedures for a Dragon. Do you have any questions?"

Answer the student's questions. If the student still does not understand, say, "Do the best you can."

Performance Measures	GO	NO-GO	NO-GO	NO-GO
NOTE: Destruction priority: First the sight, then the round.				
1. **Sight**.				
a. Destroyed sight by mechanical means. Used an entrenching tool, axe, sledgehammer, or other appropriate item.				
b. Destroyed by gunfire.				
c. Destroyed by burning. Used flammable material or gasoline.				
d. Destroyed by exploding. Used a double-primed explosive charge.				
2. **Round**. Destroyed a round of ammunition by either—				
a. *Burning it*. Used flammable material, thermite grenade, or gasoline.				
b. *Exploding it*. Used a double-primed explosive charge. (Squad leader will prepare explosives.)				
c. *Launching it*. Fired the missile, then drove over and crushed the expelled launcher.				

Signature Block

INSTRUCTOR'S SIGNATURE: ____________________

STUDENT'S INITIALS: GO __________

NO-GO __________

NO-GO __________

NO-GO __________

TRAINING OBJECTIVE 11
EXPLAIN EMERGENCY DECONTAMINATION PROCEDURES

Conditions. Given the Dragon sights and a round of ammunition (for all tasks, use the FHT instead of an actual round), M8 paper, M256 detector kit, M258A1 kit, hot, soapy water, lens tissue, camel's-hair brush, and ethyl alcohol.

Standards. Correctly explain the decontamination procedures for the Dragon sights and round of ammunition.

Evaluation Preparation.

Brief student. "When I say 'Begin,' you will correctly explain how to decontaminate the Dragon sights and a round of ammunition. Do you have any questions?"

Answer the student's questions. If the student still does not understand, say, "Do the best you can."

Instructor's Note. Decide whether the Dragon needs immediate decontamination. If the gunner is already wearing MOPP4 and the threat is imminent, decontamination must be postponed. If trainers plan to displace the gunner to a noncontaminated area, you must decide whether to fire the missile immediately or try to decontaminate it for transport. (Inform the student of the type of decontamination: nuclear, chemical, or biological.)

Performance Measures	GO	NO-GO	NO-GO	NO-GO
1. **Nuclear decontamination**.				
a. Thoroughly brushed clothing and equipment to remove fallout (away from the position).				
b. Turned over the soil in the immediate area around the position.				
c. Verified decontamination with the AN/PDR-27 radiacmeter.				
2. **Biological decontamination**.				
a. Washed the Dragon with warm, soapy water.				
b. Dried the Dragon.				
c. Used normal cleaning procedures for the lenses.				
3. **Chemical decontamination**.				
a. Checked with the M256 detector kit.				
b. Checked with M8 paper.				
c. Removed or blotted excess agent from all components.				
d. Left Dragon exposed to air and sun.				
e. Used hot, soapy water (if available). Did not submerge the Dragon or sight.				

Performance Measures	GO	NO-GO	NO-GO	NO-GO
f. Used the M258A1 kit on parts the gunner needed to touch or that touched the gunner.				
g. Used normal cleaning procedures for the lenses.				

Signature Block

INSTRUCTOR'S SIGNATURE: ______________________

STUDENT'S INITIALS: GO ________

NO-GO ________

NO-GO ________

NO-GO ________

APPENDIX H

DECONTAMINATION AND DESTRUCTION PROCEDURES

Units develop and maintain decontamination procedures so they can accomplish their missions in possible NBC environments. A unit will destroy its Dragons only in the combat zone, only on order or IAW command policy, and only to prevent the enemy from capturing and using them.

Section I. AGENTS

Decontamination means removal or neutralization of a hazardous level of NBC contamination from personnel and materials. Decide whether or not to decontaminate a Dragon contaminated by NBC agents. Remove nuclear contamination (fallout) as soon as possible. However, as long as you wear the required protective equipment, you can carry and fire a Dragon contaminated with chemical or biological agents. After doing so, however, swap your MOPP gear out IAW unit SOP. Chemical and biological contaminants soak into the exposed rubber-like and plastic parts of the Dragon. You will not be able to decontaminate these parts completely. Instead of wasting your time on them, decontaminate the other parts of the Dragon that you actually touch.

H-1. CHEMICAL CONTAMINANTS

Air out components to eliminate chemical agents.

a. Remove or blot extra liquid agents from all components.

b. Place the equipment outdoors in the sun.

c. Use M8 detector paper and an M256 detector kit to check the Dragon periodically, until it is safe to handle (FM 3-5).

d. To quickly reduce chemical contamination on Dragon surfaces, use hot, soapy water (do not submerge) or an M258A1 kit.

e. Use a camel's hair brush, ethyl alcohol, and lens paper to gently clean the lens.

f. Avoid using DS2 and STB, the standard decontaminants, because they can damage the Dragon weapon system's sensitive electronic components.

H-2. BIOLOGICAL AGENTS

Eliminate biological agents as follows:

a. Wash the Dragon with warm, soapy water.

b. Dry the Dragon.

c. Clean the lens with lens-cleaning solvents (or ethyl alcohol) and lens paper.

H-3. NUCLEAR AGENTS

Do not remove nuclear agents until all fallout has stopped. Then remove them as follows:

a. Move away from the fighting position and brush clothing and equipment thoroughly to remove fallout.

b. Decontaminate individual equipment by brushing, wiping, and scrubbing.

c. Turn over the soil to decontaminate the immediate area around the position.
d. Use the AN/PDR-27 radiacmeter to verify that the Dragon is decontaminated.

Section II. DESTRUCTION GUIDANCE

If soldiers cannot evacuate the Dragons, then the unit should destroy first the sights, then the rounds. Destroying the same component in all weapons prevents the enemy from assembling a complete Dragon. Soldiers destroy Dragons by mechanical or explosive means, gunfire, or fire. Some procedures require the use of explosives that may not be authorized. The commander must decide whether to issue these and related materials. If the situation requires the destruction of Dragons, the commander orders the unit to destroy them where the debris will obstruct the enemy but will not endanger friendly soldiers. Each organization and installation that uses, maintains, or stores Dragon missiles or sights should specify in its SOP how to destroy them. The destruction plan should allow for any situation. It should state priorities and methods of destruction; it should provide clear instructions for each method of destruction; and it should specify the quantities of explosives required.

H-4. DESTRUCTION OF SIGHT

When you destroy the sight, use one of the following methods:

a. **Explosives**. Double-prime an explosive charge (at least 1/2 pound). Place the charge on top of the sight. Detonate the explosive electrically or nonelectrically.

b. **Gunfire**. Well-aimed shots from rifles or other small arms can make the sight useless to the enemy. When using small-arms fire, aim for critical parts such as the lenses, trigger assembly, and the electronics package.

c. **Mechanical Means**. Smash the sight with axes, picks, crowbars, rocks, and so on, or drive over it with a tracked vehicle. If you have enough time and personnel, destroy the sight completely.

d. **Fire**. After doing as much damage with the other methods, use vehicle fuels and lubricants to burn whatever is left of the sight.

H-5. DESTRUCTION OF ROUND

When you destroy the missile, use the following methods:

WARNING

DO NOT **use mechanical methods to destroy live missiles.**

a. **Launching**. The easiest and best way to destroy the missile is to fire it into enemy territory. After doing this, smash the launcher with axes, picks, and so on, or drive over it with a tracked vehicle.

b. **Exploding**. Explosives not only destroy the missile, but also the launcher. Placing one-half pound of explosive material on the round near the warhead should destroy the round (Figure H-1, page H-3).

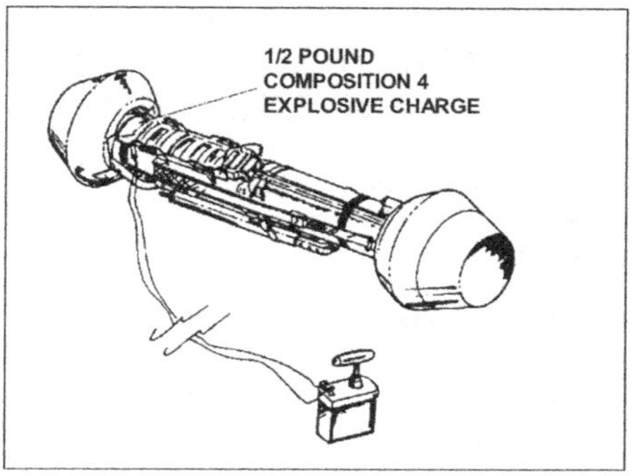

Figure H-1. Explosive charge on a round.

(1) Determine whether you can use *electric* blasting caps and wire or *nonelectric* blasting caps and safety fuzes to prime and detonate the explosive charges.

(a) If you prime the charges with electric blasting caps, take cover before firing them.

(b) If you prime the charges with nonelectric caps, crimp them onto at length of safety fuze at least 2 meters (6 1/2 feet) long. Once you prime the charges, ignite the safety fuzes and *take cover at once*.

(2) Use detonating cord to connect the charges. This produces a simultaneous detonation. Dual-prime the charges to reduce the possibility of a misfire.

c. **Burning**. If time allows after you have done as much damage as you can to the components using the other methods, burn the debris. Use vehicle fuels and lubricants to aid burning. You can also place one or more incendiary grenades on each component or part.

DANGER

1. **THE SAFETY FUZE BURNS AT THE RATE OF 1 FOOT IN 30 TO 40 SECONDS. BECAUSE SAFETY FUZE CONTAINS BLACK POWER AND BLASTING CAPS, YOU MUST PROTECT IT FROM MOISTURE AT ALL TIMES.**
2. **KEEP THE BLASTING CAPS, DETONATING CORD, AND SAFETY FUZES APART UNTIL YOU NEED TO USE THEM.**
3. **DO NOT FIRE SMALL ARMS AT THE MISSILE, AND KEEP OTHERS FROM DOING SO AS WELL. HITTING A LIVE ROCKET MOTOR OR A HIGH-EXPLOSIVE ANTITANK WARHEAD COULD INJURE OR KILL THE FIRER OR ANYONE ELSE NEAR THE MISSILE.**

APPENDIX I

THERMAL TARGET RECOGNITION, IDENTIFICATION, AND ENGAGEMENT

With the nightsight, the Dragon gunner can learn to recognize, identify, and engage targets by their unusual thermal images. On clear nights, he can recognize thermal targets at long ranges.

I-1. TEMPERATURE AND THERMAL IMAGES

An object the same temperature as its background does not show through a nightsight. However, most objects radiate some heat, and even a difference of less than one degree renders an object visible to the gunner using such a sight. Since temperatures generally drop at night, thermal targets show up better through the nightsight then.

a. Some targets, such as tanks and APCs, radiate heat in patterns recognizable to a trained, experienced Dragon gunner.

b. A soldier looking through a nightsight will not see an object of the same temperature as its background. However, the same soldier looking through the same sight will see as bright red any object or area on an object of greater heat than its background, such as engines that have recently been operated. Objects much colder than their backgrounds look black. Objects with less extreme temperatures appear as a less intense red or gray.

I-2. SOURCES OF INFRARED ENERGY

The four sources of infrared energy include the sun, fuel combustion, friction, and radiant heat.

a. **Solar Heat**. Heat from the sun affects only the outside surfaces of objects.

(1) Solar heat highlights the outlines of an object, which helps the gunner recognize targets. For example, a solar-heated M113 APC looks like a box with a sloping front. However, a solar-heated M60A1 tank looks like a small oval on top of a larger oval. Gunners can detect objects at long ranges (out to 2,000 meters), but at medium ranges (800 to 1,200 meters), they can recognize specific shape cues. Profiles are usually easier to recognize than front views.

(2) Not only do atmospheric variables and surface reflections affect the color and intensity of a thermal image, so also does an object's ability to absorb sunlight. For example, dark-colored objects normally absorb sunlight better than light-colored objects.

b. **Fuel Combustion Heat**. An operating engine produces heat that warms up the surface of its surrounding compartment. Engine temperatures can reach up to 200 degrees Fahrenheit.

(1) ***Outside Features***. A gunner can seldom clearly see the actual shape of an engine, heated personnel space, muffler, or exhaust pipes, but the transferred heat may reveal features on the surface of the engine compartment.

(2) ***Relative Locations***. From the relative locations of what he perceives to be the engine and the exhaust, a trained and experienced gunner can tell whether the vehicle has a front- or rear-mounted engine.

(3) ***Directional Movement***. If the vehicle makes evasive maneuvers, the gunner can locate the vehicle's exhaust, which provides an important cue.

c. **Frictional Heat**. Moving parts generate frictional heat, which produces less intense heat than fuel combustion heat. The transport systems of moving vehicles generate frictional heat, which appears as a dim red in the nightsight. Except when high vegetation or dust obscures its transport system, the frictional heat it generates identifies a vehicle as wheeled or tracked.

(1) ***Tracked Vehicles***. In a moving tracked vehicle, the tracks, road wheels, drive sprockets, support rollers, and shock absorbers generate heat. When shock absorbers radiate the heat they have absorbed, the gunner can identify them at longer ranges. At short to medium ranges, radiant heat from shock absorbers could help the gunner identify the vehicle.

(2) ***Wheeled Vehicles***. In a moving wheeled vehicle, the tires, shock absorbers, differentials, drive shafts, transmissions, and axles generate heat. Gunners can detect the tires, shock absorbers, and differentials of a wheeled vehicle at medium to long range.

d. **Radiant Heat**. Smooth, shiny surfaces, such as windshields and glossy, painted fenders, reflect radiant heat from other sources. These reflections can produce odd images. For example, the fenders of an enemy tank appear black due to this thermal reflection; a glossy, painted vehicle could reflect off the tank's flat surfaces. An overcast sky can cause warmer thermal reflections. Diffuse surface reflections seldom cause problems.

I-3. EFFECTS OF WEATHER AND OBSCURANTS

Variations in solar heat, fuel combustion heat, frictional heat, and thermal reflection affect infrared signatures and target recognition cues. Also, various atmospheric conditions affect the view through the nightsight, for better or worse.

a. **Precipitation**. Infrared energy transmits poorly through falling precipitation such as rain, snow, or fog, even when basic signature cues do not change. Falling rain and snow restrict visibility more than fallen rain and snow. During rain or snow, background objects and target features heated by friction and the sun lose their heat. Water, ice, or mud buildup on the transport system reduces frictional heat. The temperatures of engine compartments and exhausts remain high. Landmarks, such as trees, trails, and contour features, and background objects, such as trees and rocks, cool so much they may no longer be visible. This reduces scene clutter and can increase *target detection* capability. However, *target recognition* lessens as certain target features cool down. Rain and snow cool the target and reduce its contrast, causing a "snowy" image. The gunner adjusts the contrast control on the nightsight to compensate.

b. **Fallen Snow.** Snow reduces the temperatures of targets and background objects. Fallen snow evens out ground temperatures, which renders terrain features invisible. Without terrain features to use as a size reference, depth perception by size comparison becomes difficult.

c. **Dust, Diesel Fog, and Oil Smoke**. Dust particles created by the impact of artillery rounds reduce the nightsight's effectiveness.

I-4. TARGET IDENTIFICATION

Vehicles have distinguishing characteristics or cues by which they can be classified and identified. Changes in atmospheric and ground conditions affect identification. To help compensate for these variables, the gunner must know how to use the control settings on the nightsight.

a. **Contrast and Brightness Controls**. Gunners can set the contrast and brightness (image) controls for the greatest internal detail of the target, then reset them to enhance other target-recognition cues. To increase their understanding of how these cues affect the thermal image and target recognition cues, gunners should experiment with the contrast and brightness controls (Figure I-1).

BRIGHTNESS	CONTRAST	APPLICATION
Low	Medium to high	• Suppresses background clutter. • Use in rainy, dusty, and lightly foggy conditions. • Reveals hot objects only—cool ones will be invisible. • When possible target appears, increase brightness and lower contrast gradually to reveal more thermal detail.
Medium to low	Medium	• Best overall setting for long ranges. • Gradually reduce brightness to reveal cooler parts, such as tracks and road wheels, which will darken before the hotter parts do. • Observe carefully while adjusting from medium to low and back. Gradual changes in brightness reveal distinctive features and aid in determining the hottest ones.
Medium	Medium to high	• Ranges greater than 1,200 meters. • Best combination of settings to reveal the hull shapes and overall silhouettes of target vehicles. • Especially useful for— – Identifying small target images with few recognizable features. – Tracking long-range targets in poor visibility such as heavy fog or dust. – Searching tree lines in wet conditions. Limitation of this setting: a snowy, distorted, tough-to-interpret image does not enhance target identification.

Figure I-1. Setting combinations for contrast and brightness controls.

b. **Focus Controls**. Most nightsight focus controls are sensitive; that is, a small movement of the control knob causes a large change in focal point. If a gunner has trouble focusing, he should check the image (contrast and brightness) controls.

(1) To adjust the reticle, turn the focus control (located on the eyepiece).

(2) Adjust the range focus lever. Before focusing the objective lens, move the image controls (contrast and brightness) to the low-to-medium level. Gunners learn to focus the objective lens skillfully through trial and error. This presents a challenge, because infrared heat diffuses on objects. It seldom yields the clear, straight lines that other types of sights do. Thus, you must learn to focus by moving the control back and forth to obtain the best image. Start by aiming the nightsight at a prominent object. Once you have determined the best image, you can experiment by focusing on different objects at different ranges.

I-5. BATTLEFIELD IDENTIFICATION

Using a nightsight to identify targets on a battlefield presents a challenge also. In a target-rich environment on a dry, clear night, high-confidence identification requires a thermal image of such features as road wheels, turret shapes, gun tube, and exhaust location. In clear weather, a gunner can thermally distinguish an M60A1 main battle tank from a T-62 medium tank at a range of 800 to 1,200 meters. To identify a target, a gunner should ask himself the following questions:

- Is the target moving? In what direction?
- Where is the engine? Where is the exhaust?
- Is the target in the unit's section? Should it be there?
- Is the target in a formation?
- Is the target firing at friendly or enemy units?

I-6. PRIMARY RECOGNITION CUES

The friendly and enemy cues shown in the following figures help in training gunners to recognize vehicles.

- Figure I-2. M60A1 main battle tank (page I-5).
- Figure I-3. M551 light tank (page I-6).
- Figure I-4. M113 armored personnel carrier (page I-7).
- Figure I-5. Trucks (page I-8).
- Figure I-6. T-62 medium tank (page I-9).
- Figure I-7. BMP infantry combat vehicle (page I-10).
- Figure I-8. BTR-60 armored personnel carrier (page I-11).
- Figure I-9. BRDM-2 reconnaissance vehicle (page I-12).

M60A1 Main Battle Tank

Classification. Main characteristics of the M60A1 main battle tank include—

- A rear engine.
- An oval-shaped track and road wheel pattern.
- An overall hull, turret, and gun pattern that may be visible with the maximum setting.
- A gun tube that can be seen when the gun has recently fired.

Side-View Identification. M60A1 main battle tank characteristics visible from both sides include—

- A rear engine and a rearward exhaust.
- A high-profile track pattern with hot, taut tracks. (The gunner may see the six evenly spaced road wheels and three support rollers, especially at short ranges.)
- A large, centrally mounted turret.
- A medium-length gun tube that can be seen at short ranges. When the gun has recently fired, the gun tube can be seen at long ranges. (The bore evacuator is two-thirds of the way down the length of the barrel.)
- A high overall profile with a large turret mounted in the center.

Front View Identification. M60A1 main battle tank characteristics visible from the front include—

- Two warm tracks, separated by a cool hull. The lower front hull will appear warm if the personnel heater has operated recently.
- A cool front hull, which indicates a rear-engine vehicle.
- One track that seems wider, if the vehicle sits slightly oblique to the viewer.
- A gun tube that observers can see when the gun has recently fired.

Effects of Motion. The M60A1 main battle tank's direction of movement may indicate the location of its engine. Other characteristics that observers could see when the M60A1 main battle tank moves include—

- A transport system that becomes warmer and more visible during movement.
- A slower bounce than a lighter vehicle.
- A sometimes visible exhaust plume, whose direction indicates the direction of movement.

Distinguishing Features. The side view of the M60A1 track pattern appears higher and more uniformly warm than the side view of the T-62. The tracks of the T-62 are cooler toward the front and smaller than those of the M60A1 tank. From the front, the M60A1 main battle tank appears hotter and larger than the T-62 medium tank.

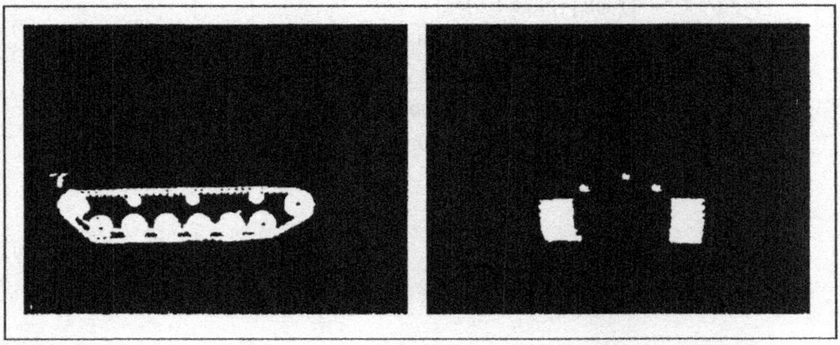

Figure I-2. M60A1 main battle tank.

M551 Light Tank

Classification. Main characteristics of the M551 light tank include—

- A rear engine.
- An oval-shaped track and road wheel pattern.
- A combined hull, turret, and gun pattern.

Side-View Identification. M551 light tank characteristics visible from both sides include—

- A rear engine and rear exhaust, whose plume may be directed upward or rearward.
- A low profile: low hull and a small, low, flat turret.
- A wedge-like shape, with the wedge pointing to the vehicle's front and visible between 800 and 1,200 meters.
- A high side decking.
- Warm, slack tracks and (at short ranges) five evenly spaced road wheels.
- A short gun tube that can be seen when the gun has recently fired.

Front-View Identification. M551 light tank characteristics visible from the front include—

- A wide, low turret, whose sides extend almost over the tracks, providing a unique front view.
- Warm tracks separated by a cool hull, visible as two red spots.

Effects of Motion. The M551 light tank's direction of movement may indicate the location of its engine. Other characteristics that observers could see when the M551 light tank moves include—

- A transport system that becomes warmer and more visible during movement.
- A slower bounce than a lighter vehicle.

Distinguishing Features. The M551 light tank has a wide, low turret (front view) and a wedge shape (side view). Otherwise, its signatures are difficult to distinguish from those of a T-62 tank.

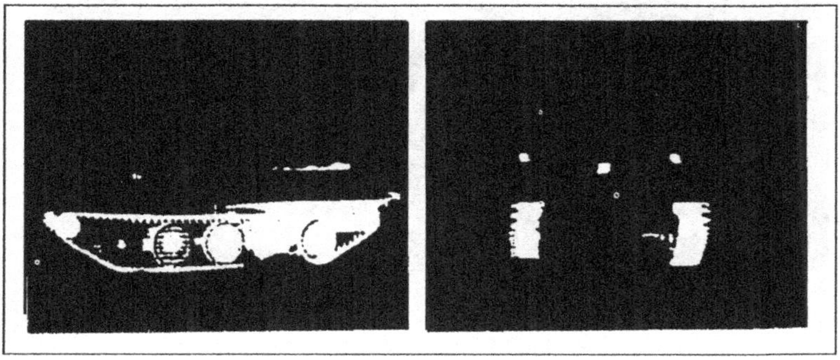

Figure I-3. M551 light tank.

M113 Armored Personnel Carrier

Classification. Main characteristics of the front-engine M113 APC include—

- An oval-shaped track and road wheel pattern.
- A box-shaped hull with no turret.

Side-View Identification. From the right, even at long ranges, you can see the M113 APC's characteristic front hot corner. From the left, you can see a cool box shape. Characteristics visible from both sides include—

- A beveled front end.
- A low track.
- A track skirt, which gives the track a low profile.
- Five evenly spaced road wheels.

Front-View Identification. The M113 APC's characteristic hot spot can be seen on its front right corner. Other characteristics visible from the front include—

- A dark surfboard across the front of the vehicle.
- An overall boxy shape (at the maximum range setting).
- On the front right corner, an upward-moving exhaust plume that can be seen at short ranges.

Effects of Motion. The M113 APC's direction of movement may indicate the location of its engine, which will get hotter and more visible as it operates. Other characteristics that observers could see when the M113 APC moves include—

- A transport system that becomes warmer and more visible during movement.
- A slower bounce than a lighter vehicle, but a faster bounce than a tank.
- A sometimes visible exhaust plume whose direction indicates the direction of movement.

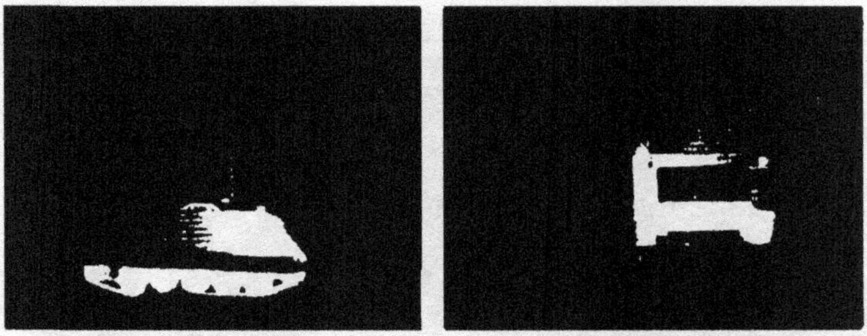

Figure I-4. M113 armored personnel carrier.

Trucks

(All truck cues are combined)

External size cues are needed to establish the size of a truck. To estimate the size of a cue, the gunner compares a nearby tree or other background feature to the size of the red spot. He can also use the reticle to gage size. This is most accurate at low to medium settings.

Side-View Identification. Truck characteristics visible from both sides include—

- A prominent square, also visible from the front, that indicates a front engine. The entire hood and radiator usually appear hot, and the front wheels may seem to merge into the image of the engine.
- Warm spots, separated from the front wheels, that indicate rear wheels. These are easily distinguished from a track pattern.
- Silhouettes of the cab and hood, which may appear at maximum settings. (Long-bed trucks often appear as other truck shapes from the side, even at long ranges.)
- A diagonally-angled drive shaft that is often visible as a bright area joining the engine and rear wheels.
- Exhaust pipes and stacks, which appear hot along their entire length. These appear as different shapes, depending upon how they are routed through the vehicle's frame.

Front-View Identification. Truck characteristics commonly visible from the front include—

- A bright red square or rectangle, indicating the engine compartment, hood, radiator, and, if a personnel heater is in use, the cab. Front views of trucks are all similar at medium and long ranges.
- Wheels and a front axle area that often merge into a continuous red spot beneath the engine. Lower settings improve this image so the gunner can see the vehicle's wheels.
- Canvas and wood parts. These parts usually appear cool at night and warm during the day.

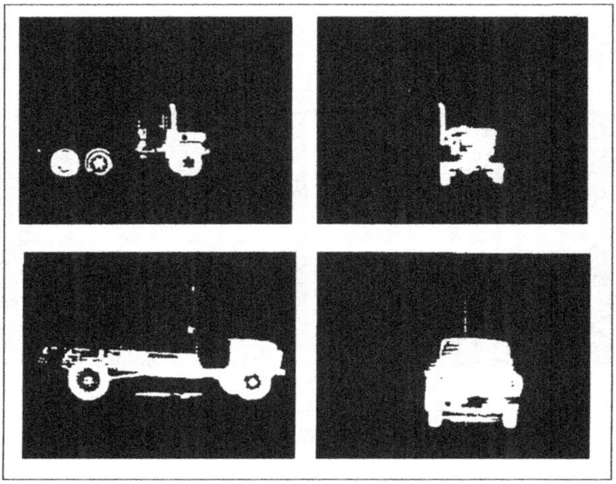

Figure I-5. Trucks.

T-62 Medium Tank

Classification. Main characteristics of the T-62 medium tank include—

- A rear engine.
- An oval-shaped track and road wheel pattern.
- A combined hull, turret, and gun pattern that may be visible at maximum range setting.
- A gun tube that can be seen when the gun has recently fired.

Side-View Identification. On the left side of the T-62 medium tank, a large heated area and exhaust flume indicate a rear engine. T-62 medium tank characteristics visible from both sides include—

- Low-profile, slack tracks and, at short ranges, five road wheels with unique (unequal) spacing.
- A low overall profile with a cool hull.
- A long gun tube that can be seen when the gun has recently fired.
- A small, centrally mounted turret.

Front-View Identification. T-62 medium tank characteristics visible from the front include—

- A cool, low overall profile, with a left side that is warmer than the right.
- Warm tracks separated by a cool hull and cool fenders that may appear black above the tracks.
- A small, dome-shaped turret.
- A long gun tube that can be seen when the gun has recently fired. When aimed straight at the viewer, it appears as a red spot. Its firing signature can be seen only at short ranges.

Effects of Motion. The direction of movement may reveal the location of the engine. Other characteristics that observers could see when the T-62 medium tank moves include—

- A transport system that becomes warmer and more visible during movement.
- A slower bounce than a lighter vehicle.
- A sometimes visible exhaust plume whose direction indicates the direction of movement.

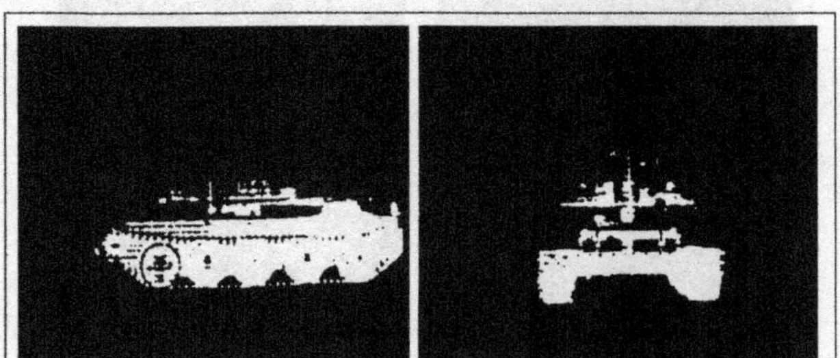

Figure I-6. T-62 medium tank.

BMP Infantry Combat Vehicle

Classification. Main characteristics of the BMP infantry combat vehicle include—

- An engine mounted in the front right of the vehicle and visible from its front and right.
- A taut track pattern and, at close ranges, visible wheels and support rollers.
- A low overall profile that can be seen at higher settings.
- A gun tube that can be seen when the gun has recently fired.

Side-View Identification. The right side of the BMP infantry combat vehicle is hotter than its left side and is usually more recognizable than other views. The engine appears as a large red spot in the forward half, and the exhaust port is hotter and brighter red than the engine. The tracks and engine area appear together to form a boat-like shape on the right side. BMP characteristics visible from both sides include—

- A track pattern that can be seen at long ranges.
- A forward end that slopes upward from the track to the front fender.
- Cool track fenders that block the view of the upper track.
- A long, low shape that can be seen at high settings.

Front-View Identification. BMP characteristics visible from the front include—

- An engine located in the right front, indicated by a square-shaped red spot.
- A cool, front deck panel, indicated by a black line across the engine's red spot.
- A sometimes visible exhaust plume which vents upward from the right side of the vehicle.
- An exhaust port that can be seen as a small red spot in the top-right corner of the engine's larger red spot.
- Visible tracks, with a small gap separating the red spot of the vehicle's left track from the engine's red spot.
- A turret that cannot be seen at long ranges.

Effects of Motion. The direction of movement may reveal the location of the engine, and changing target views may reveal other features such as—

- A transport system that becomes warmer and more visible during movement.
- A slower bounce than a lighter vehicle.
- A sometimes visible exhaust plume whose direction may indicate the direction of movement.

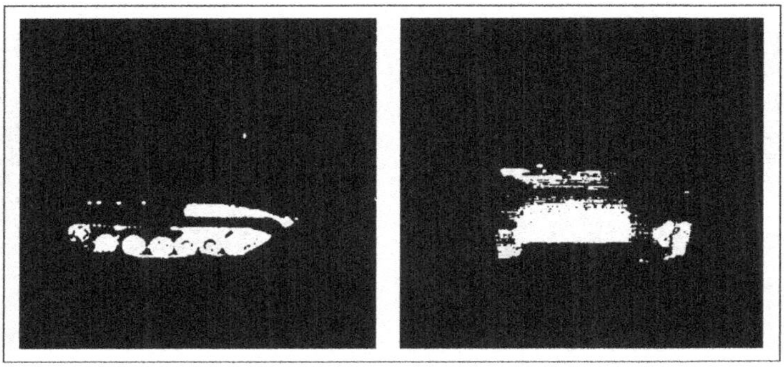

Figure I-7. BMP infantry combat vehicle.

BTR-60 Armored Personnel Carrier

Classification. Main characteristics of the BTR-60 include—

- A rear engine and multiple wheels that can be distinguished from the side at long ranges.
- Mufflers on the rear deck that intensify the rear-engine compartment cue.
- A long, high profile, apparent even at long ranges.
- A gun tube that can be seen when the gun has recently fired.

Side-View Identification. BTR-60 characteristics visible from both sides include—

- A rear engine, which creates the brightest spot on the image.
- Four large, inflatable tires on each side, with a gap between the front and rear pairs.
- Two mufflers mounted over the rear engine compartment.
- A small turret that can be seen at medium ranges and that sometimes can be seen at long ranges.
- Identical left-side and right-side views.

Front-View Identification. BTR-60 characteristics visible from the front include—

- A pattern across the front surface, caused by variations in temperature, visible as multiple shapes.
- A rounded hull.
- A cool surfboard that divides the top and bottom halves of the front hull.
- A cool front, unless the personnel heater is in use.
- Shock absorbers that can be seen at medium ranges.

Effects of Motion. The vehicle's direction of movement may indicate the location of its engine, which will get hotter as it operates and thus brighter red. Other characteristics that observers could see when the BTR-60 moves include—

- Its wheels, which will become more visible as they get warmer.
- Its mufflers, which will also become more visible as they get hot.
- A slow bounce, even on rough terrain, which identifies it as a large vehicle.
- A sometimes visible exhaust plume whose direction indicates the direction of movement.

NOTE: Some of the heat from the muffler merges with the heat from the front hull and wheels, making a front three-quarter view of this vehicle more confusing than the same view of other vehicles.

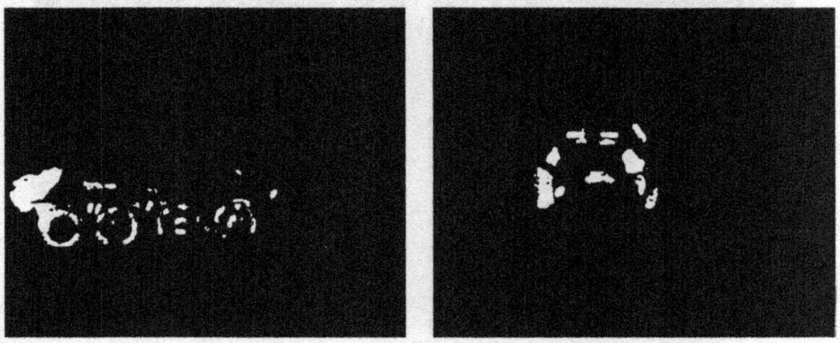

Figure I-8. BTR-60 armored personnel carrier.

BRDM-2 Reconnaissance Vehicle

Classification. Main characteristics of the BRDM-2 reconnaissance vehicle include—

- A rear engine and exhaust.
- A pattern of separate, wheel-sized red spots.
- An overall profile of a hull with (or without) turret.
- A gun tube that can be seen when the gun has recently fired.

Side-View Identification. BRDM-2 characteristics visible from both sides include —

- The vehicle's small overall size, indicated in part by the short distance between the two tires clearly visible on each side.
- A distinctive heat pattern formed by the rear mufflers and wheels. This pattern is clear even at long ranges. Left and right views are the same.
- The rear engine, which is clearly visible at long ranges.
- Two mufflers, one on each side of the top rear deck, which enlarge the hot spot created by the engine area.
- A choppy ride over rough terrain that is noticeable even at long ranges.

Front-View Identification. BRDM-2 characteristics visible from the front include —

- Two warm tires separated by a cool hull. The red spot of the differential may be visible between the front tires.
- A cool front hull, indicating a rear engine.

NOTE: Setting the brightness and contrast controls on high makes the entire front of this vehicle appear hot.

- The dark surfboard across the front of the hull can be seen.
- The tires are set closer together than the tracks of most armored vehicles.
- A greater height-to-width ratio than most armored vehicles, so that it appears narrower and taller.
- Over rough terrain, the vehicle tends to bounce more than armored vehicles.

Effects of Motion. The vehicle's direction of movement may indicate the location of its engine. Other characteristics that observers could see when the BRDM-2 reconnaissance vehicle moves include—

- A transport system that becomes warmer and more visible during movement.
- A sometimes visible exhaust plume whose direction indicates the direction of movement.

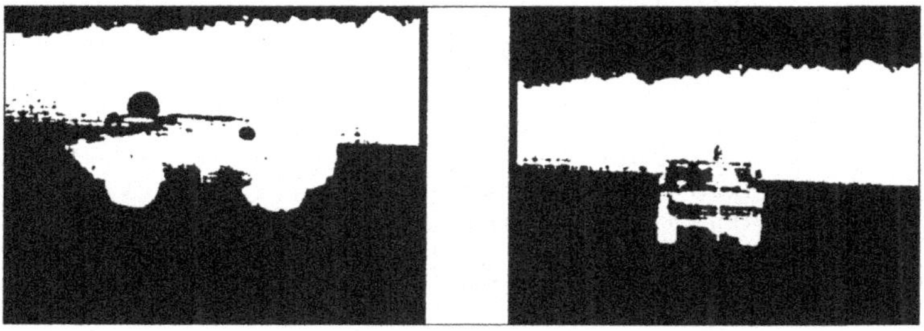

Figure I-9. BRDM-2 reconnaissance vehicle.

GLOSSARY

ACRONYMS AND ABBREVIATIONS

AAR	after-action review
amp	ampere
APC	armored personnel carrier
AR	Army regulation
ARTEP	Army Training and Evaluation Program
ASI	additional skill identifier
asst	assistant
ATGM	antitank guided missile
ATWESS	antitank weapon effect signature simulator
batt	battery
BFV	Bradley fighting vehicle
BMP	(a fighting vehicle manufactured by the former Soviet Union)
BRDM	(a scout car manufactured by the former Soviet Union)
BRT	brightness
BTR	(a personnel carrier manufactured by the former Soviet Union)
C2	command and control
CB	citizen's band (radio)
cbt	combat
CFV	cavalry fighting vehicle
CP	command post
CPX	command post exercise
CS	combat support
CTA	common table of allowances
CTRS	contrast
CVI	combat vehicle identification
DA	Department of the Army
DC	direct current
DFTT	Dragon field tactical trainer
DGT	Dragon gunnery trainer
dir	direction
DLIC	detachment left in contact
DZ	drop zone
EOD	explosive ordnance disposal
F	Fahrenheit
1SG	first sergeant
FASCAM	family of scatterable mines

FHT	field handling trainer
FIST	fire support team
FM	field manual
FOV	field of view
FPF	final protective fire
FPL	final protective line
freq	frequency
FSO	fire support officer
FTX	field training exercise
GTA	graphic training aid
HE	high explosive
headphone	An earphone held over the ear by a band worn on the head—usually used in plural, as in "*a set of headphones*."
headset	An attachment for holding an earphone and transmitter at one's head.
HEAT	high-explosive antitank
HMMWV	high-mobility multipurpose wheeled vehicle
horz	horizontal
HQ	headquarters
hr	hour
Hz	Hertz
IAW	in accordance with
KHz	kilohertz
KPH	kilometers per hour
LAW	light antitank weapon
LCE	load-carrying equipment
LD	line of departure
ldr	leader
LFX	live-fire exercise
LOS	line of sight
LP	listening post
L-R	left to right
LRA	local reproduction authorized
LTID	laser target interface device
M	monthly
mech	mechanized
METT-T	mission, enemy, terrain (and weather), troops, and time available
MILES	multiple integrated laser-engagement system
min	minute(s)
mm	millimeter

mod	model
mon	monitor
MOPP	mission-oriented protective posture
MOUT	military operations on urbanized terrain
MPH	miles per hour
MQS	military qualification standards
MTP	mission training plan
NA	not applicable
NATO	North Atlantic Treaty Organization
NAVTRADEV	Naval training development
NBC	nuclear, biological, and chemical
NCO	noncommissioned officer
NCOES	Noncommissioned Officer Education System
NCOIC	noncommissioned officer in charge
nonmech	nonmechanized
NVD	night vision device
OPFOR	opposing force
OSUT	one-station unit training
PDF	principal direction of fire
PGTS	Precision Gunnery Training System
plt	platoon
PMCS	preventive maintenance checks and services
POI	program of instruction
PSG	platoon sergeant
PSI	pounds per square inch
PSM	power supply modulator
pwr	power
Q	quarterly
qual	qualification
qtr	quarterly
RDL	Reimer Digital Library
RSTA	reconnaissance, surveillance, and target acquisition
S2	intelligence officer
S3	operations and training officer
sec	seconds
SGT	sergeant
SL	skill level
SMCT	soldier's manual of common tasks
SOP	standing operating procedure

sqd	squad
SSN	social security number
STB	supertropical bleach
STP	soldier's training publication
STX	situational training exercise
TC	training circular
tm	team
TM	technical manual
tng	training
TO	training objective
TOW	tube-launched, optically tracked, wire-guided (missile)
TRADOC	(United States Army) Training and Doctrine Command
TRP	target reference point
TSC	Training Support Center
TTP	tactics, techniques, and procedures
US	United States
USC	United States code
VAC	voltage alternating current
VDC	voltage direct current
vert	vertical
VPC	vehicle power conditioner
WP	white phosphorus
XO	executive officer

REFERENCES

SOURCES USED

These are the sources quoted or paraphrased in this publication:

AR 385-62	Regulations for Firing Guided Missiles and Heavy Rockets for Training, Target Practice and Combat. 5 January 1977; with Change 1, 1 Jun 1983.
AR 385-63	Policies and Procedures for Firing Ammunition for Training, Target Practice, and Combat. 15 October 1983.
AR 614-200	Enlisted Assignments and Utilization Management. 20 November 2000.
FM 3-5	NBC Decontamination. 28 July 2000.
FM 7-7	The Mechanized Infantry Platoon and Squad (APC). 15 March 1985.
FM 7-7J	The Mechanized Infantry Platoon and Squad (Bradley). 7 May 1993.
FM 7-8	Infantry Rifle Platoon and Squad (Infantry, Airborne, Air Assault, Ranger). 22 April 1992.
FM 7-10	The Infantry Rifle Company. 14 December 1990.
FM 21-11	First Aid for Soldiers. 27 October 1988; with Change 1, 28 August 1989 and Change 2, 4 December, 1991.
FM 71-1	Tank and Mechanized Company Team. 26 January 1998.
STP 21-1-SMCT	Soldier's Manual of Common Tasks, Skill Level 1. 1 October 1994.
STP 21-24-SMCT	Soldier's Manual of Common Tasks (SMCT), Skill Levels 2/3/4. 1 October 1992.
TM 9-1265-368-10-1	Operator's Manual for Multiple Integrated Laser Engagement System (MILES) Simulator System, Firing Laser: M62 for Dragon Weapons System. 15 July 1988; with Change 1, 16 August 1989.

TM 9-1425-484-10	Operator's Manual for Dragon Weapon Guided Missile System, Surface Attack: M47 (Medium Antitank/Assault Weapon System). 31 July 1979; with Changes 1 thru 13, 17 October 1980 thru 30 March 1993.
TM 9-5855-254-14-HR	Hand Receipt Covering Content of Components of End Item (COEI) Basic Issue Items (BII), and Additional Authorization List (AAL) for Charger, Battery PP-7382/TAS. 22 October 1987.

DOCUMENTS NEEDED

These documents must be available to the intended users of this publication:

Publications

ARTEP 7-8-MTP	Mission Training Plan for the Infantry Rifle Platoon and Squad. 3 March 2000 (RDL version).
GTA 17-2-11	Combat Vehicle Identification Training Cards (CVI). 1984.
GTA 17-2-13	Armored Vehicle Recognition. 1987.
TB Med 524	Occupational and Environmental Health: Control of Hazards to Health from Laser Radiation. 20 June 1985.

Forms

The forms prescribed by this manual can be obtained from three sources.

- *Reproducible Forms*--Copy the blank forms from the back of the manual onto 8 1/2-by 11-inch paper.
- *CD-ROM*--Find the forms on the Army Electronic Library (AEL) disks (EM0001).
- *Downloadable Forms*--Download the blank forms in PDF, HTML, BOO, or FRO from the USAPA website (http://www.usapa.army.mil).
- *Electronic Forms*--Complete FRZ and FRO forms electronically (except for signature blocks, which must be completed manually) using *FormFlow Filler*. Download the latest version of FormFlow Filler software from the same website. NOTE: If any item on the blank form prints in the wrong place, turn Enhanced Fonts on or off in the Tools, Options menu.

DA Form 4242-1-R	Dragon Performance Scorecard.
DA Form 7445-R	Dragon Gunnery Trainer (DGT) Scorecard, Firing Tables 1-4.

DA Form 5517-R	Standard Range Card (LRA). February 1986. (For a locally reproducible copy of this form, see FM 7-7J.)
DA Form 7446-R	Dragon Field Tactical Trainer (DFTT) Scorecard, Firing Tables 5-6.
DA FORM 7447-R	Dragon Gunnery Trainer (DGT) Scorecard, Firing Tables 7-8.

Internet Web Sites

U.S. Army Publishing Agency (USAPA)
http://www.usapa.army.mil

Army Doctrine and Training Digital Library (ADTDL)
http://www.adtdl.army.mil

Air Force Publishing
http://afpubs.hq.af.mil/

READINGS RECOMMENDED

These sources contain relevant supplemental information:

AR 614-200	Enlisted Assignments and Utilization Management. 20 November 2000.
CTA 50-970	Expendable/Durable Items (Except Medical, Class V, Repair Parts, and Heraldic Items), 21 September 1990.
DA Label 80	US Army Calibrated Instrument, 1 December 1977.

INDEX

DRAGON PERFORMANCE SCORECARD

For use of this form see FM 3-23.24 (FM 23-24), the proponent agency is TRADOC.

DATA REQUIRED BY PRIVACY ACT OF 1974. AUTHORITY: 10 USC 3012(g)/Executive Order 9397. PRINCIPLE PURPOSE(S): Aids in individual training on targets at various ranges. ROUTINE USE(S): To evaluate individual proficiency. SSN used for positive identification purposes only. MANDATORY OR VOLUNTARY DISCLOSURE AND EFFECT ON INDIVIDUAL NOT PROVIDING INFORMATION: Voluntary. Individuals not providing information cannot be rated or scored on a mass basis.

LAST NAME	FIRST NAME	MI	UNIT	SSN	RANK

	TEST DATE (YYYYMMDD)		RETEST DATE (YYYYMMDD)		RETEST DATE (YYYYMMDD)	
	GO	NO-GO	GO	NO-GO	GO	NO-GO
1. Maintain the M47 medium antitank weapon.						
2. Prepare the M47 for firing.						
3. Restore the M47 to carrying configuration.						
4. Demonstrate correct firing positions.						
5. Determine if targets are engageable.						
6. Perform immediate-action procedures.						
7. Prepare an antiarmor range card.						
8. Identify the six major components of an M47 firing position.						
9. Identify armored vehicles.						
10. Explain emergency destruction procedures.						
11. Explain emergency decontamination procedures.						
UNQUALIFIED (INSTRUCTOR'S INITIALS)						
QUALIFIED (INSTRUCTOR'S INITIALS)						
INSTRUCTOR'S NAME (PRINT)						
INSTRUCTOR'S SIGNATURE						

NOTES:

1. The instructor and Dragon gunners must pass all tasks.
2. A soldier eliminates himself from training if, after three tries -- the initial test and two retests -- he still fails to meet the performance standards on any task.
3. The instructor scores each test by initialing in the GO or NO-GO columns for that test.
4. After each test, the instructor initials below the GO-NO GO columns in either the "qualified" or "unqualified" block.
5. After the soldier either qualifies or fails to qualify after three tries, the instructor prints and signs his own name in the lower righthand corner of this scorecard.

REMARKS:

DA FORM 4242-1-R, DEC 2001 USAPA V1.00

DRAGON GUNNERY TRAINER (DGT) SCORECARD

MOVING AND STATIONARY TARGETS

For use of this form see FM 3-23.24 (FM 23-24); the proponent agency is TRADOC.

DATA REQUIRED BY PRIVACY ACT OF 1974. AUTHORITY: 10 USC 3012(g)/Executive Order 9397. PRINCIPLE PURPOSE(S): Aids in individual training on targets at various ranges. ROUTINE USE(S): To evaluate individual proficiency. SSN used for positive identification purposes only. MANDATORY OR VOLUNTARY DISCLOSURE AND EFFECT ON INDIVIDUAL NOT PROVIDING INFORMATION: Voluntary. Individuals not providing information cannot be rated or scored on a mass basis.

LAST NAME	FIRST NAME	MI	SSN	UNIT NO.	POINT NO.	DATE (YYYYMMDD)

FIRING TABLE 1
STATIONARY AND MOVING TARGETS
DISK: DGT #14
PERCENT HIT ZONE SETTING 50 75 100 %

EVENT A. OPTICAL SIGHT. WEIGHT LOSS OR PULL-DOWN; AIMING POINT AND TARGET ASPECT.

ENG	MSN	OBSC	HIT	MISS	TRACKING SCORE	THRUSTERS
1	1	0.0				
2	2	0.0				
3	3	0.5				
4	3	1.0				
5	4	2.0				

EVENT B. OPTICAL SIGHT. INTRODUCTION TO OBSCURATION AND MOVING.

ENG	MSN	OBSC	HIT	MISS	TRACKING SCORE	THRUSTERS
1	8	0.5				
2	9	1.0				
3	10	0.5				
4	11	0.5				
5	12	0.5				

FIRING TABLE 2
STATIONARY AND MOVING TARGETS
DISK: DGT #14
PERCENT HIT ZONE SETTING 50 75 100 %

EVENT A. OPTICAL SIGHT. STATIONARY AND MOVING TARGETS.

ENG	MSN	OBSC	HIT	MISS	TRACKING SCORE	THRUSTERS
1	8	0.5				
2	9	0.5				
3	18	0.0				
4	19	0.0				
5	20	0.0				

EVENT B. BRIGHTNESS AND CONTRAST.

ENG	MSN	OBSC	HIT	MISS	TRACKING SCORE	THRUSTERS
1	26	0.0				
2	27	0.0				
3	28	0.0				
4	29	0.0				
5	30	0.0				

LEGEND: ENG - ENGAGEMENT
MSN - MISSION
OBSC - OBSCURATION

REMARKS:

FIRING TABLE 3
MOVING TARGETS
DISK: DGT #14
PERCENT HIT ZONE SETTING 50 75 100 %

EVENT A. OPTICAL SIGHT.

ENG	MSN	OBSC	HIT	MISS	TRACKING SCORE	THRUSTERS
1	8	1.0				
2	9	1.0				
3	10	1.0				
4	11	1.0				
5	12	1.0				

EVENT B. THERMAL SIGHT.

ENG	MSN	OBSC	HIT	MISS	TRACKING SCORE	THRUSTERS
1	26	0.0				
2	27	0.0				
3	28	0.0				
4	29	0.0				
5	30	0.0				

FIRING TABLE 4
INTERMEDIATE TARGETS
DISK: DGT #14
PERCENT HIT ZONE SETTING 50 75 100 %

EVENT A. OPTICAL SIGHT.

ENG	MSN	OBSC	HIT	MISS	TRACKING SCORE	THRUSTERS
1	7	0.5				
2	9	1.0				
3	10	1.5				
4	11	0.5				
5	12	1.0				

EVENT B. OPTICAL AND THERMAL SIGHT.

ENG	MSN	OBSC	HIT	MISS	TRACKING SCORE	THRUSTERS
1	13	THERMAL				
2	26	THERMAL				
3	28	THERMAL				
4	29	THERMAL				
5	30	THERMAL				

EVENT C. OPTICAL SIGHTS, MOPP, PROTECTIVE MASK FAMILIARIZATION.

ENG	MSN	OBSC	HIT	MISS	TRACKING SCORE	THRUSTERS
1	7	0.0				
2	9	0.0				
3	10	0.0				
4	11	0.0				
5	12	0.0				

SCORER'S NAME (PRINT) ______________________

SCORER'S SIGNATURE ______________________

DRAGON FIELD TACTICAL TRAINER (DFTT) SCORECARD

MULTIPLE TARGETS

TARGET SIZE SETTING: 1 2 3

For use of this form see FM 3-23.24 (FM 23-24); the proponent agency is TRADOC.

DATA REQUIRED BY PRIVACY ACT OF 1974. AUTHORITY: 10 USC 3012(g)/Executive Order 9397. PRINCIPLE PURPOSE(S): Aids in individual training on targets at various ranges. ROUTINE USE(S): To evaluate individual proficiency. SSN used for positive identification purposes only. MANDATORY OR VOLUNTARY DISCLOSURE AND EFFECT ON INDIVIDUAL NOT PROVIDING INFORMATION: Voluntary. Individuals not providing information cannot be rated or scored on a mass basis.

LAST NAME	FIRST NAME	MI	SSN	UNIT NO.	POINT NO.	DATE (YYYYMMDD)

FIRING TABLE 5

MULTIPLE TARGETS, FAMILIARIZATION ONLY

EVENT A. OPTICAL SIGHT, SITTING.

TGT	RANGE	*OBS	HIT KILL	HIT NO KILL	MISS	MISSILE GROUNDED	EXCEEDED RANGE	THRUSTERS
1								
2								
3								
4								
5								

EVENT B. OPTICAL SIGHT, STANDING.

TGT	RANGE	*OBS	HIT KILL	HIT NO KILL	MISS	MISSILE GROUNDED	EXCEEDED RANGE	THRUSTERS
1								
2								
3								
4								
5								

EVENT C. THERMAL SIGHT, SITTING.

TGT	RANGE	*OBS	HIT KILL	HIT NO KILL	MISS	MISSILE GROUNDED	EXCEEDED RANGE	THRUSTERS
1								
2								
3								
4								
5								

EVENT D. THERMAL SIGHT, STANDING.

TGT	RANGE	*OBS	HIT KILL	HIT NO KILL	MISS	MISSILE GROUNDED	EXCEEDED RANGE	THRUSTERS
1								
2								
3								
4								
5								

FIRING TABLE 6

STATIONARY, MOVING, AND MULTIPLE TARGETS, FAMILIARIZATION ONLY

EVENT A. OPTICAL SIGHT, SITTING.

TGT	RANGE	*OBS	HIT KILL	HIT NO KILL	MISS	MISSILE GROUNDED	EXCEEDED RANGE	THRUSTERS
1								
2								
3								
4								
5								

EVENT B. OPTICAL SIGHT, STANDING.

TGT	RANGE	*OBS	HIT KILL	HIT NO KILL	MISS	MISSILE GROUNDED	EXCEEDED RANGE	THRUSTERS
1								
2								
3								
4								
5								

EVENT C. THERMAL SIGHT, SITTING.

TGT	RANGE	*OBS	HIT KILL	HIT NO KILL	MISS	MISSILE GROUNDED	EXCEEDED RANGE	THRUSTERS
1								
2								
3								
4								
5								

EVENT D. THERMAL SIGHT, STANDING.

TGT	RANGE	*OBS	HIT KILL	HIT NO KILL	MISS	MISSILE GROUNDED	EXCEEDED RANGE	THRUSTERS
1								
2								
3								
4								
5								

REMARKS

SCORER'S NAME (PRINT) ______________________

SCORER'S SIGNATURE ______________________

* OBSCURATION

DA FORM 7446-R, DEC 2001

USAPA V1.00

DRAGON GUNNERY TRAINER (DGT) SCORECARD

MULTIPLE TARGETS

For use of this form see FM 3-23.24 (FM 23-24); the proponent agency is TRADOC.

DATA REQUIRED BY PRIVACY ACT OF 1974. AUTHORITY: 10 USC 3012(g)/Executive Order 9397. PRINCIPLE PURPOSE(S): Aids in individual training on targets at various ranges. ROUTINE USE(S): To evaluate individual proficiency. SSN used for positive identification purposes only. MANDATORY OR VOLUNTARY DISCLOSURE AND EFFECT ON INDIVIDUAL NOT PROVIDING INFORMATION: Voluntary. Individuals not providing information cannot be rated or scored on a mass basis.

LAST NAME	FIRST NAME	MI	SSN	UNIT NO.	POINT NO.	DATE (YYYYMMDD)

FIRING TABLE 7

DGT PRACTICE QUALIFICATION
DISK: DGT #14
PERCENT HIT ZONE SETTING
50 % 75 % 100 %

EVENT A. OPTICAL SIGHT.

			TEST			
ENG	MSN	OBSC	GO	NO GO	TS	TH
1	1	0.5				
2	2	1.5				
3	7	0.5				
4	8	1.5				
5	9	0.5				
6	10	1.5				
7	11	0.5				
8	12	1.5				
9	13	0.5				
10	14	1.5				

EVENT B. THERMAL SIGHT.

			TEST			
ENG	MSN	OBSC	GO	NO GO	TS	TH
1	26	0.0				
2	27	0.0				
3	28	0.0				
4	29	0.0				
5	30	0.0				
6	26	0.0				
7	27	0.0				
8	28	0.0				
9	29	0.0				
10	30	0.0				

FIRING TABLE 8

DGT QUALIFICATION
DISK: DGT #15
PERCENT HIT ZONE SETTING
50 % 75 % 100 %

EVENT A. OPTICAL SIGHT.

			TEST				RETEST				RETEST			
ENG	MSN	OBSC	GO	NO GO	TS	TH	GO	NO GO	TS	TH	GO	NO GO	TS	TH
1	1	0.5												
2	6	1.5												
3	7	0.5												
4	4	0.0												
5	9	0.5												
6	11	1.5												
7	12	0.5												
8	13	1.5												
9	14	0.5												
10	15	1.5												

EVENT B. THERMAL SIGHT.

			TEST				RETEST				RETEST			
ENG	MSN	OBSC	GO	NO GO	TS	TH	GO	NO GO	TS	TH	GO	NO GO	TS	TH
1	17	0.0												
2	18	0.0												
3	19	0.0												
4	20	0.0												
5	21	0.0												
6	22	0.0												
7	23	0.0												
8	24	0.0												
9	25	0.0												
10	26	0.0												

LEGEND: ENG - ENGAGEMENT; MSN - MISSION; OBSC - OBSCURATION; TH - THRUSTERS (NARROW COLUMN); TS - TRACKING SCORE (NARROW COLUMN)

REMARKS:

SCORING (CHECK ONE):

EXPERT	☐	19 to 20
SHARPSHOOTER	☐	17 to 18
MARKSMAN	☐	16
UNQUALIFIED	☐	Less than 16

SCORER'S NAME (PRINT) ______________________ SCORER'S SIGNATURE ______________________

DA FORM 7447-R, DEC 2001 USAPA V1.00

Made in the USA
Monee, IL
05 November 2026

67808319R00187